Mastering Git

Attain expert-level proficiency with Git by mastering
distributed version control features

Jakub Narębski

‹packt›

Mastering Git

Group Product Manager: Preet Ahuja

Publishing Product Manager: Vidhi Vashisth

Book Project Manager: Ashwini Gowda

Senior Editor: Roshan Ravi Kumar

Technical Editor: Irfa Ansari

Copy Editor: Safis Editing

Proofreader: Roshan Ravi Kumar

Indexer: Rekha Nair

Production Designer: Alishon Mendonca

Senior Developer Relations Marketing Executive: Rohan Dobhal

First published: April 2016

Second edition: August 2024

Production reference: 1290724

Published by Packt Publishing Ltd.

Grosvenor House

11 St Paul's Square

Birmingham

B3 1RB, UK

ISBN 978-1-83508-607-0

www.packtpub.com

Contributors

About the author

Jakub Narębski followed Git development from the very beginning of its creation. He created, announced, and analyzed the annual Git User's Surveys from 2007 till 2012 – all except the first one (you can find his analysis of those surveys on the Git wiki). He shares his expertise in the technology on the StackOverflow question-and-answer site. He is one of the editors of Git Rev News (`https://git.github.io/rev_news/`), which is a monthly digest of all things Git.

He is an assistant professor in the Faculty of Mathematics and Computer Science at the Nicolaus Copernicus University in Toruń, Poland. He uses Git as his version control system of choice both for personal and professional work and teaches it to computer science students as a part of their coursework.

Jakub Narębski was one of the proofreaders of the *Version Control by Example* book by Eric Sink, and was the reason why it has a chapter on Git.

About the reviewer

Samuel Ng has worked as a data engineer for 3 years. He first started as a systems engineer working on military projects in Singapore before moving into the data sphere. His career has seen him gain experience in multiple industries including semiconductors, military, blockchain and now manufacturing, across Singapore, the Philippines, and South Korea. In his free time, he likes to read up on advancements in AI and his current interest is in GenAI.

I want to thank the author of this book, Jakub, for his effort in creating excellent content for Git readers. Also to the team at Packt, for the opportunity to be part of this amazing project. I hope that you, the reader, will find this content useful in your Git journey, and excel in any tech project that you undertake.

Table of Contents

3

Managing Your Worktrees   59

4

Exploring Project History   97

5

Searching Through the Repository   117

Part 2 - Working with Other Developers

6

Collaborative Development with Git   141

7

Publishing Your Changes   163

8

Advanced Branching Techniques   185

9

Merging Changes Together   219

10

Keeping History Clean   243

Part 3 - Managing, Configuring, and Extending Git

11

12

13

Customizing and Extending Git   329

14

Git Administration   367

15

Git Best Practices   391

Index   405

Other Books You May Enjoy   420

Preface

Git is the most popular open source and distributed version-control system. Version-control systems help software teams manage changes to project sources over time. Using version control is a must in any collaborative development, and it's useful even if you work alone.

Mastering Git will help novice Git professionals attain expert-level proficiency with Git, as well as understand Git concepts and the mental model behind basic and advanced Git tasks. Developers working with Git will be able to use its powerful capabilities to make their work easier. *Mastering Git* will help in various tasks during development, saving time and effort.

This book is meticulously designed to help you gain deeper insights into Git's architecture and its underlying concepts, behavior, and best practices.

You'll begin with a quick example of using Git for the collaborative development of a sample project, in order to establish a basic knowledge of Git's operational tasks and concepts. As you progress through the book, subsequent chapters provide detailed descriptions of the various areas of Git use – from managing your own work, through source code archaeology, to working with other developers. You'll learn how to examine and explore your project's history, create and manage your contributions, set up repositories and branches for collaboration in centralized and distributed workflows, integrate work sent from other developers, customize and extend Git, and recover from repository errors.

Version control topics are accompanied by detailed descriptions of the relevant parts of Git's architecture and behavior. By exploring advanced Git practices and getting to know the details of Git's workings, you will attain a deeper understanding of its behavior, allowing you to customize and extend existing recipes and write your own.

Who this book is for

If you are a Git user with a reasonable knowledge of it and you are familiar with its basic concepts, such as branching, merging, staging, and workflows, this is the book for you. If you have used Git for a long time, this book will help you understand how Git works, make full use of its power, and learn about advanced tools, techniques, and workflows.

If you are a system administrator, project lead, or operations manager, this book will help you to configure Git for better collaborative development, selecting a workflow and branching patterns that would fit best the needs of the team and a project.

A basic knowledge of installing Git and its software configuration management concepts is essential. The first chapter of the book, Git Basics in Practice, should work as a refresher and get you up to date. This book assumes that you have some skills in working from the command line, although this is not strictly necessary.

What this book covers

Chapter 1, *Git Basics in Practice*, serves as a reminder of the version-control basics with Git. The focus is on providing the practical aspects of the technology, using an example of the development of a simple project. This chapter will show and explain basic version-control operations for the development of an example project, as well as how two developers can use Git to collaborate.

Chapter 2, *Developing with Git*, shows how to selectively commit files and interactively select what to commit. You will learn how to create new revisions and new lines of development. This chapter introduces the concept of the staging area for commits (the index) and explains how to view and read differences between the working directory, the index, and the current revision. It will also teach you how to create, list, and switch branches, how to go back in history, and how to revert changes or amend the last commit.

Chapter 3, *Managing Your Worktrees*, teaches you how to manage your files in detail to prepare content for a new commit. It will explain the concept of the index and file status, teaching you how to examine the status of your working area, how to move file contents between a worktree, index, and repository, and how to change your file status. It will also show how to manage files that require special handling, introducing the concepts of ignored files and file attributes.

Chapter 4, *Exploring Project History*, introduces the concept of a graph of revisions and explains how this concept relates to the ideas of branches, tags, and the current branch in Git. You will learn how to select and view a revision or a range of revisions, as well as how to refer to them. These skills will help you focus on specific parts of project history, selecting the interesting part of it for further search.

Chapter 5, *Searching Through the Repository*, explores how to extract the information you want from selected commits. You will learn how to limit your search according to the revision metadata, such as the contents of the commit message, or look at the changes themselves. These skills will help you focus on specific parts of project history, extract information from it, examine what changed and when, and even find bugs by using history bisection.

Chapter 6, *Collaborative Development with Git*, presents a bird's-eye view of the various ways to collaborate, showing different centralized and distributed workflows, their advantages and disadvantages, and how to set them up. This chapter will focus on repository-level interactions in collaborative development. You will also learn the concept of the chain of trust and how to use signed tags, signed merges, signed commits, and signed pushes.

Chapter 7, Publishing Your Changes, examines how Git exchanges information and data between your local repository and remote repositories, describes what the choices are with respect to transport protocols, and shows how Git can help manage credentials that might be needed to access those remote repositories. This chapter also teaches you how you can send your changes upstream so that they can appear in the repository with the official history of a project.

Chapter 8, Advanced Branching Techniques, dives deeper into the details of collaboration in distributed development. It explores the relationships between local branches and branches in remote repositories and describes techniques to synchronize branches and tags. You will learn the different patterns when using branches, including a trunk-based workflow and a topic branch (also called a feature branch) workflow, their advantages and disadvantages, and when to use them.

Chapter 9, Merging Changes Together, teaches you how to merge together changes from different parallel lines of development (that is, branches) using merge and rebase (and squash merge). This chapter will also explain the different types of merge conflicts, how to examine them, and how to resolve them. You will learn how to copy changes with cherry-pick and how to apply a single patch and a patch series.

Chapter 10, Keeping History Clean, explains why you might want to keep a clean history, when it can and should be done, and how it can be done. You will find step-by-step instructions on how to reorder, squash, and split commits. This chapter also demonstrates how you can recover from a history rewrite and explains what to do if you cannot rewrite history, how to revert the effect of commit, how to add a note to it, and how to change the view of a project's history with a replacement mechanism.

Chapter 11, Managing Subprojects, explains and shows different ways to connect different projects in a single repository of a framework superproject, from a strong inclusion by embedding the code of one project in another (monorepos and subtrees) to a light connection between projects by nesting repositories (submodules and similar solutions).

Chapter 12, Managing Large Repositories, presents various solutions to the problem of large Git repositories, whether they are large because of a long history, contain a large number of files, or contain a project that includes some large files.

Chapter 13, Customizing and Extending Git, covers configuring and extending Git to fit your needs. You will find here details on how to set up the command line for easier use and a short introduction to graphical interfaces. This chapter explains how to automate Git with hooks (focusing on client-side hooks) – for example, how to make Git check whether a commit being created passes specific coding guidelines.

Chapter 14, Git Administration, focuses on the administrative side of Git. It briefly touches on the topic of serving Git repositories. Here, you will learn how to use server-side hooks for logging, access control, enforcing a development policy, and other purposes.

Chapter 15, *Git Best Practices*, presents a collection of version-control, generic and Git-specific recommendations and best practices. These cover issues such as managing the working directory, creating commits and a series of commits (pull requests), submitting changes for inclusion, and a peer review of changes.

To get the most out of this book

To follow the examples used in this book and run the provided commands, you will need the Git software (`https://git-scm.com/`), preferably version 2.41.0 or later. Git is available for free on every platform (such as Linux, Windows, and macOS). All examples use the textual Git interface, using the bash shell (which is provided with Git for Microsoft Windows, where it is not present by default).

Software covered in the book	Operating system requirements
Git	Windows, macOS, or Linux

To follow the development of a sample program, which is tracked in *Chapter 1*, *Git Basics in Practice*, as a demonstration of using version control, you would also need a web browser and a text editor (although a programmers' editor or IDE is preferred).

Download the example code files

You can download the example code files for this book from GitHub at `https://github.com/PacktPublishing/Mastering-Git---Second-Edition`. If there's an update to the code, it will be updated in the GitHub repository.

We also have other code bundles from our rich catalog of books and videos available at `https://github.com/PacktPublishing/`. Check them out!

Conventions used

There are a number of text conventions used throughout this book.

`Code in text`: Indicates code words in text, database table names, folder names, filenames, file extensions, pathnames, dummy URLs, user input, and Twitter handles. Here is an example: "Then, Bob writes JavaScript source code (`random.js`) that is responsible for web application behavior."

A block of code is set as follows:

```
function getRandomInt(max) {
    return Math.floor(Math.random() * max) + 1;
}
```

```
function generateRandom() {
    let max = document.getElementById('max').value;
    alert(getRandomInt(max));
}
```

When we wish to draw your attention to a particular part of a code block, the relevant lines or items are set in bold:

```
<body>
<button disabled>Generate number</button>
<label for="max">up to</label>
<input type="number" id="max" name="rand_max" value="10" />
<div id="result"></div>
</body>
</html>
```

Any command-line input or output is written as follows:

```
$ mkdir css
$ cd css
```

Bold: Indicates a new term, an important word, or words that you see on screen. For instance, words in menus or dialog boxes appear in **bold**. Here is an example: "Select **System info** from the **Administration** panel."

> **Tips or important notes**
> Appear like this.

Get in touch

Feedback from our readers is always welcome.

General feedback: If you have questions about any aspect of this book, email us at customercare@ packtpub.com and mention the book title in the subject of your message.

Errata: Although we have taken every care to ensure the accuracy of our content, mistakes do happen. If you have found a mistake in this book, we would be grateful if you would report this to us. Please visit www.packtpub.com/support/errata and fill in the form.

Piracy: If you come across any illegal copies of our works in any form on the internet, we would be grateful if you would provide us with the location address or website name. Please contact us at copyright@packt.com with a link to the material.

If you are interested in becoming an author: If there is a topic that you have expertise in and you are interested in either writing or contributing to a book, please visit `authors.packtpub.com`.

Share Your Thoughts

Once you've read *Mastering Git*, we'd love to hear your thoughts! Scan the QR code below to go straight to the Amazon review page for this book and share your feedback.

`https://packt.link/r/1-835-08607-1`

Your review is important to us and the tech community and will help us make sure we're delivering excellent quality content.

Download a free PDF copy of this book

Thanks for purchasing this book!

Do you like to read on the go but are unable to carry your print books everywhere?

Is your eBook purchase not compatible with the device of your choice?

Don't worry, now with every Packt book you get a DRM-free PDF version of that book at no cost.

Read anywhere, any place, on any device. Search, copy, and paste code from your favorite technical books directly into your application.

The perks don't stop there, you can get exclusive access to discounts, newsletters, and great free content in your inbox daily

Follow these simple steps to get the benefits:

1. Scan the QR code or visit the link below

https://packt.link/free-ebook/978-1-83508-607-0

2. Submit your proof of purchase

3. That's it! We'll send your free PDF and other benefits to your email directly

Part 1 -
Exploring Project History and
Managing Your Own Work

In this part, you will start by understanding the basics of using Git in a simple example. You will then learn how to use it to answer questions about the project and its history. You will also learn how to examine the state of your worktree, manage changes, and create a good commit.

This part has the following chapters:

Git Basics in Practice

1

This book is intended for intermediate and advanced Git users to help them on their road to mastering Git. Therefore, the chapters following this one will assume you know the basics of Git, and have advanced past the beginner stage.

This chapter will serve as a reminder of version control basics with Git. The focus will be on providing practical aspects of the technology, showing and explaining basic version control operations on the example of the development of an example project, and the collaboration between two developers.

In this chapter, we will cover the following:

- A brief introduction to version control and Git
- Setting up a Git environment and Git repository (`init` and `clone`)
- Adding files, checking status, creating commits, and examining the history
- Interacting with other Git repositories (`pull` and `push`)
- Creating and listing branches, switching to a branch, and merging changes
- Resolving a simple merge conflict
- Creating and publishing a tag

Technical requirements

To follow the examples shown in this chapter, you will need Git: `https://git-scm.com/`. You will also need an interactive shell (for example, Git Bash if you are using MS Windows), a text editor or an IDE for web development (for editing JavaScript and HTML), and a web browser.

You can access the code of an example project used in this chapter at the following URLs: `https://github.com/PacktPublishing/Mastering-Git---Second-Edition/tree/main/chapter01` and `https://github.com/jnareb/Mastering-Git---Second-Edition---chapter01-sample_project`.

A brief introduction to version control and Git

A **version control system** (sometimes called **revision control**) is a tool that lets you track the history and attribution of your project files over time (stored in a **repository**) and helps the developers in the team to work together. Modern version control systems give each developer their own sandbox, preventing their work in progress from conflicting, and all the while providing a mechanism to merge changes and synchronize work. They also allow us to switch between different lines of development, called **branches**; this mechanism allows the developer to change, for example, from working on introducing a new feature step by step to fixing the bug in an older, released version of the project.

Distributed version control systems (such as Git) give each developer their own copy of the project's history, which is called a **clone** of a repository. This is what makes Git fast, because nearly all operations are performed locally. It is also what makes Git flexible because you can set up repositories in many ways. Repositories meant for development also provide a separate **working area** (or a **worktree**) with project files for each developer. Git's branching model enables cheap local branching, allowing the use of branches for context switching by creating sandboxes for different tasks. It also makes it possible to use a very flexible *topic branch* workflow for collaboration.

The fact that the whole history is accessible allows for a *long-term undo*, rewinding to the last working version, and so on. Tracking ownership of changes automatically makes it possible to find out who was responsible for any given area of code, and when each change was done. You can compare different revisions, go back to the revision a user is sending a bug report against, and even automatically find out which revision introduced a regression bug (with `git bisect`). Tracking changes to the tips of branches with **reflog** allows for easy undo and recovery.

A unique feature of Git is that it enables explicit access to the **staging area** for creating **commits** (new revisions—that is, new versions of a project). This brings additional flexibility to managing your working area and deciding on the shape of a future commit.

All this flexibility and power come at a cost. It is not easy to master using Git, even though it is quite easy to learn its basic use. This book will help you attain this expertise, but let us start with a reminder about the basics of Git.

Git by example

Let's follow a step-by-step, section-by-section, simple example of two developers using Git to work together on a simple project. You can find all three repositories (for two developers, and the bare server repository) with the example code files for this chapter, where you can examine the code, history, and reflog, at `https://github.com/PacktPublishing/Mastering-Git---Second-Edition`, in a `sample_project.zip` archive.

Following the example

To follow this example of the team development process on a single computer, you can simply create three folders called, for example, `alice/`, `bob/`, and `server/`, and switch to the appropriate folder when following work done by Alice, Bob, and Carol, respectively.

There are a few simple changes you need to make for this simulation to work. When creating a repository as Carol, you don't need to create and switch to the `/srv/git` directory, so you can simply skip these commands. In Alice or Bob's role, you need to create separate identities in the repository's *local* configuration, either with the `git config` command without the `--user` option or by editing the `.git/config` file in the appropriate repository. In place of the `https://git.company.com/random` repository URL, which does not exist, simply use the path to the server repository: `../server/random.git`.

Additionally, if you plan on moving the directory with `alice/`, `bob/`, and `server/` subdirectories, you will need to edit the "origin" repository URL that is stored in repository config files by changing it from the absolute path to a relative path—namely, `../../server/random.git`.

Setup and initialization

A company has begun work on a new product. This product calculates a random number—an integer value of a specified range.

The company has assigned two developers to work on this new project, Alice and Bob. Both developers are telecommuting to the company's corporate headquarters. After a bit of discussion, they have decided to implement their product as a simple web application in JavaScript and HTML and to use Git 2.41.0 (`git-scm.com`) for version control.

Note

This project and the code are intended for demonstration purposes only and will be much simplified. The details of code are not important here—what is important is how the code changes, and how Git is used to help with the development.

Repository setup

With a small team, they have decided on the setup shown in the following diagram.

Important note

This is only one possible setup, with the **central canonical repository**, and without a dedicated maintainer responsible for this repository (all developers are equal in this setup). It is not the only possibility; other ways of configuring repositories will be shown in *Chapter 6, Collaborative Development with Git*.

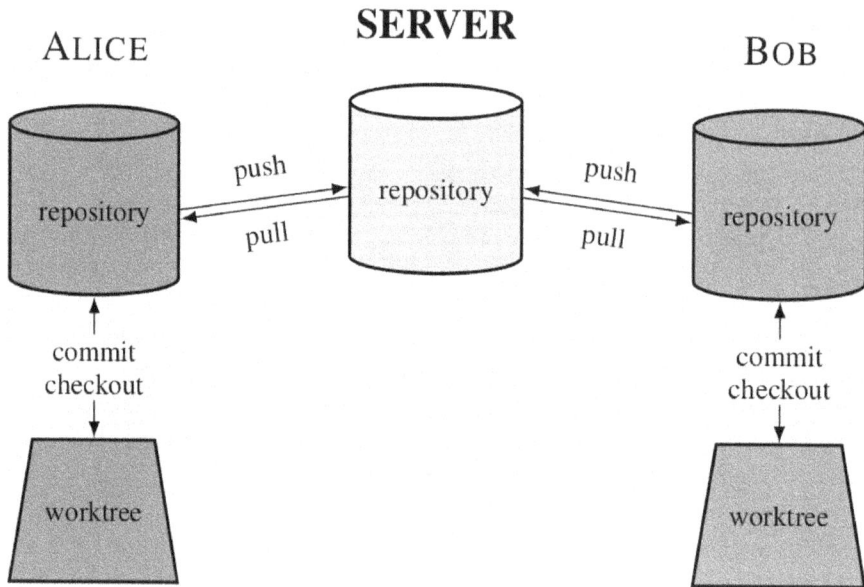

Figure 1.1 – Repository setup for the sample project (using a centralized workflow)

Creating a Git repository

Alice gets the project started by asking Carol, an administrator, to create a new repository specifically for collaborating on a project, to share work with the whole team:

```
carol@server:~$ mkdir -p /srv/git
carol@server:~$ cd /srv/git
carol@server:/srv/git$ git init --bare random.git
Initialized empty Git repository in /srv/git/random.git/
```

> **Important note**
>
> Command-line examples follow the Unix convention of having user@host:directory at the beginning of the command prompt, to make it easier to guess from first glance who performs a command, on what computer, and in which directory (here, the tilde, ~, denotes the user's home directory). This is the usual command prompt setup on Linux; a similar-looking prompt is used by Git Bash.

You can configure your command prompt to show Git-specific information, such as the name of the repository, the subdirectory within the repository, the current branch, and even the worktree status (see *Chapter 13, Customizing and Extending Git*).

I consider the details of server configuration to be too much for this chapter. Just imagine that it happened, and nothing went wrong, or look at *Chapter 14, Git Administration*.

You can also use a tool to manage Git repositories (for example, `gitolite`); creating a public repository on a server would then, of course, look different. Often, though, it involves creating a Git repository with `git init` (without `--bare`) in your own home directory and then pushing it with an explicit URI to the server, which would then automatically create the public repository.

Or perhaps the repository was created through the web interface of tools such as GitHub, Bitbucket, or GitLab (either hosted in the cloud, or installed on-premises).

Cloning the repository and creating the first commit

Bob gets the information that the project repository is ready, and he can start coding.

Since this is Bob's first time using Git, he first sets up his `~/.gitconfig` file with information that will be used to identify his commits (for example, with `git config --global --edit`):

```
[user]
    name = Bob Hacker
    email = bob@company.com
```

Now, he needs to get his own repository instance (which currently is empty):

```
bob@hostB:~$ git clone https://git.company.com/random
Cloning into 'random'...
warning: You appear to have cloned an empty repository.
done.
bob@hostB:~$ cd random
bob@hostB:~/random$
```

> **Tip**
>
> All examples in this chapter use the command-line interface. Those commands might be given using a Git GUI or IDE integration, as explained in the *Graphical interfaces* section in *Chapter 13, Customizing and Extending Git*. The book *Git: Version Control for Everyone*, published by Packt Publishing, shows GUI equivalents for the command line.

Bob notices that Git said that it is an empty repository with no source code yet, and starts coding. He opens his text editor (or IDE of choice) and creates the starting point for their product.

First, he creates an HTML file (`index.html`) with the simplest possible interface for the web application being created, just a button and an input field:

```
<!DOCTYPE html>
<html lang="en">
<head>
    <meta charset="UTF-8">
    <title>Random Number Generator</title>
```

```
    <script src="random.js" defer></script>
</head>
<body>
<button disabled>Generate number</button>
<label for="max">up to</label>
<input type="number" id="max" name="rand_max" value="10" />
</body>
</html>
```

Then, Bob writes JavaScript source code (`random.js`) that is responsible for web application behavior—in this case, generating and displaying a random integer number within a given range from 1 to a configurable maximum, inclusive:

```
function getRandomInt(max) {
    return Math.floor(Math.random() * max) + 1;
}


function generateRandom() {
    let max = document.getElementById('max').value;
    alert(getRandomInt(max));
}


let button = document.querySelector('button');
button.addEventListener('click', generateRandom);
button.disabled = false;
```

Typically, for most initial implementations, this version is missing a lot of features but it is a good place to begin. Before committing his code, Bob wants to make sure that it looks all right and that it works correctly. He opens the `index.html` file in a web browser or uses a live preview feature of his IDE, as shown in *Figure 1.2*.

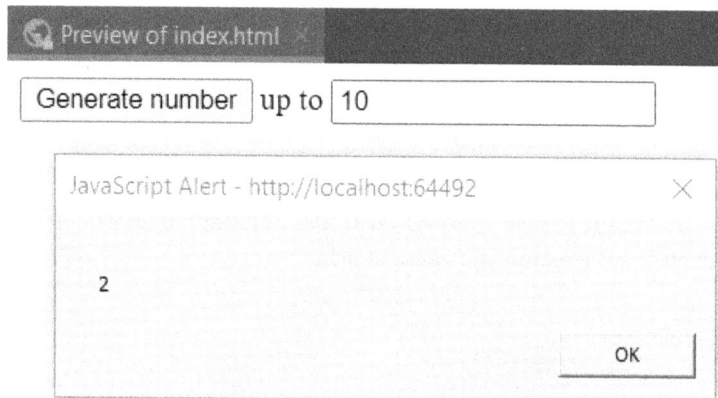

Figure 1.2 – Preview of the first version of the example application

Alright! It's time to *add* both files to the repository:

```
bob@hostB:~/random$ git add index.html random.js
```

Bob uses the `status` operation to make sure that the pending changeset (the future commit) looks proper.

We use a short form of `git status` here to reduce the amount of space taken by examples; you can find an example of a full `status` output further in the chapter:

```
bob@hostB:~/random$ git status --short
A  index.html
A  random.js
```

Now it's time to *commit* to the current version:

```
bob@hostB:~/random$ git commit -a -m "Initial implementation"
[master (root-commit) 961e72b] Initial implementation
 2 files changed, 25 insertions(+)
 create mode 100644 index.html
 create mode 100644 random.js
```

> **Important note**
>
> Normally, you would create a **commit message** not by using the `-m <message>` command-line option but by letting Git open an editor. We use this form here to make the examples more compact. In actual practice, it is recommended to provide a more detailed description of changes.
>
> The `-a`/`--all` option in the `git commit -a` command means to take *all* changes to the tracked files. This is not the only possible way of creating revisions; you can separate manipulating the staging area from creating a commit—this is, however, a separate issue, left for *Chapter 3, Managing Your Worktrees*.

Now it's time to make those changes visible to Alice.

Collaborative development

One of the main goals of a version control system is to help developers work together on a common project. With a distributed version control system, such as Git, this involves an explicit step of publishing changes to be visible to others.

Publishing changes

After finishing working on the initial version of the project, Bob decides that it is ready to be published (to be made available for other developers). He *pushes* the changes as follows:

```
bob@hostB:~/random$ git push
Enumerating objects: 4, done.
Counting objects: 100% (4/4), done.
Delta compression using up to 4 threads
Compressing objects: 100% (4/4), done.
Writing objects: 100% (4/4), 670 bytes | 22.00 KiB/s, done.
Total 4 (delta 0), reused 0 (delta 0)
To https://git.company.com/random.git
 * [new branch]      master -> master
```

> Tip
>
> Note that, depending on the speed of the network, Git could show progress information during remote operations such as `clone`, `push`, and `fetch`. Such information is omitted from examples in this book, except where that information is actually discussed while examining the history and viewing changes.
>
> Also, if you use an older Git version, it could require setting a `push.default` configuration variable.

Since it is Alice's first time using Git on her desktop machine, she first tells Git how her commits should be identified:

```
alice@hostA:~$ git config --global user.name "Alice Developer"
alice@hostA:~$ git config --global user.email alice@company.com
```

Now, Alice needs to set up her own repository instance:

```
alice@hostA:~$ git clone https://git.company.com/random
Cloning into 'random'...
done.
```

Alice examines the working directory:

```
alice@hostA:~$ cd random
alice@hostA:~/random$ ls -alF
total 6
drwxr-xr-x   1 alice staff       0 May 2 16:44 ./
drwxr-xr-x   4 alice staff       0 May 2 16:39 ../
drwxr-xr-x   1 alice staff       0 May 2 16:39 .git/
-rw-r--r--   1 alice staff     331 May 2 16:39 index.html
-rw-r--r--   1 alice staff     327 May 2 16:39 random.js
```

> **Tip**
>
> The `.git` directory contains Alice's whole copy (clone) of the repository in Git internal format and some repository-specific administrative information. See the **gitrepository-layout(5)** man page for details of the file layout, which can be done, for example, with the `git help repository-layout` command.

She wants to check the log to see the details (to examine the project history):

```
alice@hostA:~/random$ git log
commit 961e72b31b0d2dacc0584cbe8953c3aed1042e9b (HEAD -> master)
Author: Bob Hacker bob@company.com
Date:   Sun May 2 22:34:40 2021 +0200

    Initial implementation
```

> **Naming revisions**
>
> At the lowest level, a Git version identifier is a SHA-1 hash, for example, `2b953b4e80`. Git supports various forms of referring to revisions, including unambiguously shortened SHA-1 (with a minimum of four characters)—see *Chapter 4, Exploring Project History*, for more ways.

When Alice decides to take a look at the application, she decides that using `alert()` to show the result is not a good user interface. To generate a new random number, the user needs to first close the window. It would be better if it was possible to generate a new result immediately.

She decides that a better solution would be to put the result on the page, below the form. She adds a single line to `index.html` to make a place for it:

```html
<!DOCTYPE html>
<html lang="en">
<head>
    <meta charset="UTF-8">
    <title>Random Number Generator</title>
    <script src="random.js" defer></script>
</head>
<body>
<button disabled>Generate number</button>
<label for="max">up to</label>
<input type="number" id="max" name="rand_max" value="10" />
<div id="result"></div>
</body>
</html>
```

She then replaces the use of `alert()` in the JavaScript code by showing the result directly on the application page, using just the added `<div id="result"></div>` placeholder:

```
function getRandomInt(max) {
    return Math.floor(Math.random() * max) + 1;
}

function generateRandom() {
    let max = document.getElementById('max').value;
    let res = document.getElementById('result');
    res.textContent = 'Result: ' + getRandomInt(max);
}

let button = document.querySelector('button');
button.addEventListener('click', generateRandom);
button.disabled = false;
```

She then opens her web browser to check that it works correctly. She clicks the **Generate number** button a few times to check that it really generates random numbers:

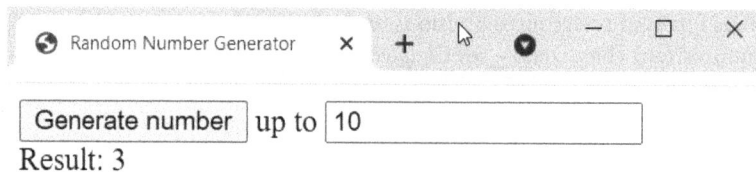

Figure 1.3 – Application after Alice's changes, with the result on the page itself

Everything looks alright, so she uses the `status` operation to see the pending changes:

```
alice@hostA:~/random$ git status -s
 M index.html
 M random.js
```

No surprise here. Git knows that `index.html` and `random.js` have been modified. She wants to double-check by reviewing the actual changes with the `diff` command:

```
alice@hostA:~/random$ $ git diff
diff --git a/index.html b/index.html
index 1e79bb1..3021b9d 100644
--- a/index.html
+++ b/index.html
@@ -9,5 +9,6 @@
 <button disabled>Generate number</button>
 <label for="max">up to</label>
```

```
 <input type="number" id="max" name="rand_max" value="10" />
+<div id="result"></div>
 </body>
 </html>
diff --git a/random.js b/random.js
index 3533d15..b036fa1 100644
--- a/random.js
+++ b/random.js
@@ -4,7 +4,8 @@ function getRandomInt(max) {

 function generateRandom() {
     let max = document.getElementById('max').value;
-    alert(getRandomInt(max));
+    let res = document.getElementById('result');
+    res.textContent = 'Result: ' + getRandomInt(max);
 }

 let button = document.querySelector('button');
```

Now, it's time to commit the changes and push them to the public repository:

```
alice@hostA:~/random$ git commit -a -m "Show result on the page
instead of using alert()"
[master a030d99] Show result on the page instead of using alert()
 2 files changed, 14 insertions(+), 12 deletions(-)
alice@hostA:~/random$ git push
To https://git.company.com/random.git
   961e72b..a030d99  master -> master
```

Renaming and moving files

While this is happening, Bob moves on to his next task, which is to restructure the tree a bit. He doesn't want the top level of the repository to get too cluttered, so he decides to follow one of the established conventions for the directory structure, and to move all the JavaScript source code files into the scripts/ subdirectory:

```
bob@hostB:~/random$ mkdir scripts
bob@hostB:~/random$ git mv *.js scripts/
```

He then checks that everything works correctly, and it turns out that he needs to update the path to the JavaScript code in the index.html file, so he does that:

```
<!DOCTYPE html>
<html lang="en">
<head>
```

```
        <meta charset="UTF-8">
        <title>Random Number Generator</title>
        <script src="scripts/random.js" defer></script>
</head>
<body>
<button disabled>Generate number</button>
<label for="max">up to</label>
<input type="number" id="max" name="rand_max" value="10" />
</body>
</html>
```

He checks that everything works fine now, examines the status, and commits changes:

```
bob@hostB:~/random$ git status --short
 M index.html
R  random.js -> scripts/random.js
bob@hostB:~/random$ git commit -a -m "Directory structure"
[master 1b58e54] Directory structure
 2 files changed, 1 insertion(+), 1 deletion(-)
 rename random.js => scripts/random.js (100%)
```

While he's at it, to minimize the impact of reorganization on the `diff` output, he configures Git to always use `rename` and copy detection:

```
bob@hostB:~/random$ git config --global diff.renames copies
```

Bob then decides the time has come to add a README.md file for the project:

```
bob@hostB:~/random$ git status -s
?? README.md
bob@hostB:~/random$ git add README.md
bob@hostB:~/random$ git status -s
A  README.md
bob@hostB:~/random$ git commit -a -m "Added README.md"
[master 6789f76] Added README.md
 1 file changed, 3 insertions(+)
 create mode 100644 README.md
```

Bob decides to rename random.js to gen_random.js:

```
bob@hostA:~/random$ git mv scripts/random.js scripts/gen_random.js
```

This, of course, also requires changes to index.html:

```
bob@hostB:~/random$ git status -s
 M index.html
R  scripts/random.js -> scripts/gen_random.js
```

He then commits those changes.

```
bob@hostB:~/random$ git commit -a -m "Rename random.js to gen_random.js"
```

Updating your repository (with merge)

Reorganization done, now Bob tries to publish those changes:

```
bob@hostA random$ git push
To https://git.company.com/random
 ! [rejected]        master -> master (fetch first)
error: failed to push some refs to 'https://git.company.com/random'
hint: Updates were rejected because the remote contains work that you
do
hint: not have locally. This is usually caused by another repository
pushing
hint: to the same ref. You may want to first integrate the remote
changes
hint: (e.g., 'git pull ...') before pushing again.
hint: See the 'Note about fast-forwards' in 'git push --help' for
details.
```

But Alice was working at the same time, and she had her change ready to commit and push first. Git is not allowing Bob to publish his changes because Alice has already pushed something to the master branch, and Git is preserving her changes.

> **Important note**
> Hints and pieces of advice in Git command output will be skipped from here on for the sake of brevity.

Bob uses pull to bring in changes (as described in hint in the command output):

```
bob@hostB:~/random$ git pull
From https://git.company.com/random
 + 3b16f17...db23d0e master       -> origin/master
Auto-merging scripts/gen_random.c
Merge made by the 'recursive' strategy.
 index.html            | 1 +
 scripts/gen_random.js | 3 ++-
 2 files changed, 3 insertions(+), 1 deletion(-)
```

The git pull command fetched the changes, automatically merged them with Bob's changes, and created a merge commit—opening editor to confirm committing the merge.

> **Important note**
>
> From version 2.31 onward, Git asks the user to set the `pull.rebase` configuration variable; we assume that Alice and Bob set it to `false`. See *Chapter 9, Merging Changes Together*, the *Methods of combining changes* section, for a more detailed description of the difference between using merge commits and using rebasing to combine changes.

Everything now seems to be good. The merge commit is done. Apparently, Git was able to merge Alice's changes directly into Bob's moved and renamed copy of a file without any problems. Marvelous!

```
bob@hostB:~/random$ git show
commit df9132d4482dfd66d6d9843db205d4e775c76509 (HEAD -> master)
Merge: eabf309 a030d99
Author: Bob Hacker bob@company.com
Date:   Mon May 3 02:31:23 2021 +0200


    Merge branch 'master' of https://git.company.com/random
```

Bob checks that it works correctly (because automatically merging does not necessarily mean that the merge output is correct). It works fine and he is ready to push the merge:

```
bob@hostB random$ git push
To https://git.company.com/random
   a030d99..df9132d  master -> master
```

Creating a tag – a symbolic name for the revision

Alice and Bob decide that the project is ready for wider distribution. Bob creates a **tag** so they can more easily access and refer to the released version. He uses an **annotated tag** for this; an often-used alternative is to use a **signed tag**, where the annotation contains a PGP signature (which can later be verified):

```
bob@hostB:~/random$ git tag -a -m "random v0.1" v0.1
bob@hostB:~/random$ git tag --list
v0.1
bob@hostB:~/random$ git log -1 --oneline --decorate
df9132d (HEAD -> master, tag: v0.1, origin/master) Merge branch
'master' of https://git.company.com/random
```

Of course, the v0.1 tag wouldn't help if it was only in Bob's local repository. He, therefore, pushes the just-created tag:

```
bob@hostB random$ git push origin tag v0.1
To https://git.company.com/random
 * [new tag]         v0.1 -> v0.1
```

Alice updates her repository to get the v0.1 tag, and to start with up-to-date work:

```
alice@hostA:~/random$ git pull
From https://git.company.com/random
   a030d99..df9132d  master      -> origin/master
 * [new tag]          v0.1        -> v0.1
Updating a030d99..df9132d
Fast-forward
 README.md                           | 3 +++
 index.html                          | 2 +-
 random.js => scripts/gen_random.js | 0
 3 files changed, 4 insertions(+), 1 deletion(-)
 create mode 100644 README.md
 rename random.js => scripts/gen_random.js (100%)
```

Resolving a merge conflict

Alice decides that it would be a good idea to add a piece of information about where the result of the random number generator would appear:

```
<!DOCTYPE html>
<html lang="en">
<head>
    <meta charset="UTF-8">
    <title>Random Number Generator</title>
    <script src="scripts/gen_random.js" defer></script>
</head>
<body>
<button disabled>Generate number</button>
<input type="number" id="max" name="rand_max" value="10" />
<div id="result">Result:</div>
</body>
</html>
```

Grand! Let's see that it works correctly.

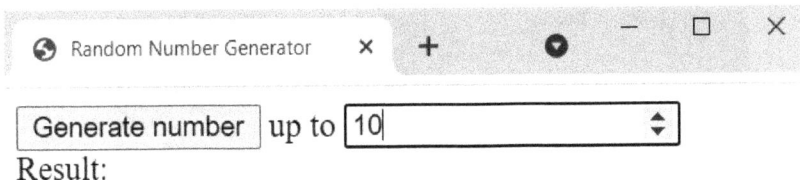

Figure 1.4 – After adding information about where the result would appear

Good. Time to commit the change:

```
alice@hostA:~/random$ git status -s
 M index.html
alice@hostA:~/random$ git commit -a -m "index.html: Show where result
goes"
[master e04655f] index.html: Show where result goes
 1 file changed, 1 insertion(+), 1 deletion(-)
```

No problems here.

Meanwhile, Bob notices that if one has JavaScript disabled in the web browser or uses a text browser without support for JavaScript, the web application as it is now does not work , without explaining why. It would be a good idea to notify the user about this issue:

```
bob@hostB:~/random$ git pull
Already up-to-date.
```

He decides to add a <noscript> tag to explain that JavaScript is required for the application to work:

```
bob@hostB:~/random$ $ git status -s
 M index.html
bob@hostB:~/random$ git diff
diff --git a/index.html b/index.html
index 108885f..80348b7 100644
--- a/index.html
+++ b/index.html
@@ -10,5 +10,6 @@
 <label for="max">up to</label>
 <input type="number" id="max" name="rand_max" value="10" />
 <div id="result"></div>
+<noscript>To use this web app, please enable JavaScript.</noscript>
 </body>
 </html>
```

Bob uses the w3m text-based web browser to check that <noscript> works as intended:

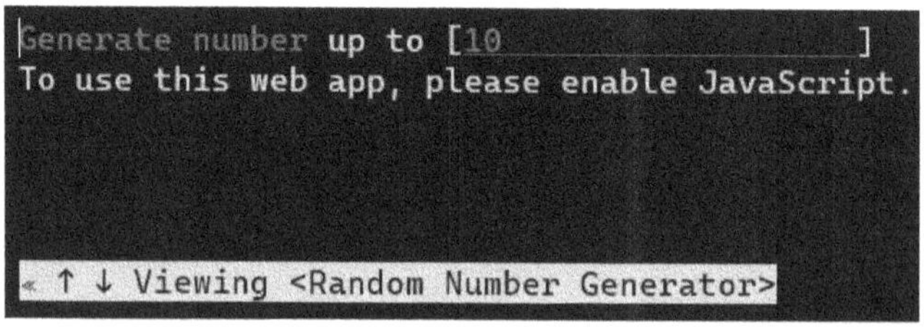

Figure 1.5 – Testing the application in w3m, a text-based web browser without JavaScript support

He then checks in a graphical web browser (or a live preview) that nothing changed for JavaScript-capable clients. He has his change ready to commit and push first:

```
bob@hostB:~/random$ git commit -a -m "Add <noscript> tag"
[master a808ecf] Add <noscript> tag
 1 file changed, 1 insertion(+)
bob@hostB:~/random$ git push
To https://git.company.com/random
   df9132d..a808ecf  master -> master
```

So, when Alice is ready to push her changes, Git rejects them:

```
alice@hostA:~/random$ git push
To https://git.company.com/random
 ! [rejected]          master -> master (non-fast-forward)
error: failed to push some refs to 'https://git.company.com/random'
```

Ah. Bob must have pushed a new changeset already. Alice once again needs to pull and merge to combine Bob's changes with her own:

```
alice@hostA:~/random$ git pull
Auto-merging index.html
CONFLICT (content): Merge conflict in index.html
Automatic merge failed; fix conflicts and then commit the result.
```

The merge didn't go quite as smoothly this time. Git wasn't able to automatically merge Alice's and Bob's changes. Apparently, there was a conflict. Alice decides to open the index.html file in her editor to examine the situation (she could have used a graphical merge tool via git mergetool instead):

```
<!DOCTYPE html>
<html lang="en">
<head>
    <meta charset="UTF-8">
    <title>Random Number Generator</title>
    <script src="scripts/gen_random.js" defer></script>
</head>
<body>
<button disabled>Generate number</button>
<label for="max">up to</label>
<input type="number" id="max" name="rand_max" value="10" />
<<<<<<< HEAD
<div id="result">Result:</div>
=======
<div id="result"></div>
```

```
<noscript>To use this web app, please enable JavaScript.</noscript>
>>>>>>> a808ecfb89919fd05cf50fbf879b493c83499002
</body>
</html>
```

Git has included both Bob's code (between the <<<<<<< HEAD and ======== conflict markers) and Alice's code (between ======== and >>>>>>>). What we want as a final result is to include both blocks of code. Git couldn't merge it automatically because those blocks were not separated. Alice work adding Result: can be simply included right before <noscript> added by Bob. After resolving the conflict, the changes look like this:

```
alice@hostA:~/random$ $ git diff
diff --cc index.html
index ea1a830,80348b7..0000000
--- a/index.html
+++ b/index.html
@@@ -9,7 -9,6 +9,6 @@@
   <button disabled>Generate number</button>
   <label for="max">up to</label>
   <input type="number" id="max" name="rand_max" value="10" />
  -<div id="result"></div>
  +<noscript>To use this web app, please enable JavaScript.</noscript>
 + <div id="result">Result:</div>
   </body>
   </html>
```

That should take care of the problem. Alice refreshes the web application in a web browser to check that it works correctly. She marks the conflict as resolved and commits changes:

```
alice@hostA:~/random$ git status -s
UU index.html
alice@hostA:~/random$ git commit -a -m 'Merge: mark output + noscript'
[master 919f0f7] Merge: mark output + noscript
```

And then she retries the push:

```
alice@hostA:~/random$ git push
To https://git.company.com/random
   a808ecf..919f0f7  master -> master
```

And... done.

Adding files in bulk and removing files

Bob decides to add a COPYRIGHT file with a copyright notice for the project. There was also a NEWS file planned (but not created), so he uses a bulk add to add all the files:

```
bob@hostB:~/random$ git add -v
Nothing specified, nothing added.
Maybe you wanted to say 'git add .'?
bob@hostB:~/random$ git add -v .
add 'COPYRIGHT'
add 'COPYRIGHT~'
```

Oops! Because Bob didn't configure his **ignore patterns**, the backup file, COPYRIGHT~, was caught too (such a system-specific pattern should go to the repository's .git/info/exclude or personal ignore file, ~/.config/git/ignore, as described in *Chapter 3, Managing Your Worktrees*, in the *Ignoring files* section). Let's remove this file:

```
bob@hostB:~/random$ git status -s
A   COPYRIGHT
A   COPYRIGHT~
bob@hostB:~/random$ git rm COPYRIGHT~
error: the following file has changes staged in the index:
    COPYRIGHT~
(use --cached to keep the file, or -f to force removal)
bob@hostB:~/random$ git rm -f COPYRIGHT~
rm 'COPYRIGHT~'
```

Let's check the status and commit the changes:

```
bob@hostB:~/random$ git status -s
A   COPYRIGHT
bob@hostB:~/random$ git commit -a -m 'Added COPYRIGHT'
[master ca3cdd6] Added COPYRIGHT
 1 files changed, 2 insertions(+), 0 deletions(-)
 create mode 100644 COPYRIGHT
```

Undoing changes to a file

A bit bored, Bob decides that their web application looks bland, and adds the Bootstrap CSS library (https://getbootstrap.com) to the index.html header:

```
<link rel="stylesheet" href="https://cdn.jsdelivr.net/npm/
bootstrap@5.3.0/dist/css/bootstrap.min.css" integrity="sha384-9nd
CyUaIbzAi2FUVXJi0CjmCapSmO7SnpJef0486qhLnuZ2cdeRhO02iuK6FUUVM"
crossorigin="anonymous">
```

He checks how much source code it changed:

```
bob@hostB:~/random$ git diff --stat
index.html | 4 ++++
 1 file changed, 4 insertions(+)
```

That looks all right; however, the application doesn't look that much better without further changes, and now it requires access to the internet. Bob decides that it is not the time to move to the Bootstrap CSS framework, and undoes the changes to index.html:

```
bob@hostB:~/random$ git status -s
 M index.html
bob@hostB:~/random$ git restore index.html
bob@hostB:~/random$ git status -s
```

If you can't remember how to revert a particular type of change or to update what is to be committed (using git commit without -a), the output of git status (without -s) contains information about what commands to use. This is shown in the following example:

```
bob@hostB:~/random$ git status
On branch master
Your branch is ahead of 'origin/master' by 1 commit.
  (use "git push" to publish your local commits)

Changes not staged for commit:
  (use "git add <file>..." to update what will be committed)
  (use "git restore <file>..." to discard changes in working
directory)
        modified:   index.html

no changes added to commit (use "git add" and/or "git commit -a")
```

Branching and merging

Developers often need to isolate a specific set of changes that is expected to not be ready for some time, to create another line of development: a branch. Usually, when the mentioned set of changes is ready, you would then want to join those branches, which can be done with a merge operation.

Creating a new branch

Alice decides that it would be a good idea to provide a way for the user to configure the lower bound of the range the random number is chosen from (currently set to 1)—that is, make both the minimum and maximum of the generated number configurable.

She needs to add a new input to the `index.html` file. Alice notices that labels for inputs need to be adjusted, too:

```
<!DOCTYPE html>
<html lang="en">
<head>
    <meta charset="UTF-8">
    <title>Random Number Generator</title>
    <script src="scripts/gen_random.js" defer></script>
</head>
<body>
<button disabled>Generate number</button>
<label for="min">between</label>
<input type="number" id="min" name="rand_min" value="1" />
<label for="max">and</label>
<input type="number" id="max" name="rand_max" value="10" />
<div id="result">Result:</div>
<noscript>To use this web app, please enable JavaScript.</noscript>
</body>
</html>
```

Then, Alice needs to adjust the JavaScript code to read another input and to generate a random integer between two given values, inclusive:

```
function getRandomIntInclusive(min, max) {
    min = Math.ceil(min);
    max = Math.floor(max);
    return Math.floor(Math.random() * (max - min + 1) + min);
}

function generateRandom() {
    let min = document.getElementById('min').value;
    let max = document.getElementById('max').value;
    let res = document.getElementById('result');
    res.textContent = 'Result: ' + getRandomIntInclusive(min, max);
}

let button = document.querySelector('button');
button.addEventListener('click', generateRandom);
button.disabled = false;
```

Alice then checks that everything works correctly:

| Generate number | between | 50 | and | 100 |

Result: 100

Figure 1.6 – Upper and lower bounds are made configurable

However, during testing, she notices that the application does not ensure that the minimum is smaller than or equal to the maximum value, and does not behave correctly if the input's order is switched.

She decides to try to fix this issue. However, to make each commit small and self-contained, ensuring that the application works sanely in such cases (when, for example, the user provides 10 and 5 as the minimum and maximum, respectively) will be done as a separate change.

To isolate this line of development from other changes and prevent integrating the feature that is not fully ready, she decides to create her own branch named 'min-max' (see also *Chapter 8, Advanced Branching Techniques*), and switch to it:

```
alice@hostA:~/random$ git checkout -b min-max
Switched to a new branch 'min-max'
alice@hostA:~/random$ git branch
  master
* min-max
```

> **Tip**
>
> Instead of using the `git checkout -b min-max` or `git switch --create min-max` shortcut to create a new branch and switch to it in one command invocation, Alice could have first created a branch with `git branch min-max`, then switched to it with `git switch min-max`.

She commits her changes and pushes them, knowing that the push will succeed because she is working on her private branch:

```
alice@hostA:~/random$ git commit -a -m 'Make lower bound configurable'
[min-max 2361cfc] Make lower bound configurable
 2 files changed, 9 insertions(+), 4 deletions(-)
alice@hostA:~/random$ git push
fatal: The current branch min-max has no upstream branch.
To push the current branch and set the remote as upstream, use

    git push --set-upstream origin min-max
```

Alright! Git just wants Alice to set up a remote origin as the **upstream** for the newly created branch (it is using a simple push strategy); this will also push this branch explicitly:

```
alice@hostA:~/random$ git push --set-upstream origin min-max
To https://git.company.com/random
* [new branch]      min-max -> min-max
Branch 'min-max' set up to track remote branch 'min-max' from
'origin'.
```

> **Tip**
>
> If she wants to make her branch visible but private (so nobody but her can push to it), she needs to configure the server with **hooks** or use Git repository management software such as gitolite to manage it for her.

Merging a branch (no conflicts)

Meanwhile, over in the default branch, Bob decides to push his changes by adding the COPYRIGHT file:

```
bob@hostB random$ git push
To https://git.company.com/random
 ! [rejected]        master -> master (fetch first)
[…]
```

OK. Alice was busy working at making the minimum value of the range configurable to choose random integers from (and resolving a merge conflict), and she pushed her changes first:

```
bob@hostB:~/random$ git pull
From https://git.company.com/random
   a808ecf..919f0f7  master      -> origin/master
 * [new branch]      min-max     -> origin/min-max
```

Git then opens the editor with the commit message for the merge. Bob exits the editor to confirm the default description:

```
Merge made by the 'recursive' strategy.
 index.html | 2 +-
 1 file changed, 1 insertion(+), 1 deletion(-)
```

Well, Git has merged Alice's changes cleanly, but there is a new branch present. Let's take a look at what is in it, showing only those changes exclusive to the min-max branch (the double-dot syntax is described in *Chapter 4, Exploring Project History*):

```
bob@hostB:~/random$ git log HEAD..origin/min-max
commit 2361cfc062809d96b9a04d8032b9c433cae5c350 (origin/min-max)
Author: Alice Developer <alice@company.com>
```

```
Date:    Mon May 3 14:35:33 2021 +0200

    Make lower bound configurable
```

Interesting! Bob decides he wants that. So, he asks Git to merge stuff from Alice's branch (which is available in the respective remote tracking branch) into the default branch:

```
bob@hostB:~/random$ git merge origin/min-max
Merge made by the 'recursive' strategy.
  index.html              | 4 +++-
  scripts/gen_random.js | 9 ++++++---
  2 files changed, 9 insertions(+), 4 deletions(-)
```

Undoing an unpublished merge

Bob realizes that it should be up to Alice to decide when the feature is ready for inclusion (and hears that it is not ready yet). He decides to undo a merge. Because it is not published, it is as simple as **rewinding** to the previous state of the current branch:

```
bob@hostB:~/random$ git reset --hard @{1}
HEAD is now at 02ad67e Merge branch 'master' of https://git.company.
com/random
```

> **Important note**
>
> This example demonstrates the use of the **reflog** for undoing operations; another solution would be to go to a previous (pre-merge) commit following the first parent, with HEAD^ instead of @{1}.

Bob then pushes his changes.

Summary

This chapter walked us through the process of working on a simple example project by a small development team.

We have recalled how to start working with Git, either by creating a new repository or by cloning an existing one. We have seen how to prepare a commit by adding, editing, moving, and renaming files, how to revert changes to the file, how to examine the current status and view changes to be committed, and how to tag a new release.

We have recalled how to use Git to work at the same time on the same project, how to make our work public, and how to get changes from other developers. Though using a version control system helps with simultaneous work, sometimes Git needs user input to resolve conflicts in work done by different developers. We have seen how to resolve a merge conflict.

We have recalled how to create a tag marking a release, and how to create a branch starting an independent line of development. Git requires tags and new branches to be pushed explicitly, but it fetches them automatically. We have seen how to merge a branch.

The next chapter will cover creating new revisions and new lines of development in much more detail, and it will introduce and explain the concept of the staging area for commits.

Questions

Answer the following questions to test your knowledge of this chapter:

1. Describe how to create a repository from existing files and how to get your own copy of an existing repository.

2. Describe how to create a new version of the project locally, and how to publish those changes.

3. Explain how to get changes from other developers, and how to combine those changes.

4. What do merge conflict markers look like, and how can you resolve a merge conflict?

5. What can you do to make Git not show temporary backup files as unknown files in the status output? What about the products and byproducts of the build system?

6. Where can you find information about how to undo adding a file, or how to undo changes to a file?

7. How can you abandon the commit? What are the dangers of doing so?

8. Explain how Git manages moving, copying, and renaming files.

Answers

Here are the answers to the questions given above:

1. Use `git init`, `git add .`, and `git commit` to create a repository from existing files. Use `git clone` to get your own copy of the existing repository.

2. Use `git commit` or `git commit -a` to create a new revision, and use `git push` to publish changes.

3. Use `git fetch` to get updates from other developers, or `git pull` to get updates and merge them together. Use `git merge` (or, as mentioned in later chapters, `git rebase`) to combine changes.

4. Merge conflicts are presented using the <<<<<<<, =======, and >>>>>>> markers; you can also find the ||||||| marker used, depending on the configuration. To resolve the conflicts, you need to edit files marked as conflicting into shape, use `git add` on them when finished, and then finalize the merge with `git commit` or `git merge --continue` (or rebase with `git rebase --continue`).

5. To make Git ignore specific types of files, you need to add appropriate glob patterns to one of the `ignore` files. It is a good practice to ignore byproducts of the build system and other generated files using the `.gitignore` file and add patterns for temporary files specific to one's individual choice to a per-repository (`.git/info/ignore`) or per-user `ignore` file.

6. All information about how to undo adding, removing, or staging a file can be found in the `git status` output.

7. You can abandon a commit with `git reset --hard HEAD^`, but it may lead to losing your changes (you can recover committed changes with the help of reflog if it did not expire; uncommitted changes are lost forever).

8. Git handles code movement, such as renaming, moving, and copying files, by using rename detection during merging and `diff` generation.

Further reading

If you need a reminder about Git basics, the following references might help you.

- *Everyday Git With 20 Commands or So*, part of the Git documentation as `giteveryday(7)`: `https://git-scm.com/docs/giteveryday`

- *A tutorial introduction to Git*, part of the Git documentation as `gittutorial(7)`: `https://git-scm.com/docs/gittutorial`

- *The Git User's Manual*, part of the Git documentation: `https://git-scm.com/docs/user-manual`

- Eric Sink, *Version Control by Example*, Pyrenean Gold Press (2011): `https://ericsink.com/vcbe/index.html`

- Scott Chacon and Ben Straub, *Pro Git, 2nd Edition*, Apress (2014): `https://git-scm.com/book/en/v2`

2
Developing with Git

This chapter will describe how to create new revisions and new lines of development (new branches) with Git.

Here, we will focus on committing one's own work on the solo development. The description of working as one of the contributors is left for *Chapter 6, Collaborative Development with Git*, while *Chapter 9, Merging Changes Together*, will show how to join created lines of development and how Git can help in maintainer duties.

This chapter will introduce the very important Git concept of the **staging area** (also called the index), while more advanced techniques for manipulating it will be described in *Chapter 3, Managing Your Worktrees*. It will also explain, in detail, the idea of a **detached HEAD** — that is, an anonymous, unnamed branch. Here, you can also find how Git describes differences between two versions of the project, or changes to the project, including a detailed description of the so-called extended **unified diff format**.

The following is the list of the topics we will cover in this chapter:

- The index — a staging area for commits
- Examining the status of the working area, and changes in it
- How to read the extended unified diff that is used to describe changes
- Selective and interactive commit, and amending a commit
- Creating, listing, renaming, and switching to branches, and listing tags
- What can prevent switching branches, and what you can do then
- Rewinding a branch with `git reset`
- Detached HEAD — that is, the unnamed branch (for example, a result of checking out a tag)

Creating a new commit

Before starting to develop with Git, you should introduce yourself with a name and an email, as shown in *Chapter 1, Git Basics in Practice*. This information will be used to identify your work, either as an author or as a committer. The setup can be global for all your repositories (with `git config --global`, or by editing the `~/.gitconfig` file directly), or local to a repository (with `git config`, or by editing the `.git/config` file inside the given repository). The per-repository configuration overrides the per-user one (you will learn more about this in *Chapter 13, Customizing and Extending Git*).

> **Multiple identities**
>
> You might want to use your company email for *work* repositories, but your own, non-work email for public repositories you work on. This can be done by setting one identity globally (for the user) and using the local repository config for setting an alternate identity for exceptions. Another possible solution would be to use **conditional includes** with the `includeIf` section, using it to include appropriate configuration files with per-directory identities.

The relevant fragment of the appropriate config file could look like the following example:

```
[user]
    name = Joe R. Hacker
    email = joe@company.com
```

How a new commit extends a project's history

Contributing to the development of a project usually consists of creating new revisions of said project. To mark the current state of the project as a new version, you use the `git commit` command. Git will then ask for a description of changes (**commit message**), and then extend the project history with the newly created revision. Here's what is happening behind the curtain — it's useful to understand this to better use advanced Git techniques.

In Git, the history of the project is stored as a graph of revisions (versions), where each revision points to the previous version it was based on. The `git commit` command simply creates a new node in this graph (a **commit** node), extending it.

To know where each branch is, Git uses **branch HEAD** as a reference to the graph of revisions. The **HEAD** denotes which branch is the current branch — that is, on which branch to create new commits at a given point in time.

You can find out more about the concept of the **Directed Acyclic Graph** (**DAG**) of revisions in *Chapter 4, Exploring Project History*. Creating a new commit adds a new node to the graph of revisions, and adjusts the position of branch tips (heads), as shown on the following figure.

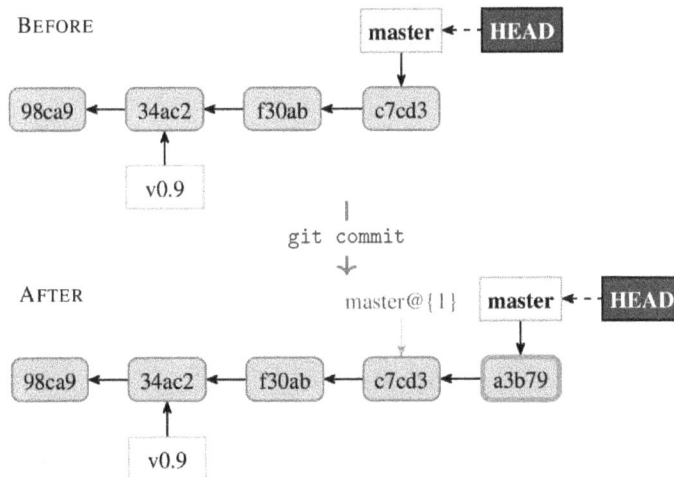

Figure 2.1 – The graph of revisions for an example project, before and after creating a new commit on the "master" branch

Let's assume that we are on the master branch and that we want to create a new version (the details of this operation will be described in more detail later). The git commit command will create a new commit object — a new revision node. This commit will have the checked-out revision (**c7cd3** in the example in *Figure 2.1*) as the previous node (as a **parent**).

That revision is found by following references starting from HEAD; here, it is a chain starting at HEAD, then following to master, and finally arriving at c7cd3.

Then, Git will create a new commit node, a3b79, and then move the master pointer to that new node. In *Figure 2.1*, the new commit is marked with a thick red outline. Note that the HEAD pointer doesn't change; all the time, it points to master. The performed commit operation is logged in the **reflog** for the master branch and for HEAD (current branch); one can examine this log with the git reflog master or git reflog HEAD command.

The index — a staging area for commits

Each of the files inside the working area of the Git repository can at a given point in time be either known or unknown to Git — that is, version-controlled or not. Any file known to Git is also known as a **tracked file**. The files unknown to Git can be either untracked or **ignored** (you can find more information about ignoring files in *Chapter 3, Managing Your Worktrees*). You can make an unknown file become tracked with the git add command.

Files tracked by Git are usually in either of the two states: committed (or unchanged) or modified. The **committed** state means that the file contents in the **working directory** are the same as in the last release (HEAD), which is safely stored in the repository. The file is **modified** if it has changed compared to the last committed version, which means it is different than in HEAD.

However, in Git, there are other states possible. Let's consider what happens when we use the `git add` command to add a file that was previously unknown to Git (an untracked file), but before creating a new commit that adds this file. A version control system needs to store somewhere the information that the given file is to be included in the next commit. Git uses something called the **index** for this; it is the **staging area** that stores information that will go into the next commit. The `git add <file>` command **stages** the current contents (current version) of the file, adding it to the index.

> **Important note**
>
> If you want to only *mark a file for the addition*, you can use `git add -N <file>` or `git add --intent-to-add <file>`; these commands simply stage the empty contents for a file (`<file>` here is a placeholder for the file's name).

The staging area stores the state of the project. It is the third such section, after a working directory (which contains your own copy of the project files and is used as a private isolated workspace to make changes) and a local repository (which stores your own copy of the project history and is used to synchronize changes with other developers). *Figure 2.2* shows how you can interact with these three sections, specifically in the context of creating a new commit:

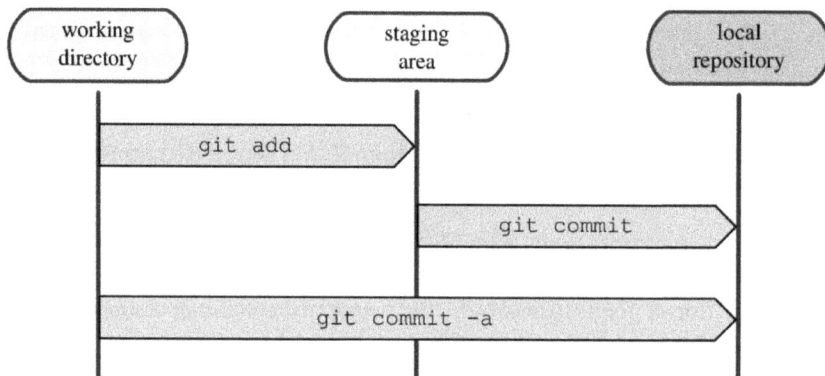

Figure 2.2 – The working directory, the staging area, and the local git repository, creating a new commit

The arrows on this diagram show how the Git commands copy contents. For example, `git add` takes the content of the file from the working directory and puts it into the staging area. Creating a new commit requires, explicitly or implicitly, the following steps:

1. You make changes to files in your working directory, usually modifying them using your favorite editor.

2. You stage the files, adding snapshots of them (their current contents) to your staging area, usually with the `git add` command.

3. You create a new revision with the `git commit` command, which takes the files as they are in the staging area and stores that snapshot permanently in your **local repository**.

In the beginning, and usually just after the commit (unless it was a selective commit), the tracked files are identical in the working directory, in the staging area, and in the last commit (in the committed version, that is HEAD).

> **Examining the staged file contents and the committed file contents**
>
> To examine the working directory state of a file, you can simply open it in an editor, or (on Linux or in Git Bash) simply use `cat <filename>`; examining other stages is more involved. To see the state in the staging area, you can use the `git show :<filename>` command. To see the committed version, use the `git show <revision>:<filename>` command (where `<revision>` may be HEAD). Here, a filename starting with `./` or `../` denotes that the path is relative to the current directory; otherwise, it is taken to be relative to the top-level directory of the repository you are in.

Often, however, one would use a special shortcut, the `git commits -a` command (spelled as `git commit --all` in the long form), which would take *all the changed tracked files*, add them to the staging area, and create a new commit (see *Figure 2.2*). This command gives the same result as running `git add --update`, followed by a `git commit` command. Note that the new files still need to be explicitly added using `git add` to be tracked and included in new commits.

Examining the changes to be committed

Before committing to the changes and creating a new revision (a new commit), you would want to see what you have done.

Git adds the information about the changes to be committed to the commit message template, which is then passed to the editor, and you will see this when writing the commit message. This is, of course, unless you specify the commit message on the command line — for example, with `git commit -m "Short description"`. The commit message template in Git is configurable (refer to *Chapter 13, Customizing and Extending Git,* for more information).

> **Important note**
>
> You can always abort creating a commit by exiting the editor without any changes, or with an empty commit message (comment lines — that is, lines beginning with # — do not count). If you want to create a commit with an empty commit message, you need to use the `--allow-empty-message` option.

In most cases, you would want to examine pending changes for correctness before creating a commit.

The status of the working directory

The main tool you use to examine which files are in which state — that is, which files have changed, whether there are any new files, and so on — is the `git status` command.

The default output is explanatory and quite verbose. If there are no changes, for example, directly after cloning, you could see something like this:

```
$ git status
On branch master
nothing to commit, working tree clean
```

If the current branch (which, in this example, is the master branch) is a local branch intended to create changes that are to be published and to appear in the public repository, and is configured to track its upstream branch, origin/master, you would also see the information about the tracked branch:

```
Your branch is up to date with 'origin/master'.
```

In further examples in this chapter, we will ignore it and not include the information about branches and tracking branches.

Let's say you have added two new files to your project: a COPYING file with the copyright and license, and a NEWS file, which is currently empty. In order to begin tracking a new COPYING file, you used git add COPYING. Accidentally, you removed the README file from the working directory with rm README. You also modified the Makefile and renamed rand.c to random.c with git mv without modifying the file.

The default long format is designed to be human-readable, verbose, and descriptive:

```
$ git status
On branch master
Changes to be committed:
  (use "git restore --staged <file>..." to unstage)
        new file:   COPYING
        renamed:    src/rand.c -> src/random.c

Changes not staged for commit:
  (use "git add <file>..." to update what will be committed)
  (use "git restore <file>..." to discard changes in working
directory)
        modified:   Makefile
        deleted:    README

Untracked files:
  (use "git add <file>..." to include in what will be committed)
        NEWS
```

Older versions of Git will suggest using different commands than git restore.

As you can see, Git not only describes which files have changed but also explains how to change their status — either include it in the commit or remove it from the set of pending changes (more information about commands shown in the git status output can be found in *Chapter 3*, *Managing Your Worktrees*). There are up to three sections present in the output:

- **Changes to be committed**: This section is about the staged changes that would be committed with git commit (without the -a/--all option). It lists files whose snapshot in the staging area is different from the version from the last commit (HEAD).

- **Changes not staged for commit**: This section lists the files whose working directory contents are different from their snapshots in the staging area. Those changes would not be committed with git commit, but would be committed with git commit -a as changes in the tracked files.

- **Untracked files**: This lists the files, unknown to Git, that are not ignored (refer to *Chapter 3*, *Managing Your Worktrees* for how to use **gitignores** to make files be ignored). These files would be added with the bulk add command, git add ., if run in the top directory of the project. You can skip this section with the --untracked-files=no (-uno for short) option.

If the section does not contain any files, it will be not shown. Note also that the file may appear in more than one section. For example, a new file that got added with git add and then modified would appear in both **Changes to be committed** and **Changes not staged for commit**.

One does not need to make use of the flexibility that the explicit staging area gives; one can simply use git add just to add new files and git commit -a to create the commit from changes to all tracked files. In this case, you would create a commit from both the **Changes to be committed** and **Changes not staged for commit** sections.

There is also a terse --short output format for git status. Its --porcelain version is suitable for scripting because it is promised to remain stable, while --short is intended for user output, uses color if possible, and could change in the future. For the same set of changes, this output format would look something like the following:

```
$ git status --short
A  COPYING
 M Makefile
 D README
R  src/rand.c -> src/random.c
?? NEWS
```

In this format, the status of each path is shown using a *two-letter status code*. The first letter shows the status of the index (the difference between the staging area and the last commit), and the second letter shows the status of the worktree (the difference between the working area and the staging area):

Symbol	Meaning
(a space)	Not updated/unchanged
M	Modified (updated)
A	Added
D	Deleted
R	Renamed
C	Copied

Table 2.1 – Letter status codes used in the short format of the git-status command

Not all combinations are possible. Status letters *A*, *R*, and *C* are possible only in the first column, for the status of the index.

A special case, ??, is used for the unknown (untracked) files, and !! for ignored files (when using `git status --short --ignored`).

> **Note about status codes**
>
> All the possible outputs are described here; the case where we have just done a merge that resulted in merge conflicts is not shown in *Table 2.1* but is left to be described in *Chapter 9, Merging Changes Together.*

Examining differences from the last revision

If you want to know not only which files were changed (which you get with `git status`), but also what exactly you have changed, you can use the `git diff` command:

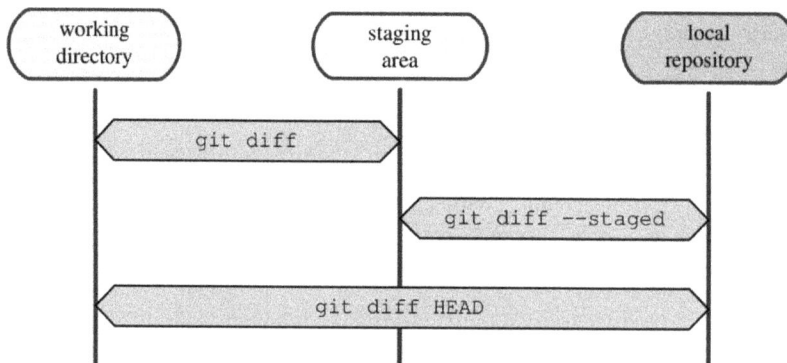

Figure 2.3 – Examining the differences between the working directory, the staging area, and the local Git repository

In the last section, we learned that in Git, there are three stages: the working directory, the staging area, and the repository (usually the last commit). Therefore, we have not one set of differences, but three, as shown in *Figure 2.3*. You can ask Git the following questions:

- What have you changed but not yet staged — that is, what are the differences between the staging area and the working directory?

- What have you staged that you are about to commit — that is, what are the differences between the last commit (HEAD) and the staging area?

- What changes have you made to the files in the working directory since the last commit (HEAD)?

To see what you've changed but not yet staged, type `git diff` with no other arguments. This command compares what is in your working directory to what is in your staging area. These are the changes that could be added but wouldn't be present if we created a commit with `git commit` (without `-a`); those changes are listed in the **Changes not staged for commit** section in the `git status` output.

To see what you've staged that will go into your next commit, use `git diff --staged` (or `git diff --cached`). This command compares what is in your staging area to the content of your last commit. These are the changes that *would* be added with `git commit` (without `-a`) — **Changes to be committed** in the `git status` output. You can compare your staging area to any commit with `git diff --staged <commit>`; HEAD (the last commit) is just the default.

You can use `git diff HEAD` to compare what is in your working directory to the last commit (or arbitrary commit with `git diff <commit>`). These are the changes that would be added with the `git commit -a` shortcut.

If you are using `git commit -a` and not making use of the staging area, usually, it is enough to use `git diff` to examine the changes that will land in the next commit. The only issue is the new files that are added with bare `git add`; they won't show in the `git diff` output unless you use `git add --intent-to-add` (or its equivalent `git add -N`) to add new files.

Unified Git diff format

Git, by default, and in most cases, will show the changes in the **unified diff output format**. Understanding this output is very important, not only when examining changes to be committed, but also when reviewing and examining changes. This may happen, for example, during the code review, or when finding bugs after `git bisect` has found the suspected commit.

> **Different ways of examining differences**
>
> You can request only *statistics of changes* with the `--stat` or `--dirstat` option, just *names of the changed files* with `--name-only`, filenames with the type of changes with `--name-status`, a tree-level view of changes with `--raw`, or a *condensed summary* of extended header information with `--summary` (see later for an explanation of what "extended header" means and what information it contains). You can also request **word diff** rather than line diff with `--word-diff`, though this changes only the description of changes; headers and hunk headers remain similar.
>
> Diff generation can also be configured for specific files or types of files with appropriate **gitattributes**. You can specify an external *diff helper* — that is, the command that describes the changes — or you can specify a *text conversion* filter for binary files (you will learn more about this in *Chapter 3, Managing Your Worktrees*).
>
> If you prefer to examine the changes in a graphical tool (which usually provides *side-by-side diff*), you can do it by using `git difftool` instead of `git diff`. You can specify the tool with the `--tool=<tool>` option or with the `diff.tool` configuration variable. If the tool you use is not supported by Git or is not in PATH, this may require some configuration. Using external tools with Git like this will be explained in more detail in *Chapter 13, Customizing and Extending Git*.

Let's look at an example of an advanced diff from Git project history, using the diff from the `1088261f` commit from the `git.git` repository. You can view these changes in a web browser — for example, on GitHub; this is the third patch in this commit, `https://github.com/git/git/commit/1088261f6fc90324014b5306cca4171987da85`:

```
diff --git a/builtin-http-fetch.c b/http-fetch.c
similarity index 95%
rename from builtin-http-fetch.c
rename to http-fetch.c
index f3e63d7206..e8f44babd9 100644
--- a/builtin-http-fetch.c
+++ b/http-fetch.c
@@ -1,8 +1,9 @@
 #include "cache.h"
 #include "walker.h"

-int cmd_http_fetch(int argc, const char **argv, const char *prefix)
+int main(int argc, const char **argv)
 {
+        const char *prefix;
         struct walker *walker;
         int commits_on_stdin = 0;
         int commits;
@@ -18,6 +19,8 @@ int cmd_http_fetch(int argc, const char **argv,
```

```
         int get_verbosely = 0;
         int get_recover = 0;

+        prefix = setup_git_directory();
+
         git_config(git_default_config, NULL);

         while (arg < argc && argv[arg][0] == '-') {
```

Let's analyze this patch line by line. The first line, diff --git a/builtin-http-fetch.c b/http-fetch.c, is a **git diff header** in the form diff --git a/file1 b/file2. The a/ and b/ filenames are the same unless rename or copy operation is involved (such as in our case), even if the file is added or deleted. The --git option means that the diff is in the git diff output format.

The next lines are one or more **extended header lines**. The first three lines in this example tell us that the file was renamed from builtin-http-fetch.c to http-fetch.c and that these two files are 95% identical (information that was used to detect this rename):

```
similarity index 95%
rename from builtin-http-fetch.c
rename to http-fetch.c
```

> **Important note**
>
> Extended header lines describe information that cannot be represented in an ordinary unified diff, except for information that the file was renamed. Besides a similarity or dissimilarity score, like in this example, those lines can also describe the changes in file type (such as from a non-executable to an executable file).

The last line in the extended diff header, in this example, is as follows:

```
index f3e63d7206..e8f44babd9 100644
```

The preceding code tells us about the mode (permissions) of a given file. Here, 100644 means that it is an ordinary file and not a symbolic link, and that it hasn't set the executable permission bit (these three are the only file permissions tracked by Git). This line also tells us about the shortened hash of the **pre-image** (the version of the file before the given change, f3e63d7206 here) and **post-image** (the version of the file after the change, e8f44babd9 here). This line is used by git am --3way to try to do a three-way merge if the patch cannot be applied itself. For the new files, the pre-image hash is 0000000000; it's the same for the deleted files with the post-image hash.

Next is the **unified diff header**, which consists of two lines:

```
--- a/builtin-http-fetch.c
+++ b/http-fetch.c
```

Compared to the `diff -U` result, it doesn't have a from-file modification time or a to-file modification time. Those should be present after the space just after the source (pre-image) and the destination or the target (post-image) filenames. If the file was created, the source would be `/dev/null`; if the file was deleted, the target would be `/dev/null`.

> **Tip**
>
> If you set the `diff.mnemonicPrefix` configuration variable to `true` in place of the `a/` prefix for the pre-image and `b/` for the post-image in this two-line header, you would instead have the `c/` prefix for commit, `i/` for index, `w/` for worktree, and `o/` for object, respectively, to show what you're comparing This makes it easy to distinguish sides in `git diff`, `git diff --cached`, `git diff HEAD` output, and so on.

Next comes one or more **change hunks**, or hunks of differences; each hunk shows one area where the files differ. Unified format hunks start with the line describing where the changes were in the file, called the **hunk header**, as follows:

```
@@ -1,8 +1,9 @@
```

This line matches the following format pattern: `@@ from-file-range to-file-range @@`. The from-file range is in the form `-<start line>,<number of lines>`, and the *to-file range* is `+<start line>,<number of lines>`. Both `start line` and `number of lines` refer to the position and length of the hunk in the pre-image and post-image, respectively. If `number of lines` is not shown, it means that it is `0`.

In this example, we can see that the changes begin at the first line of the file, both in the pre-image (file before the changes) and post-image (file after the changes). We also see that the fragment of code corresponding to this hunk of diff has eight lines in the pre-image and nine lines in the post-image. This difference in the number of lines means that one line is added. By default, Git will also show three unchanged lines surrounding changes (three so-called **context lines**).

Git will also show the "*function name*" where each change occurs (or equivalent, if any, for other types of files; this can be configured with `.gitattributes` via diff driver —see *Chapter 3*, *Managing Your Worktrees*, in the *Configuring diff output* section in *File attributes*); it is like the `-p` option in GNU diff:

```
@@ -18,6 +19,8 @@ int cmd_http_fetch(int argc, const char
```

Git includes many builds in patterns for extracting the "*function*" heading for the hunk for various programming languages.

Next is the description of where and how files differ. The lines common to both files are prefixed with a space " (" ") " indicator character. The lines that differ between the two files have one of the following indicator characters in the left print column:

- +: A line was added here to the second file

- -: A line was removed here from the first file

> **Note**
>
> In the **plain word-diff** format, instead of comparing file contents line by line, added words are surrounded by { + and + } and removed words by [- and -] , as in the following example:

```
int [-cmd_http_fetch-]{+main+}(int argc, const char **argv[-, const
char *prefix-])
```

If the last hunk includes, among its lines, the very last line of either version of the file, and that last line is **incomplete line** (which means that the file does not end with the end-of-line character at the end of the hunk), you will find the following:

```
\ No newline at end of file
```

This situation is not present in the example used.

So, for the example used here, the first hunk means that cmd_http_fetch was replaced by main and the const char *prefix; line was added:

```
#include "cache.h"
 #include "walker.h"

-int cmd_http_fetch(int argc, const char **argv, const char *prefix)
+int main(int argc, const char **argv)
 {
+       const char *prefix;
        struct walker *walker;
        int commits_on_stdin = 0;
        int commits;
```

See how for the *replaced line*, the old version of the line appears as removed (-) and the new version as added (+).

In other words, before the change, the appropriate fragment of the file, which was then named builtin-http-fetch.c, looked similar to the following code fragment:

```
#include "cache.h"
#include "walker.h"

int cmd_http_fetch(int argc, const char **argv, const char *prefix)
{
        struct walker *walker;
        int commits_on_stdin = 0;
        int commits;
```

After the change, this fragment of the file, which is now named `http-fetch.c`, looks similar to the following instead:

```
#include "cache.h"
#include "walker.h"

int main(int argc, const char **argv)
{
        const char *prefix;
        struct walker *walker;
        int commits_on_stdin = 0;
        int commits;
```

Selective commit

Sometimes, after examining the pending changes as explained, you realize that you have two (or more) unrelated changes in your working directory that should belong to two different logical changes; such a problem is sometimes called the **tangled working copy problem**. You need to put those unrelated changes into separate commits as separate changesets. This is the type of situation that can occur even when trying to follow best practices (see *Chapter 15, Git Best Practices*).

One solution is to create the commit as-is and fix it later (split it in two). You can read about how to do this in *Chapter 10, Keeping History Clean*.

Sometimes, however, some of the changes are needed now and must be shipped immediately (for example, a bugfix to a live website), while the rest of the changes are not ready yet (they are a work in progress). You need to tease those changes apart into two separate commits.

Selecting files to commit

The simplest situation is when these unrelated changes touch different files. For example, if the bug was in the `view/entry.tmpl` file and the bugfix changes only this file (and there were no other changes to this file, unrelated to fixing the bug), you can create a bugfix commit with the following command:

```
$ git commit view/entry.tmpl
```

This command will ignore changes staged in the index (what was in the staging area), and instead record the current contents of a given file or files (what is in the working directory).

Interactively selecting changes

Sometimes, however, the changes cannot be separated in this simple way; the changes to the file are tangled together. You can try to tease them apart by giving the --interactive option to the git commit command:

```
$ git commit --interactive
           staged     unstaged path
   1:    unchanged      +3/-2 Makefile
```

```
  2:    unchanged        +64/-1 src/rand.c

*** Commands ***
  1: status         2: update        3: revert        4: add untracked
  5: patch          6: diff          7: quit          8: help
What now>
```

Here, Git shows us the status and the summary of changes to the working area (unstaged) and the staging area (staged), which is also the output of the status subcommand. The changes are described by the number of added and deleted lines — for example, +3/-2 here (this is similar to what the git diff --numstat command would show).

> **Tip**
>
> It might be easier to use a graphical tool such as git gui with this capability. In GUIs, such as the one mentioned, one can use the mouse to select which lines of changes to include or exclude.

You can use the help subcommand, accessed by pressing *h*, to check what those listed operations mean:

```
What now> h
status         - show paths with changes
update         - add working tree state to the staged set of changes
revert         - revert staged set of changes back to the HEAD version
patch          - pick hunks and update selectively
diff           - view diff between HEAD and index
add untracked - add contents of untracked files to the staged set of
changes
```

To tease apart changes, you need to choose the patch subcommand (for example, with *5* or *p*). Git will then ask for the files with the Update>> prompt; you then need to select the files to selectively update with their numeric identifiers, as shown in the status, and type return. You can type * to select all the files possible. After making the selection, end it by answering with an empty line:

```
What now> p
              staged       unstaged path
    1:     unchanged        +3/-2 Makefile
    2:     unchanged       +64/-1 src/rand.c
Patch update>> 1
              staged       unstaged path
*   1:     unchanged        +3/-2 Makefile
    2:     unchanged       +64/-1 src/rand.c
Patch update>>
```

You can skip directly to patching files by using git commit --patch instead of git commit --interactive.

Git will then display all the changes to the specified files on a hunk-by-hunk basis, and let you choose, among others, one of the following options for each hunk:

```
y - stage this hunk
n - do not stage this hunk
q - quit; do not stage this hunk or any of the remaining ones
a - stage this hunk and all later hunks in the file
...
s - split the current hunk into smaller hunks
e - manually edit the current hunk
? - print help
```

The hunk output and the prompt look similar to the following:

```
@@ -16,7 +15,6 @@ int main(int argc, char *argv[])

        int max = atoi(argv[1]);

+       srand(time(NULL));
        int result = random_int(max);
        printf("%d\n", result);

Stage this hunk [y,n,q,a,d,/,j,J,g,e,?]? y
```

In many cases, it is enough to simply select which of those hunks of changes you want to have in the commit. In extreme cases, you can split a chunk into smaller pieces, or even manually edit the difference.

Many graphical tools, including git gui, also allow for the interactive selection of changes going to the next commit. You can find out more in *Chapter 13, Customizing and Extending Git*, in the *Graphical interfaces* section.

Creating a commit step by step

Using git commit --interactive to select changes to a commit doesn't, unfortunately, allow you to test the changes to be committed. You can always check that everything works after creating a commit (that is, compile the code and/or run tests), and then amend it if there are any errors (see the next section, *Amending a commit*). There is, however, an alternative solution.

Instead of using the interactive commit feature, you can prepare to commit by putting the pending changes into the staging area with git add --interactive or an equivalent solution (such as a graphical commit tool for Git — for example, git gui). The interactive commit is just a shortcut for **interactive add** followed by commit. Then, you should examine these changes with git diff --cached, modifying them as appropriate with git add <file>, git checkout <file>, and git reset <file>.

In theory, you should also test whether these changes are correct, checking that at least they do not break the build. To do this, first use `git stash save --keep-index` to save the current state and bring the working directory to the state prepared in the staging area (the index). After this command, you can run tests (or at least check whether the program compiles and doesn't crash). If tests pass, you can then run `git commit` to create a new revision. If tests fail, you should restore the working directory while keeping the staging area state with the `git stash pop --index` command; it might be required to precede it with `git reset --hard`. The latter might be needed because Git is overly conservative when preserving your work and does not know that you have just stashed. First, there are uncommitted changes in the index that prevent Git from applying the stash, and second, the changes to the working directory are the same as those stashed, so, of course, they would conflict.

You can find more information about **stashes**, including how they work, in *Chapter 3, Managing Your Worktrees*, in the *Stashing away your changes* section.

Amending a commit

One of the better things about Git is that you can undo almost anything; you only need to know how to. This is because no matter how carefully you craft your commits, sooner or later, you'll forget to add a change or mistype the commit message. That's when the `--amend` flag of the `git commit` command comes in handy; it allows you to change the very last commit really easily. Note that with `git commit --amend`, you can also amend the merge commits (for example, to fix a merging error). *Figure 2.4* shows how this amend operation changes the graph of revisions which represents the history of the project.

> **Tip**
>
> If you want to change a commit deeper in the history (assuming that it was not published, or, at least, there isn't anyone who based their work on the old version of the said commit), you need to use **interactive rebase**, or some specialized tool such as **StGit** (a **patch stack management interface** on top of Git). Refer to *Chapter 10, Keeping History Clean*, for more information.

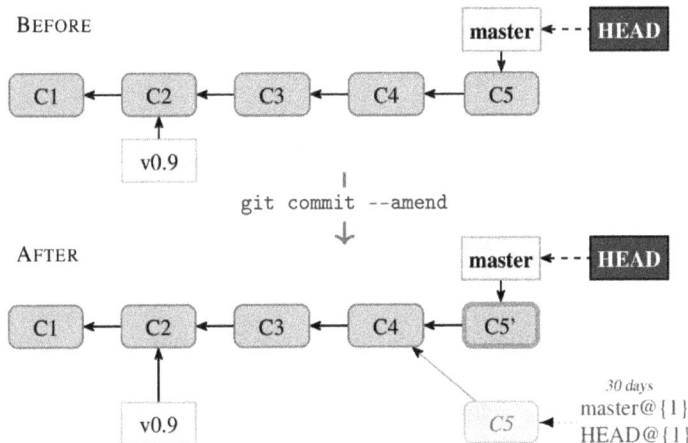

Figure 2.4 – The graph of revisions before and after amending the last commit

If you just want to correct the commit message, assuming you don't have any staged changes, you simply run `git commit --amend` and fix it (note that we use `git commit` without the `-a` / `--all` flag):

```
$ git commit --amend
```

If you want to add some more changes to that last commit, you can simply stage them as normal with `git add` and then *amend* the last commit, as shown in the preceding example, or make the changes and then use `git commit -a --amend`:

> **Important note**
>
> There is a very important caveat: you should *never* amend a commit that has already been published! This is because amending effectively produces a completely new commit object that replaces the old one, as can be seen in *Figure 2.4*. In this figure, you can see that the most recent commit before the operation, denoted by **C5**, is replaced in the project history by the commit **C5'**, with amended changes.
>
> If you're the only person who had this commit, doing this is safe. However, after publishing the original commit to a remote repository, other people might already have based their new work on that version of the commit. In this case, replacing the original with an amended version will cause problems downstream. You will find out more about this issue in *Chapter 10, Keeping History Clean.*
>
> That is why, if you try to push (publish) a branch with the published commit amended, Git prevents overwriting the published history and asks you to **force push** if you really want to replace the old version (unless you configure it to force push by default). More about that in *Chapter 6, Collaborative Development with Git.*

The old version of the commit before amending would be available in the branch reflog and in the HEAD reflog; for example, just after amending, the amended version would be available as `@{1}`. This means that you can undo the amend operation with, for example, `git reset --keep HEAD@{1}`, as described in the *Rewinding or resetting a branch* section. Git would keep the old version for a month (30 days) by default if not configured otherwise, unless the reflog is manually purged.

You can always check the log of operations in the reflog by using the `git reflog` command. Just after amending a commit, that command output would look like the following:

```
$ git reflog --no-decorate
94d3e03 HEAD@{0}: commit (amend): After amending
d69a0a9 HEAD@{1}: commit: Before amending
```

Here, `HEAD@{1}` means the position of the current branch 1 operation back. Besides the HEAD reflog, there is also a reflog for each branch, as described later. Note that you can read more about using reflog to refer to commits in *Chapter 4, Exploring Project History.*

Working with branches and tags

In version control, **branches** are separate lines of development, a way of separating different ideas and different parts of changes. You can use branches in different ways, which are described in *Chapter 8, Advanced Branching Techniques*.

Tags are a way to give a meaningful name to mark a specific version of a project. They are used to make it possible to return to a given point in history — for example, with the `v1.0` tag, you will be able to go to exactly version 1.0 of the code. Additionally, with **annotated tags**, you can give a longer description of the tagged revision, and **signed tags** also help ensure that it was you who created it.

In Git, each branch is realized as a named "*pointer*" (reference) to some commit in the graph of revisions, the so-called branch head. The same is true for **lightweight tags**; for annotated and signed tags, the "*pointer*" refers to the tag object (with annotation or signature), which points to a commit.

> **Representation of branches in Git**
>
> Git currently uses two different on-disk representations of branches: the "*loose*" format (which takes precedence) and the "*packed*" format.
>
> Take, for example, the `master` branch (which is currently the default name of the branch in Git; you start on this branch in a newly created repository, unless configured otherwise). In the "*loose*" format (which takes precedence), the branch is represented as the one-line `.git/refs/heads/master` file with a textual hexadecimal representation of SHA-1 at the tip of the branch. In the "*packed*" format, a branch is represented as a line in the `.git/packed-refs` file, connecting the SHA-1 identifier of the top commit with the fully qualified branch name.

The (named) **line of development** is then a set of all the revisions that are reachable from the branch head. It is not necessarily a straight line of revisions — it can fork and join.

Creating a new branch

When creating a new branch, you can just create it and switch to it later, or you can create it and switch to it with a single command. This is explained in *Figure 2.5*.

You can create a new branch with the `git branch` command. For example, to create a new `testing` branch starting from the current branch (see the top-right part of *Figure 2.5*), run the following:

```
$ git branch testing
```

What happens here? Well, this command creates a new pointer (a new reference) for you to move around. You can give an optional parameter to this command if you want to create the new branch pointing to some other commit, like in the following example:

```
$ git branch testing HEAD^^^
```

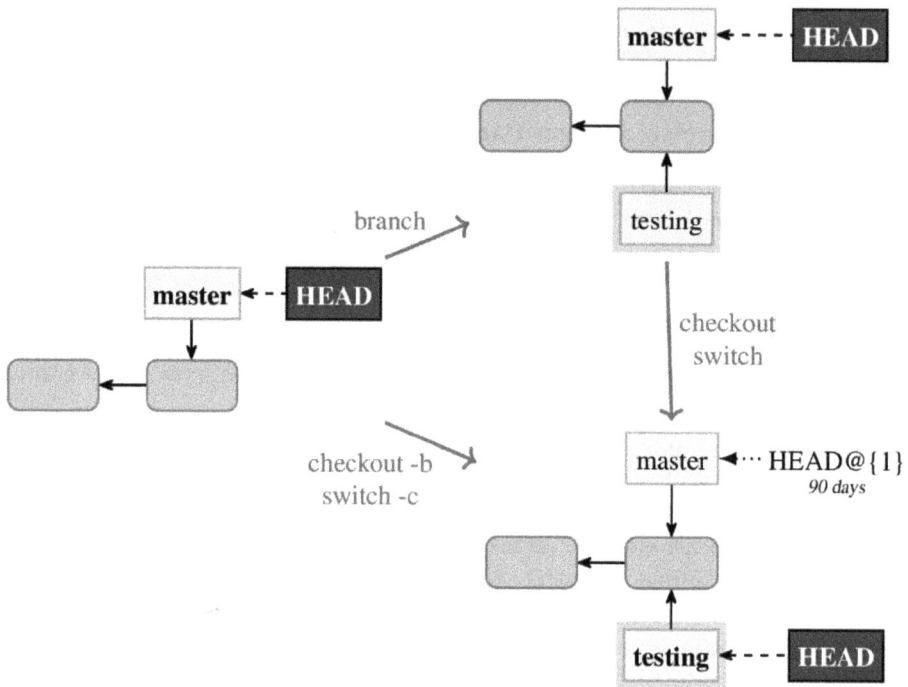

Figure 2.5 – Creating a new branch named "testing" and switching to this branch,
or creating a new branch and switching to it at once, with one command

> **Note**
>
> The HEAD^^^ notation will be explained in *Chapter 4, Exploring Project History*.

However, the git branch <new branch> command would not change which branch is the current branch; it does not switch to the just-created branch. It would not change the position of the HEAD (the symbolic reference pointing to the current branch) and would not change the contents of the working directory.

If you want to create a new branch and switch to it (to start working on a new branch immediately), you can use the following shortcut:

```
$ git switch -c testing
```

Here, the short -c option stands for --create. You can also use the following alternative command, which is the only option for older Git:

```
$ git checkout -b testing
```

If you want to forcibly create a branch with a name that already exists, effectively deleting the existing branch, you will need to use the -C and -B options instead of the -c and -b options, respectively.

If we create a new branch at the current state of the repository, the switch -c and checkout -b commands differ only in that they also move the HEAD pointer; see the transition from the left-hand side to the bottom right in *Figure 2.5*.

Creating orphan branches

Rarely, you might want to create a new unconnected **orphan branch** in your repository that doesn't share any history with other branches. Perhaps you want to store the generated documentation for each release to make it easy for users to get readable documentation (for example, as man pages or HTML help) without requiring the installation of conversion tools or renderers (for example, AsciiDoc or a Markdown parser). Or, you might want to store web pages for a project in the same repository as the project itself; that is what GitHub project pages can use. Perhaps you want to open source your code, but you need to clean up the code first (for example, because of copyrights and licensing).

One solution is to create a separate repository for the contents of an orphan branch and fetch from it into some remote-tracking branch. You can then create a local branch based on it. This has the advantage of having unconnected contents separately, but on the other hand, it is one more repository to manage.

You can also do this with either the git switch or git checkout command by using the --orphan option:

```
$ git switch --orphan gh-pages
Switched to a new branch 'gh-pages'
```

This reproduces the state similar to just after git init: the HEAD symref points to the gh-pages branch, which does not exist yet; it will be created on the first commit.

If you want to start with a clean state, such as with GitHub Pages, you will also need to remove the contents of the start point of the branch (which defaults to HEAD — that is, to the current branch and the current state of the working directory) — for example, with the following:

```
$ git rm -rf .
```

In the case of open sourcing code with proprietary parts to be excluded, where the orphan branch is used to make sure not to bring the proprietary code accidentally to the open source version on merging, you would want to carefully edit the working directory instead.

Selecting and switching to a branch

To switch to an existing local branch, you need to run the `git switch` command. For example, after creating the `testing` branch, you can switch to it with the following command:

```
$ git switch testing
```

This is shown in *Figure 2.6* as the vertical transition from the top-right to the bottom-right state; this figure also shows that you can use `git checkout` to switch branches.

Obstacles to switching to a branch

When switching to a branch, Git also checks out its contents in the working directory. What happens then if you have uncommitted changes (that are not considered by Git to be on any branch)?

> **Tip**
>
> It is a good practice to switch branches in a clean state, stashing away changes or creating a commit if necessary. Checking out a branch with uncommitted changes is useful only in a few rare cases, some of which are described in the following section.

If the difference between the current branch and the branch you want to switch to does not touch the changed files, the uncommitted changes are moved to the new branch. This is very useful if you started working on something and only later realized that it would be better to do this work in a separate feature branch.

If uncommitted changes conflict with changes on the given branch, Git will refuse to switch to the said branch to prevent you from losing your work:

```
$ git checkout other-branch
error: Your local changes to the following files would be overwritten
by checkout:
        file-with-local-changes
Please commit your changes or stash them before you switch branches.
Aborting
```

In such a situation, you have a few possible different solutions:

- You can *stash away* your changes with the `git stash` command and restore them when you come back to the branch you were on. This is usually the preferred solution.

  Alternatively, you can simply create a temporary commit of the work in progress with those changes, and then either amend the commit or rewind the branch when you get back to it.

- You can try to move your changes to the new branch by *merging*, either with `git switch --merge` (which would do the three-way merge between the current branch, the contents of your working directory with unsaved changes, and the new branch), or by stashing away your changes before checkout and then unstashing them after a switch.

- You can also *throw away* your changes with `git switch --discard-changes` or `git checkout --force`.

Anonymous branches

What happens if you try to check out (switch to) something that is not a local branch — for example, an arbitrary revision (such as `HEAD^`), a tag (such as `v0.9`), or a remote-tracking branch (for example, `origin/master`)? Git assumes that you need to be able to create commits on top of the current state of the working directory.

Older Git refused to switch to a non-branch. Nowadays, Git will create an **anonymous branch** by detaching the **HEAD** pointer and making it refer directly to a commit (that's why it is also called a detached HEAD state) rather than being a symbolic reference to a branch; see *Figure 2.6* for an example.

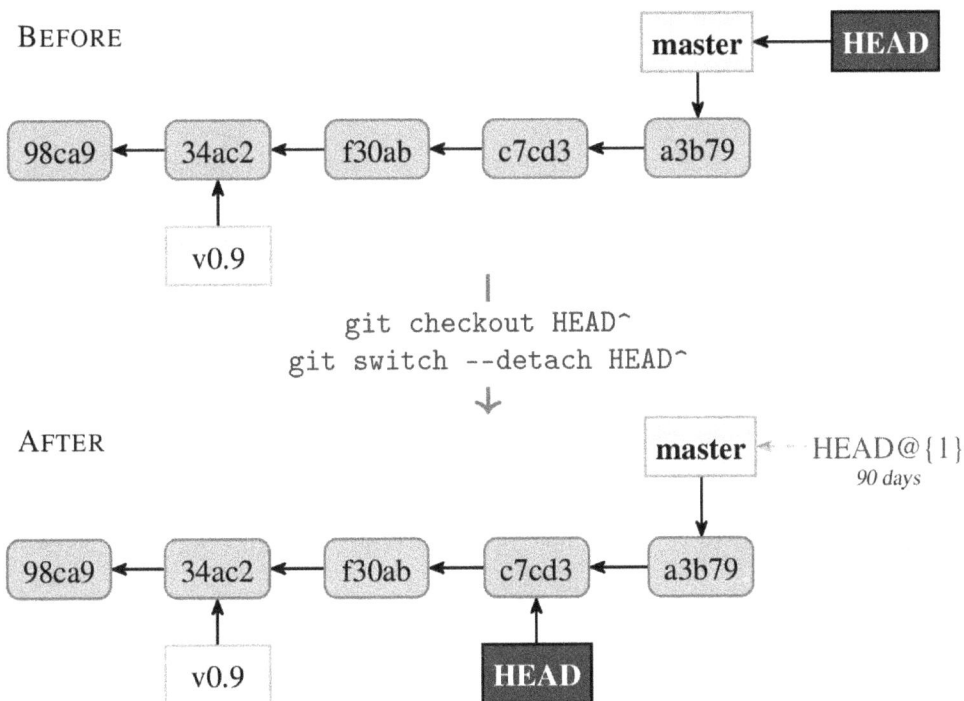

BEFORE

master ← HEAD

98ca9 ← 34ac2 ← f30ab ← c7cd3 ← a3b79

v0.9

```
git checkout HEAD^
git switch --detach HEAD^
```

AFTER

master ← HEAD@{1}
90 days

98ca9 ← 34ac2 ← f30ab ← c7cd3 ← a3b79

v0.9   HEAD

Figure 2.6 – The result of checking out a non-branch is a detached HEAD
state (which is like being on the anonymous branch)

Because this feature is used only rarely, to avoid landing in such a state explicitly, the `git switch` command requires the use of the `--detach` option; for backward compatibility, `git checkout` does not require using this option to detach the **HEAD** pointer. This option is also useful to explicitly create an anonymous branch at the current position. The detached HEAD state is shown in the branch listing as (**no branch**) in older versions of Git, or (**detached from HEAD**) or (**HEAD detached at**...) in newer versions.

If you did detach HEAD by mistake, you can always *go back to the previous branch* with the following command (where - means the name of the previous branch):

```
$ git switch -
Previous HEAD position was a3b19 <Some commit message>
Switched to branch 'master'
```

> **Important note**
>
> The git switch - command uses the HEAD reflog to switch to a previous branch. This may not work correctly after renaming the branch.

As Git informs you, when creating a detached branch without using the --detach option, you can always give a name to the anonymous branch with git switch -c <new-branch-name>.

Because tags are meant to be immutable, trying to check one out (or switch to it) also creates a detached HEAD — tags are not branches.

The switch command DWIM-mery

There is a special case of checking out something that is not a branch. If you check out a remote-tracking branch (for example, origin/next) by its short name (in this case, next) as if it were a local branch, Git would assume that you meant to create new contents on top of the remote-tracking branch state and will do what it thinks you need. This **do what I mean** (**DWIM**) feature will create a new local branch, tracking the remote-tracking branch. This behavior can be turned off with the --no-guess option, or the accompanying checkout.guess configuration variable.

This means that:

```
$ git switch next
```

is equivalent to:

```
$ git switch -c next --track origin/next
```

Git will do it only if there are no ambiguities; the local branch must not exist (otherwise the command would simply switch to the local branch given) and there can be only one remote-tracking branch that matches. The latter condition can be checked by running git show-ref next (using the short name) and verifying that it returns only one line, with remote-tracking branch information:

```
$ git show-ref --abbrev next
4936735 refs/remotes/origin/next
```

Listing branches and tags

If you use the `git branch` command without any other arguments, it will list all the branches, marking the current branch with an asterisk — that is, `*`.

> **Programmatically determining the current branch**
>
> The `git branch` command is intended for the end user; its output may change in the future version of Git. To find out programmatically, in a shell script, the name of the current branch, uses `git symbolic-ref HEAD` (or `git branch --show-current`). To find the SHA-1 function of the current commit, use `git rev-parse HEAD`. To list all the branches, use `git show-ref` or `git for-each-ref`; this also works for tags and remote-tracking branches.
>
> The `git symbolic-ref`, `git rev-parse`, `git show-ref`, and `git for-each-ref` commands are all **plumbing** — that is, commands intended for use in scripts.

You can request more information with `-v` (`--verbose`) or `-vv`. You can also limit branches shown to only those matching the given shell wildcard with `git branch --list <pattern>` (quoting the pattern to prevent its expansion by the shell, if necessary).

Querying information about remotes, which includes the list of remote branches, by using `git remote show` is described in *Chapter 8*, *Advanced Branching Techniques*.

To list all tags, you can use the `git tag` command without any arguments, or `git tag --list`; with `git tag --list <pattern>`, you can select which tags to show (such as for branches), as in the following example:

```
$ git tag --list "v0.9*"
v0.99
v0.99.1
v0.99.2
```

Rewinding or resetting a branch

What do you do if you want to abandon the last commit and **rewind** (reset) the current branch to its previous position? For this, you need to use the **reset command**. It would change where the current branch points to. Note that, unlike the *checkout* command, the *reset* command does not change the working directory by default; you need to use `git reset --keep` (to try to keep the uncommitted changes) or `git reset --hard` (to drop them). The result of such reset operation is shown in *Figure 2.7*.

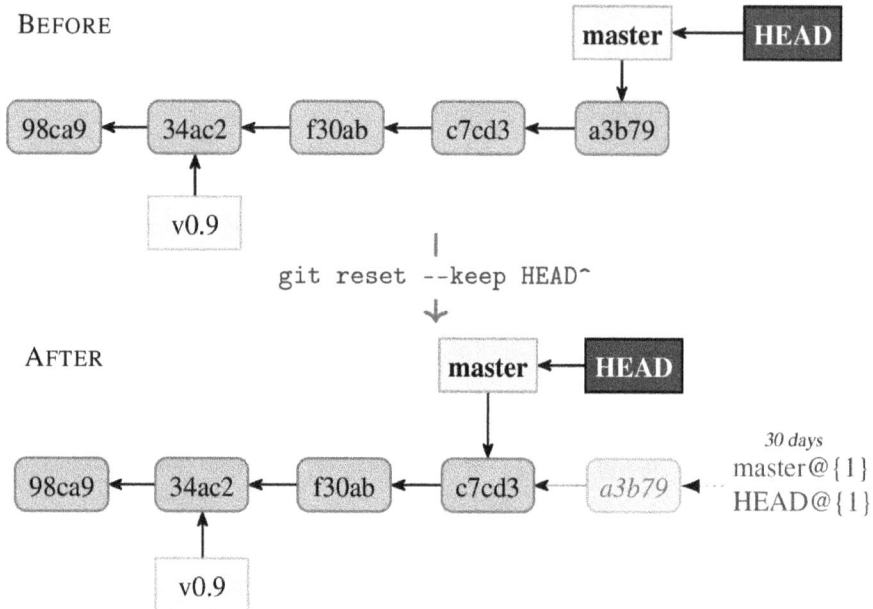

Figure 2.7 – Rewinding a branch one commit back, to HEAD^, with the reset command

The reset command and its effects on the working area will be explained in more detail in *Chapter 3, Managing Your Worktrees*.

> **Note**
>
> The `git reset <commit>` command always changes where the current branch points to (moves the ref), while `git switch` always modifies where the HEAD points to, either changing the current branch or detaching it.

Deleting a branch

Because in Git, a branch is just a pointer and an external reference to the node in the DAG of revisions, deleting a branch is just deleting a pointer. This means that deleting the branch does not immediately delete the history, but it might make it not accessible except via `reflog`. It is not kept forever, though; the **garbage collection** process would remove unreferenced parts of the commit-graph file after the `reflog` entries expire.

> **Important note**
>
> Actually, deleting a branch also removes, irretrievably (at least, in the current Git version), the `reflog` for the branch being deleted — that is, the log of its history of local operations.

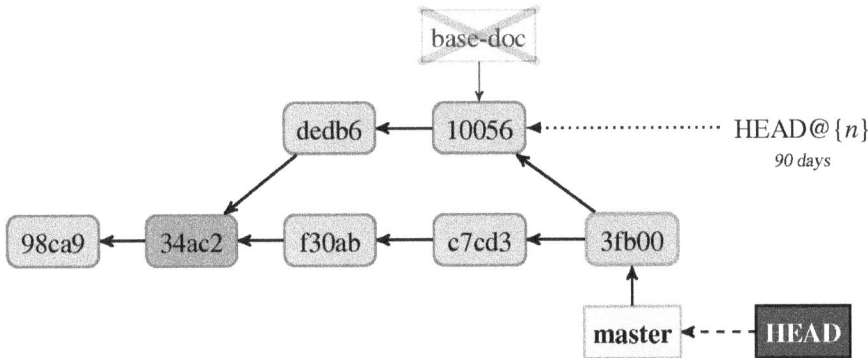

Figure 2.8 – Deleting the 'based-doc' branch that just got merged into
'master', while being on the 'master' branch that includes it

You can delete a branch with `git branch --delete <branch-name>`, or with `branch -d`, provided that the branch is not checked out anywhere.

There is, however, one issue to consider — what happens if you delete a branch, and there is no other reference to the part of the project history it pointed to? Those revisions will become unreachable, and Git would delete them after the HEAD reflog expires (which, with default configuration, is after 30 days).

That is why Git would allow you to delete only the completely merged-in branch, whose commits are all reachable from HEAD, as in the example in *Figure 2.8* (or if the branch deleted is reachable from its upstream branch, if it exists).

To delete a branch that was not merged in, which risks parts of the DAG becoming unreachable, as seen in *Figure 2.9*, you need a stronger command — namely, `git branch -D` or `git branch --delete --force`. Git will suggest this operation when refusing to delete an unmerged branch.

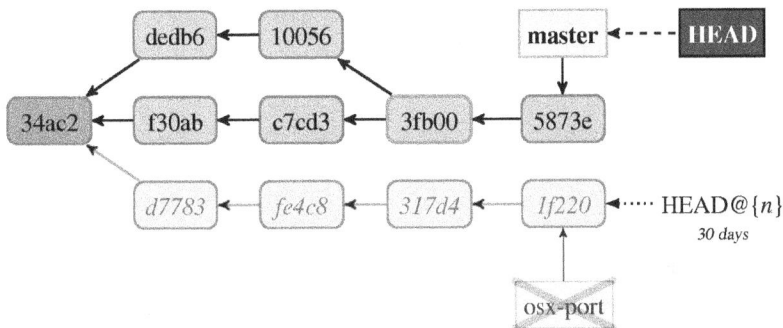

Figure 2.9 – Deleting the unmerged branch, which results in parts of the history being unreachable

You can check whether the branch was merged into any other branch by checking whether `git branch --contains <branch>` shows anything. You cannot delete the current branch.

If you ever switched to the branch that got deleted, this event and the switch away from the branch will be kept in the `reflog` for HEAD. This information can then be used to undelete that branch, or rather, to recreate it:

```
$ git reflog --no-decorate HEAD
...
3a59408 HEAD@{3}: checkout: moving from base-doc to master
```

Changing the branch name

Sometimes, the name chosen for a branch needs to be changed. This can happen, for example, if the scope of the branch changes during the development and the old name no longer fits it. Names of branches will appear and be kept forever, by default, in commit messages for merge commits; that's why you want them to be meaningful.

You can rename a branch with `git branch -m` (use `-M` if the target name exists and you want to overwrite it); it will rename a branch and move the corresponding reflog. This will also change the name of the branch in all of its configuration (its description, its upstream, and so on).

The renaming event is stored in the reflog, where you can find the previous name and use it to undo the operation (to rename the branch back to the old name):

```
$ git reflog --no-decorate new-name
3a59408 new-name@{0}: branch: renamed refs/heads/old-name to refs/
heads/new-name
```

Summary

In this chapter, we have learned how to develop with Git and extend the project history by creating new commits and new lines of development (branches). We know what it means to create a commit, amend a commit, create a branch, switch a branch, rewind a branch, and delete a branch from the point of view of the graph of revisions.

This chapter showed a very important Git feature — the staging area for creating commits, also known as the index. This is what makes it possible to untangle the changes to the working directory by selectively and interactively choosing what to commit.

We learned how to examine the changes to the working area before creating a commit. This chapter described, in detail, the extended unified diff format that Git uses to describe the changes.

We also learned about the concept of detached HEAD (or anonymous branch) and orphan branches.

In *Chapter 3, Managing Your Worktrees*, we will learn how to use Git to prepare new commits and how to configure it to make our work easier. We will also learn how to examine, search, and study the contents of the working directory, the staging area, and the project history. We will also see how to use Git to deal with interruptions and recover from mistakes.

Questions

Answer the following questions to test your knowledge of this chapter:

1. How does creating a new commit change the history stored in the repository — that is, how does it change the graph of revisions and where branch heads point to?

2. What is the difference between `git commit` and `git commit --all` (or `git commit -a`)?

3. How do you check what changes you have made in the local repository? How do you undo them?

4. What is the simplest way to fix an error in the commit message of the last commit on the current branch?

5. What do you do when you realize that the changes you started to write (but didn't commit) should be made on a separate new branch?

6. What is the simplest way to switch to the previous branch, and when can it fail?

Answers

Here are the answers to the questions given above:

1. Creating a new commit makes a new node in the graph of revisions that has a previous commit as a parent, advances the branch head ref for the current branch to the freshly created node, and keeps HEAD unchanged.

2. The `git commit` operation creates the new commit out of the staging area contents, while the `git commit --all` creates it out of the changes to all tracked files.

3. You can use `git status` to examine what files have changed and `git diff` or `git diff HEAD` to examine the changes. You can find the explanation of how to undo changes you have made in the full `git status` output.

4. To change the commit message (that is, the description of the changes) of the last commit, you can use `git commit --amend`.

5. Because uncommitted changes do not belong to a branch, you can simply create a new branch and switch to it with `git switch -c <branch-name>` or `git checkout -b <branch-name>`.

6. To switch to the previous branch, you can use `git switch -` (with - in place of the branch name). Git finds what the previous branch was by searching reflogs. This operation can fail if the branch was deleted or rename

Further reading

To learn more about the topics that were covered in this chapter, take a look at the following resources:

- Scott Chacon, Ben Straub. *Pro Git*, 2nd Edition (2014), Apress, *Chapter 2.2, Git Basics - Recording Changes to the Repository*: `https://git-scm.com/book/en/v2/Git-Basics-Recording-Changes-to-the-Repository`

- Jakub Narębski. *How to read the output from git diff?* (the answer to the question on StackOverflow): `https://stackoverflow.com/questions/2529441/how-to-read-the-output-from-git-diff/2530012#2530012`

- Dragos Barosan. *New in Git: switch and restore* (2021): `https://www.banterly.net/2021/07/31/new-in-git-switch-and-restore/`

- Junio C Hamano. *Fun with a new feature in recent Git* (2016), about the `--sort` option of the `git branch` command: `https://git-blame.blogspot.com/2016/05/fun-with-new-feature-in-recent-git.html`

- Tobias Günther. *A look under the hood: how branches work in Git* (2021): `https://stackoverflow.blog/2021/04/05/a-look-under-the-hood-how-branches-work-in-git/`

- Ryan Tomayko. *The Thing About Git* (2008), about the tangled working copy problem, and how to resolve it in Git: `https://tomayko.com/blog/2008/the-thing-about-git`

- Nick Quaranto. *Reflog, your safety net* (2009), on Gitready: `http://gitready.com/intermediate/2009/02/09/reflog-your-safety-net.html`

3

Managing Your Worktrees

The previous chapter, *Developing with Git*, described how you can use Git for project development, including how to create new revisions. In this chapter, we will focus on learning how to manage a working directory (worktree) so that you can prepare content for a new commit. This chapter will teach you how to manage your files in detail. It will also show you how to care for files that require special handling while introducing the concepts of ignored files and file attributes. Then, you will learn how to fix mistakes in handling files, both in the working directory and in the staging area, as well as how to fix or split the latest commit. Finally, you will learn how to safely handle interruptions in the workflow with stashes and multiple working directories.

The previous chapter also taught you how to examine changes. In this chapter, you will learn how to undo and redo those changes selectively, as well as how to view different versions of a file.

This chapter will cover the following topics:

- Ignoring files – marking files as intentionally not being under version control
- File attributes – path-specific configuration
- Using various modes of the `git reset` command
- Stashing away your changes to handle interruptions
- Managing the working directory's contents and the staging area
- Multiple working directories (worktrees)

Ignoring files

The files inside your **working area** (also known as the **worktree**) can be *tracked* or *untracked* by Git. **Tracked files**, as the name suggests, are those files whose changes Git will follow. For Git, if a file is present in the **staging area** (also known as **the index**), it will be tracked, and – unless specified otherwise – it will be a part of the next revision. You *add* files to be tracked, to have them as a part of the project history.

> **The purpose of the staging area**
>
> The **index**, or the **staging area**, is used not only for Git to know which files to track, but also as a kind of scratchpad to create new **commits**, as described in *Chapter 2, Developing with Git*, and to help resolve merge conflicts, as shown in *Chapter 9, Merging Changes Together*.

Often, you will have some individual files or a class of files that you never want to be a part of the project history, and never want to track. These can be your editor backup files, or automatically generated files that are produced by the project's build system (executables, object files, minified sources, source maps, and so on).

You don't want Git to automatically add such files, for example, when doing **bulk add** with `git add :/` (adding the entire working tree) or with `git add .` (adding the current directory), or when updating the index to the worktree's state with `git add --all`.

Quite the opposite: you want Git to actively prevent you from accidentally adding them. You also want such files to be absent from the `git status` output as there can be many of them. They could otherwise drown out legitimate new *unknown* files there. You want such files to be intentionally untracked – that is, *ignored*.

> **Un-tracking and re-tracking files**
>
> If you want to start ignoring a file that was formerly tracked, such as when you're moving from a hand-generated HTML file to using a lightweight markup language such as **Markdown** instead, you usually need to **un-track** the file without removing it from the working directory while adding it to the list of ignored files. You can do this with `git rm --cached <file>` (as shown in the output of `git status`). This command removes the named file from the staging area.
>
> To add (start tracking) an intentionally untracked (that is, ignored) file, you need to use `git add --force <file>`, as Git will tell you.

Marking files as intentionally untracked (ignored)

If you want to mark a file or a set of files as intentionally ignored, you need to add a **shell glob pattern** that matches files that you want to have ignored by Git to one of the following **gitignore** files, one pattern per line:

- The per-user file, which can be specified by the `core.excludesFile` configuration variable. If this configuration variable is not set, then the default value of `$XDG_CONFIG_HOME/git/ignore` is used. This, in turn, defaults to `$HOME/.config/git/ignore` if the `$XDG_CONFIG_HOME` environment variable is not set or empty (where `$HOME` is the current user's home directory).

- The per-local repository $GIT_DIR/info/exclude file in the administrative area of the local clone of the repository (in most cases, $GIT_DIR points to the .git/ directory in the top-level directory of the project).

- The .gitignore files in the working directories of a project. These are usually tracked, and in this case, they are shared among all developers.

Some commands, such as git clean, also allow us to specify ignore patterns from a command line with the --exclude=<pattern> option.

When deciding whether to ignore a path, Git checks all those sources in the order specified in the preceding list, with the last matching pattern deciding the outcome. The .gitignore files are checked in order, starting from the top directory of the project down to the directory of files to be examined.

To make .gitignore files more readable, you can use blank lines to separate groups of files (a blank line matches no files). You can also describe patterns or groups of patterns with comments; a line starting with the hash character, #, serves as one (to ignore a pattern beginning with #, escape the first hash character with a backslash, \ – for example, \#*#). Trailing spaces (at the end of the line) are ignored unless escaped with a backslash, \.

Each line in the .gitignore file specifies a Unix glob pattern, a shell wildcard. The * wildcard matches zero or more characters (any string), while the ? wildcard matches any single character. You can also use character classes with brackets, [...]. Take, for example, the following list of patterns:

```
*.[oa]
*~
```

Here, the first line tells Git to ignore all files with the .a or .o extension – *.a files are archive files (for example, a static library), and *.o files are object files that may be the products of compiling your code. The second line tells Git to ignore all files ending with a tilde, ~; this is used by many Unix text editors to mark temporary backup files.

If the pattern does not contain a slash, /, which is a path component separator, Git treats it as a **shell glob** and checks the filename or directory name for a match, starting at the appropriate depth (this means the .gitignore file location if the pattern is in such a file, or the top level of the repository otherwise). The exception is patterns ending with a slash, /, which is used to have the pattern only matched against directories but otherwise treated as if the trailing slash was removed.

A leading slash matches the beginning of the pathname. This means the following:

- Patterns not containing a slash match everywhere in the repository; we can say that the pattern is recursive.

  For example, the *.o pattern matches object files anywhere, both at the .gitignore file level and in subdirectories such as file.o, obj/file.o, and others.

- Patterns ending with a slash only match directories but are otherwise recursive (unless they contain other slashes).

  For example, the `auto/` pattern will match both the top-level `auto` directory and the `src/auto` directory but will not match the `auto` file (or a symbolic link either).

- To *anchor* a pattern and make it non-recursive, add a leading slash.

  For example, the `/TODO` pattern will match and make Git ignore the current-level `TODO` file, but not files in subdirectories, such as `src/TODO`.

- Patterns containing a slash inside are anchored and non-recursive, and wildcard characters (`*`, `?`, or a character class such as `[ao]`) do not match the directory separator that is the slash, `/`. If you want the pattern to match any number of directories, use two consecutive asterisks, `**`, in place of the path component (which means `**/foo`, `foo/**`, and `foo/**/bar`).

  For example, `doc/*.html` matches the `doc/index.html` file but not `doc/api/index.html`; to match HTML files anywhere inside the `doc` directory, you can use the `doc/**/*.html` pattern (or put the `*.html` pattern in the `doc/.gitignore` file).

You can also negate a pattern by prefixing it with an exclamation mark, `!`; any matching file excluded by the earlier rule is then included (non-ignored) again. For example, to ignore all generated HTML files, but include the one HTML file generated by hand, you can put the following in the `.gitignore` file:

```
# ignore html files, generated from AsciiDoc sources
*.html
# except for the files below which are generated by hand
!welcome.html
```

> **Note**
>
> For performance reasons, Git doesn't go into excluded directories, and (up until *Git 2.7*) this means that you cannot re-include a file if a parent directory is excluded.

This means that to ignore everything except for the subdirectory, you need to write the following:

```
# exclude everything except directory t0001/bin
/*
!/t0001
/t0001/*
!/t0001/bin
```

To match a pattern beginning with `!`, escape it with a backslash, similar to what you need to do for the `#` character – for example, use the `\!important!.md` pattern to match the file named `!important!.md`.

Which types of files should be ignored?

Now that we know how to mark files as intentionally untracked (ignored), there is the question of *which* files (or classes of files) should be marked as such. Another issue is *where* we should add a pattern for ignoring specific types of files – that is, in which of the three types of .gitignore files.

The first rule is that you should never track *automatically generated files* (usually generated by the build system of a project). If you add such files to the repository and if you track them, there is a high chance that they will get out of sync with their source. Besides, they are not necessary, as you can always re-generate them. The only possible exception is generated files where the source rarely changes and generating them requires extra tools that developers might not have (if the source changes more often, you can use an orphan branch to store these generated files and refresh this branch only at release time; see *Chapter 2, Developing With Git*, the *Creating orphan branches* section for more information).

Those automatically generated files are the files that *all developers* will want to ignore. Therefore, they should go into a tracked .gitignore file. This list of patterns will be version-controlled and distributed to other developers via a clone; this way, all developers will get it. You can find a collection of useful .gitignore templates for different programming languages at https://github.com/github/gitignore, or you can use the web app at https://gitignore.io.

Second, there are *temporary files* and byproducts specific to one user's toolchain; those should usually not be shared with other developers. If the pattern is specific to both the repository and the user – for example, auxiliary files that live inside the repository but are specific to the workflow of a user (for example, to the IDE used for the project) – it should go into the per-clone $GIT_DIR/info/exclude file.

Patterns that the user wants to ignore in all situations and are not specific to the repository (or to the project) should generally go into a per-user .gitignore file specified by the core.excludesFile config variable, set in the per-user (global) ~/.gitconfig config file (or ~/.config/git/config). This is usually ~/.config/git/ignore by default.

> **Important note about the per-user .gitignore file**
>
> The per-user ignore file cannot be ~/.gitignore as this would be the in-repository .gitignore file for the versioned user's home directory if the user wants to keep the ~/ directory ($HOME) under version control.

This is the place where you can put patterns that match the backup or temporary files generated by your editor or IDE of choice.

> **Ignored files are considered expendable**
>
> Warning: Do not add *precious files* – that is, those you do not want to track in a given repository but whose contents are important – to the list of ignored files! The types of files that are ignored (excluded) by Git are either easy to regenerate (build products and other generated files) or not important to the user (temporary or backup files).
>
> Therefore, Git considers ignored files *expendable* and will remove them without warning when required to do a requested command – for example, if the ignored file conflicts with the contents of the revision being checked out.

Listing ignored files

You can list untracked ignored files by appending the `--ignored` option to the `git status` command:

```
$ git status --ignored
On branch master
Ignored files:
  (use "git add -f <file>..." to include in what will be committed)

        rand.c~

no changes added to commit (use "git add" and/or "git commit -a")

$ git status --short --branch --ignored
## master
!! rand.c~
```

Instead of using `git status --ignored`, you can use the dry-run option of cleaning ignored files, `git clean -Xnd`, or the low-level (plumbing) `git ls-files` command:

```
$ git ls-files --others --ignored --exclude-standard
rand.c~
```

The latter command can also be used to list *tracked files* that match *ignore patterns*. If there are any such files, it might mean that some files need to be un-tracked (perhaps because what was once a source file is now generated), or that ignore patterns are too broad. Since Git uses the existence of a file in the staging area (*cache*) to know which files to track, this can be done with the following command:

```
$ git ls-files --cached --ignored --exclude-standard
```

An empty result, like what's shown here, means that everything is fine.

Plumbing versus porcelain commands

Git commands can be divided into two sets: high-level **porcelain** commands intended for interactive usage by the end user and low-level **plumbing** commands intended mainly for shell scripting. The major difference is that high-level commands have outputs that can change and are constantly improving. For example, the output of the `git branch` command in the detached HEAD case changed from (`no branch`) to (`detached from HEAD`). Their output and behavior are also subject to the configuration. Note that some porcelain commands have the option to switch to unchanging output via `--porcelain`.

Another important difference is that plumbing commands try to guess what you meant, they have default parameters, use the default configuration, and so on. This isn't the case with plumbing commands. You need to pass the `--exclude-standard` option to the `git ls-files` command to make it respect the default set of ignore files.

You can find more on this topic in *Chapter 13, Customizing and Extending Git*.

Trick – ignoring changes in tracked files

You might have files in your repository that are changed but rarely committed. These can be various local configuration files that are edited to match the local setup but should never be committed upstream. This can be a file containing the proposed name for a new release, to be committed later when tagging the next released version.

You would want to keep such files in a *dirty* state most of the time, but you would like Git not to tell you about their changes all the time in case you miss other changes because you're used to ignoring such messages.

Dirty working directory

The working directory is considered **clean** if it is the same as the committed and staged version and **dirty** if any modifications or changes have been made.

Git can be configured – or rather tricked in this case – to skip checking the worktree (to assume that it is always up to date), and to use the staged version of the file instead. This can be done by setting the aptly named `skip-worktree` flag for a file. For this, you would need to use the low-level `git update-index` command, which is the plumbing equivalent of the user-facing `git add` porcelain. You can check file status and flags with `git ls-files`, which will use the letter S for files with this flag set:

```
$ git update-index --skip-worktree GIT-VERSION-NAME
$ git ls-files -v
S GIT-VERSION-NAME
H ...
```

Note that this elision of the worktree also includes the `git stash` command; to stash away your changes and make the working directory *clean*, you need to disable this flag (at least temporarily). To make Git look at the working directory version and start tracking changes to the file, use the following command:

```
$ git update-index --no-skip-worktree GIT-VERSION-NAME
```

This problem is caused by the fact that this use of the `skip-worktree` flag is not intended use; this flag was created to manage so-called sparse checkout – more on that in *Chapter 12, Managing Large Repositories.*

> **Important note**
>
> There is a similar `assume-unchanged` flag that can be used to make Git completely ignore any changes to the file, or rather *assume that it is unchanged*. Files marked with this flag never show as changed in the output of the `git status` or `git diff` command. The changes to such files will not be staged, nor committed.
>
> This is sometimes useful when you're working with a big project on a filesystem that's very slow at checking for changes. However, do not use `assume-unchanged` to *ignore* changes to tracked files. You are promising that the file didn't change, lying to Git. This means, for example, that with `git stash save` believing what you stated, you would lose your precious local changes.

File attributes

There are some settings and options in Git that can be specified on a per-path basis, similar to how ignoring files (marking files as intentionally untracked) works. These path-specific settings are called **attributes**.

To specify attributes for files matching a given pattern, you need to add a line with a pattern, separated by a space and followed by a whitespace-separated list of attributes, to one of the **.gitattributes files** (similar to how `.gitignore` files work):

- The per-user file, for attributes that should affect all repositories for a single user, specified by the `core.attributesFile` configuration variable. By default, this is `~/.config/git/attributes`

- The per-repository `.git/info/attributes` file in the administrative area of the local clone of the repository, for attributes that should only affect a single specific clone of the repository (for one user's workflow).

- The `.gitattributes` files in the working directories of a project, for those attributes that should be shared among developers.

The rules for how patterns are used to match files are the same as for the .gitignore files, as described previously, except that there is no support for negative patterns, and that patterns matching the directory do not recursively match paths inside that directory.

Each attribute can be in one of the following states for a given path: set (special value true), unset (special value false), set to a given value, or unspecified:

```
pattern*  set -unset set-to=value !unspecified
```

> **Note**
> There can be no whitespace around the equals sign, =, when setting an attribute to a string value!

When more than one pattern matches the path, a later line overrides an earlier line on a per-attribute basis. .gitattributes files are used in order, from the per-user, through per-repository, to the .gitattributes file in a given directory, like for .gitignore files.

Identifying binary files and end-of-line conversions

Different operating systems and different applications can differ in how they represent newlines in text files. Unix and Unix-like systems (including Mac OS X) use a single control character LF (\n), while Windows uses CRLF – that is, CR followed by LF (\n\r); macOS up to version 9 used CR alone (\r).

That might be a problem for developing portable applications if different developers use different operating systems. We don't want to have spurious changes because of different end-of-line conventions. Therefore, Git makes it possible to automatically normalize end-of-line characters to be LF in the repository on commit (check-in), and optionally to convert them to CR + LF in the working directory on checkout.

You can control whether a file should be considered for end-of-line conversion with the text attribute. Setting it enables end-of-line conversion, and unsetting it disables it. Setting it to the auto value makes Git guess if the given file is a text file; if it is, end-of-line conversion is enabled. For files where the text attribute is unspecified, Git uses core.autocrlf to decide whether to treat them as text=auto case.

> **How Git detects if a file contains binary data**
> To decide whether a file contains binary data, Git examines the beginning of the file for an occurrence of a zero byte (the null/NUL character or \0). When deciding whether to convert a file (as in end-of-line conversion), the criterion is stricter: for a file to be considered text, it must have no nulls, and no more than around 1% of it should be non-printable characters.
>
> However, this means that Git usually considers files saved in the UTF-16 encoding to be binary.

To decide what line ending type Git should use in the working directory for text files, you need to set up the `core.eol` configuration variable. This can be set to `crlf`, `lf`, or `native` (the last is the default). You can also force a specific line ending for a given file with the `eol=lf` or `eol=crlf` attribute:

Old crlf attribute	New text and eol attributes
`crlf`	`text`
`-crlf`	`-text`
`crlf=input`	`eol=lf`

Table 3.1 – Backward compatibility of the text and eof attributes with the crlf attribute

End-of-line conversion bears a slight chance of corrupting data. If you want Git to warn or prevent conversion for files with a mixture of LF and CRLF line endings, use the `core.safecrlf` configuration variable.

Sometimes, Git might not detect that a file is binary correctly, or there may be some type of file that is nominally text, but which is opaque to a human reader. Examples include PostScript documents (`*.ps`) and Xcode build settings (`*.pbxproj`). Such files should not be normalized and using textual `diff` for them doesn't make sense. You can mark such files explicitly as binary with the `binary` attribute macro (which is equivalent to `-text -diff`):

```
*.ps binary
*.pbxproj binary
```

> **What to do if files start without end-of-line normalization**
>
> When the normalization of line endings is turned on in the repository (by editing the `.gitattributes` file), you should also force the **normalization** of files. Otherwise, the change in newline representation will be misattributed to the next change to the file. This can be done, for example, with the `git add --renormalize` command. This should also be done when changing which files have the `text` attribute.

Diff and merge configuration

In Git, you can use the attributes functionality to configure how to show differences between different versions of a file, and how to do a three-way merge of its contents. This can be used to enhance that operation, making `diff` more attractive and `merge` less likely to conflict. It can even be used to make it possible to effectively `diff` binary files, or to describe differences in a specific way.

In both cases, we would usually need to set up the `diff` and/or `merge` driver. The attributes file only tells us which driver to use; the rest of the information is contained in the configuration file, and this configuration is not automatically shared among developers, unlike the `.gitattributes` file (though you can create a shared configuration fragment, add it to the repository, and have developers include it in their local per-repository config via the relative `include.path`). The reason for this behavior is easy to understand – the tool's configuration may be different on different computers, and some tools may be not available for the developer's operating system of choice. But this means that some information needs to be distributed out-of-band.

There are, however, a few built-in **diff drivers** and **merge drivers** that anyone can use without further configuration.

Generating diffs and binary files

Diffs that are generated for particular files are affected by the `diff` attribute. If this attribute is unset, Git will treat files as binary concerning generating diffs and show just **binary files differ** (or show a binary diff). Setting it will force Git to treat a file as text, even if it contains byte sequences that normally mark the file as binary, such as the null (`\0`) character.

You can use the `diff` attribute to make Git more effectively describe the differences between two versions of a binary file via a **diff driver**. In this case, you have two options: the easier one is to tell Git how to convert a binary file into a text format, or how to extract text information (for example metadata) from the binary data. This text representation is then compared using the ordinary textual `diff` command. Even though conversion to text usually loses some information, the resulting differences is useful for human viewing (even though it is not information about all the changes).

This can be done with the `textconv` config key for a `diff` driver, where you specify a program that takes the name of the file as an argument and returns a text representation on its output.

For example, you might want to see the difference in the contents of Microsoft Word documents and see the difference in metadata for JPEG images. First, you need to put something like this in your `.gitattributes` file:

```
*.doc   diff=word2text
*.jpg   diff=exif
```

For example, you can use the `catdoc` program to extract text from binary Microsoft Word documents and `exiftool` to extract EXIF metadata from JPEG images.

Because conversion can be slow, Git provides a mechanism to cache the output in the form of the Boolean `cachetextconv` attribute; the cached data is stored using **notes** (this mechanism will be explained in *Chapter 10*, *Keeping History Clean*). The part of the configuration file that's responsible for this setup looks like this:

```
[diff "word2text"]
    textconv = catdoc

# cached data will be stored in refs/notes/textconv/exif
```

```
[diff "exif"]
    textconv = exiftool
    cachetextconv = true
```

You can see what the output of the `textconv` filter looks like with `git show`, or with `git cat-file -p` with the `--textconv` option.

The more complicated but also more powerful option is to use an **external diff driver** (an attribute version of the global driver that can be specified with the `GIT_EXTERNAL_DIFF` environment variable or the `diff.external` configuration variable) with the `command` option of the `diff` driver. However, when choosing to use this option, you lose some options that Git `diff` gives for free: colorization, word diff, and combined diff for merges.

Such a program will be called with seven parameters: `path`, `old-file`, `old-hex`, `old-mode`, `new-file`, `new-hex`, and `new-mode`. Here, `old-file` and `new-file` are files that the `diff` driver can use to read the contents of two versions of the differing file, `old-hex` and `new-hex` are SHA-1 identifiers of file contents, and `old-mode` and `new-mode` are octal representations of file modes. The command is expected to generate a `diff`-like output. For example, you might want to use the XML-aware `diff` tool to compare XML files:

```
$ echo "*.xml diff=xmldiff" >>.gitattributes
$ git config diff.xmldiff.command xmldiff-wrapper.sh
```

This example assumes that you have written the `xmldiff-wrapper.sh` shell script to reorder options so that they fit the expectations of the XML `diff` tool.

Configuring diff output

The `diff` format that Git uses to show changes for users was described in detail in *Chapter 2, Developing with Git*. Each group of changes (called a hunk) in textual `diff` output is preceded by the hunk header line, as shown here:

```
@@ -18,6 +19,8 @@ int cmd_http_fetch(int argc, const char **argv,
```

The text after the second `@@` is meant to describe the section of the file where the chunk is; for C source files, it is the start of the function. The decision on how to detect the beginning of such a section depends on the type of file. Git allows you to configure this by setting the `xfuncname` configuration option of the `diff` driver to the regular expression, which matches the description of the section of the file. For example, for LaTeX documents, you might want to use the following configuration for the `tex` `diff` driver (you don't need to as `tex` is one of the pre-defined, built-in `diff` drivers):

```
[diff "tex"]
    xfuncname = "^(\\\\(sub)*section\\{.*)$"
    wordRegex = "\\\\[a-zA-Z]+|[{}]|\\\\.|[^\\{}[:space:]]+"
```

The `wordRegex` configuration defines what `word` is to define it for the `git diff --word-diff` command (described in *Chapter 2, Developing with Git*, near the end of the *Unified diff output* section). Here, it's being used for LaTeX documents.

> **Note**
>
> You would need to double the backslashes: \\ matches the literal backslash, \, in a regexp, so you need to use \\\\ here (which is typical for storing regexps in strings).

Performing a three-way merge

You can also use the `merge` attribute to tell Git to use specific merge strategies for specific files or classes of files in your project. By default, Git will use the thee-way merge driver (similar to `rcsmerge`) for text files, and it will take our (being merged) version and mark the result as a conflicted merge for binary files. You can force a three-way merge by setting the `merge` attribute (or by using `merge=text`); you can force binary-like merging by unsetting this attribute (with `-merge`, which is equivalent to `merge=binary`).

You can also write your **merge driver** or configure Git to use a third-party external merge driver. For example, if you keep a GNU-style `ChangeLog` file in your repository (with a curated list of changes with their description), you can use the `git-merge-changelog` command from the **GNU Portability Library (Gnulib)**. You need to add the following to the appropriate Git config file:

```
[merge "merge-changelog"]
    name = GNU-style ChangeLog merge driver
    driver = git-merge-changelog %O %A %B
```

Here, the token, `%O`, in `merge.merge-changelog.driver` will be expanded to the name of the temporary file holding the contents of the merge ancestor's (old) version. The `%A` and `%B` tokens expand to the names of temporary files holding contents being merged – that is, the current (ours, merged into) version and the other branches' (theirs, merged) version, respectively. The `merge` driver is expected to leave the merged version in the `%A` file, exiting with a non-zero status if there is a merge conflict. You can also use `%L` to denote the conflict marker size and `%P` to find a pathname where the merged results will be stored.

> **Note**
>
> You can use a different driver for an internal merge between common ancestors (when there is more than one). You can do this by setting the `merge.*.recursive` configuration variable for a given driver. For example, here, you can use the predefined `binary` driver.

Of course, you will also need to tell Git to use this driver for `ChangeLog` files, adding the following line to `.gitattributes`:

```
ChangeLog merge=merge-changelog
```

Transforming files (content filtering)

Sometimes, the format of the content you want to put in a version control system may depend on where it is stored, be it on disk or in the repository, with different shapes in different places that are more convenient for Git, the platform (operating system), the filesystem, and the user to use. End-of-line conversion can be considered a special case for such an operation.

To do this, you need to set the `filter` attribute for appropriate paths and configure the `clean` and `smudge` commands of the specified filter driver (either command can be left unspecified for a pass-through filter). When checking out the file matching the given pattern, the `smudge` command is fed file contents from the repository in its standard input, and its standard output is used to update the file in the working directory. See *Figure 3.1* for details:

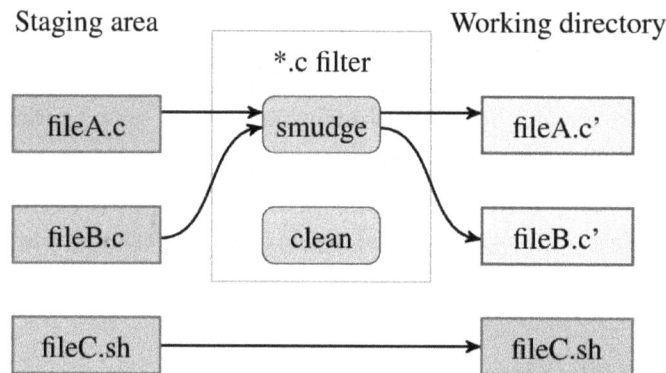

Figure 3.1 – The "smudge" filter is run on checkout (when writing files to the working directory)

Similarly, the `clean` command of a filter is used to convert the contents of the worktree file into a shape suitable to be stored in the repository; see *Figure 3.2*:

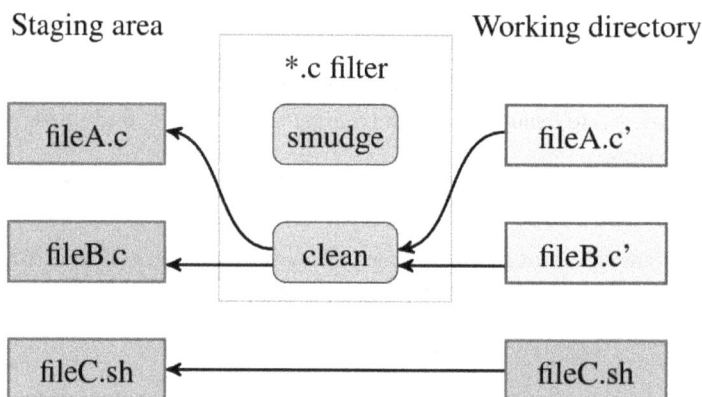

Figure 3.2 – The "clean" filter is run when files are staged (added
to the index, also known as the staging area)

When specifying a command, you can use the %f token, which will be replaced by the name of the file the filter is working on.

One simple example of how you can use this feature is to use the rezip script for **OpenDocument Format (ODF)** files. ODF documents are ZIP archives of mainly XML files. Git uses compression itself and also does deltaification (but cannot do it on already compressed files); the idea is to store uncompressed files in the repository but to check out compressed files:

```
[filter "opendocument"]
    clean = "rezip -p ODF_UNCOMPRESS"
    smudge = "rezip -p ODF_COMPRESS"
```

Of course, you also need to tell Git to use this filter for all kinds of ODF files:

```
*.odt filter=opendocument
*.ods filter=opendocument
*.odp filter=opendocument
```

Another example of an *advisory* filter is to use the indent program to force a code formatting convention, as shown in the following example, or gofmt for the Go programming language. A similar example would be to replace tabs with spaces on check-in:

```
[filter "indent"]
    clean = indent
```

Yet another example is **nbdev**, where you can install a filter that uses the nbdev_clean command to strip metadata and cell output from **Jupyter Notebook** files. This is done to reduce the number of merge conflicts and to avoid storing generated data in the repository.

Obligatory file transformations

Another use of content filtering is to store the content that cannot be directly used in the repository and turn it into a usable form upon checkout.

One such example might be to use .gitattributes files to configure Git so that it stores large binary files outside the Git repository (such files are often only used by a subset of developers); inside the repository, there is only an identifier that allows us to get file contents from external storage. That's how git-media works:

```
$ git config filter.media.clean  "git-media filter-clean"
$ git config filter.media.smudge "git-media filter-smudge"
$ echo "*.mov filter=media -crlf" >> .gitattributes
```

> **Tip**
>
> You can find the git-media tool at https://github.com/alebedev/git-media. Other similar tools will be mentioned in *Chapter 12, Managing Large Repositories*, as one of the possible solutions to the problem of handling large files.

Another example of obligatory transformations would be encrypting sensitive content or replacing a local sensitive program configuration that is required for an application to work (for example, a database password) with a placeholder. Because running such a filter is, like in the preceding example, *required* to get useful content, you can mark it as such:

```
[[filter "clean-password"]
    clean = sed -e 's/^pass = .*$/pass = @PASSWORD@/'
    smudge = sed -e 's/^pass = @PASSWORD@/pass = passw0rd/'
    required
```

> **Important note**
>
> This is only a simplified example; in real use, you would have to consider the security of the config file itself if you do this or store the real password in an external smudge script. In such a case, you should also set up pre-commit, pre-push, and update hooks to ensure that the password won't make it to the public repository (see *Chapter 13, Customizing and Extending Git*, for details).

If many files need to be processed, and the time it takes to invoke and run the clean and smudge scripts becomes a problem, you can configure Git to use a program that will process all files with a single filter invocation for the entire lifetime of a Git command. You can define such a filter with a process key in place of clean and smudge.

Keyword expansion and substitution

Sometimes, though rare, there is a need to have a piece of dynamic information about the versioned file in the contents of the file itself. To keep such information up to date, you can request the version control system to perform **keyword expansion**: replace the **keyword anchor** in the form of a string of text (in the file contents) formatted as $Keyword$, with the keyword inside dollar characters (keyword anchor). This is usually replaced by a version-control system with $Keyword: value$, which is a keyword followed by its expansion.

The main problem with doing this in Git is that you cannot modify the file contents stored in the repository with information about the commit after you've committed because of the way Git works (more information about this can be found in *Chapter 10, Keeping History Clean*). This means that keyword anchors must be stored in the repository as-is, and only expanded in the worktree on checkout. However, this is also an advantage; you would get no spurious differences due to keyword expansion when examining the history.

The only built-in keyword that Git supports is Id: its value is the SHA-1 identifier of the file contents (the SHA-1 checksum of the blob object representing the file contents, which is not the same as the SHA-1 of the file; see *Chapter 10, Keeping History Clean*, to learn how objects are constructed). You need to request this keyword expansion by setting the ident attribute for a file.

However, you can write your keyword expansion support with an appropriate filter while defining the smudge command, which would expand the keyword, and the clean command, which would replace the expanded keyword with its keyword anchor.

With this mechanism you can, for example, implement support for the $Date$ keyword, expanding it on checkout to the date when the file was last modified:

```
[filter "dater"]
    clean = sed -e 's/\\\$Date[^\\\$]*\\\$/\\\$Date\\\$/'
    smudge = expand_date %f
```

The expand_date script, which is passed the name of the file as an argument, could run the git log --pretty=format:"%ad" "$1" command to get the substitution value, for example.

However, you need to remember another limitation: for better performance, Git does not touch files that did not change, be it on commit, on switching the branch (on checkout), or on rewinding the branch (on reset). This means that this trick cannot support keyword expansion for the date of the last revision of a project (as opposed to the last revision that changed the file).

If you need to have such information in distributed sources (for example, the description of the current commit, or how long it was since the tagged release), you can either make it a part of the build system, or use **keyword substitution** for the git archive command. The latter is quite a generic feature: if the export-subst attribute is set for a file, Git will expand the $Format:<PLACEHOLDERS>$ generalized keyword when adding the file to an archive.

> **Limitation of the keyword expansion with export-subst**
>
> The expansion of the $Format$ meta-keyword depends on the availability of the revision identifier; it cannot be done if you, for example, pass the SHA-1 identifier of a tree object to the git archive command.

The placeholders are the same as for the --pretty=format: custom formats for git log, which are described in *Chapter 4, Exploring Project History*. For example, the $Format:%H$ string will be *replaced* (not expanded) by the commit hash. It is an irreversible keyword substitution; there is no trace of the keyword in the result of the archive (export) operation.

Other built-in attributes

You can also tell Git not to add certain files or directories when generating an archive. For example, in the user-facing archive, you may not want to include the directory with distribution tests, which are useful for the developer but not for end users (those tests may require additional tools or checking the quality of the program and processing it rather than checking the correctness of the application behavior). This can be done by setting the `export-ignore` attribute – for example, by adding the following line to the `.gitattributes` file:

```
# Do not include extra tests in the archive
xt/   export-ignore
```

Another thing that can be configured with file attributes is defining what `diff` and `apply` should consider a **whitespace error** for specific types of files; this is a fine-grained version of the `core.whitespace` configuration variable. Note that the list of common whitespace problems to take notice of should use commas as an element separator, without any surrounding whitespace, when put in the `.gitattributes` file. See the following example (taken from the Git project):

```
*   whitespace=!indent,trail,space
*.[ch] whitespace=indent,trail,space
*.sh whitespace=indent,trail,space
```

With file attributes, you can also specify the **character encoding** that is used by a particular file by providing it as a value of the `encoding` attribute. Git can use it to select how to display the file in GUI tools (for example, `gitk` and `git gui`). This is a fine-grained version of the `gui.encoding` configuration variable and is only used when explicitly asked for due to performance considerations. For example, GNU gettext **Portable Object** (**.po**) files holding translations should use the UTF-8 encoding:

```
/po/*.po encoding=UTF-8
```

To have Git convert between UTF-8 encoding in the staging area and the repository, as well as specify the encoding of a file in the working directory on checkout, you can use the `working-tree-encoding` attribute. For example, **PowerShell script files** (***.ps1**) are sometimes encoded in UTF-16; files encoded using this encoding are interpreted by Git as binary. For `diff` and other commands to work correctly, you might want to use the following command:

```
*.ps1 text working-tree-encoding=UTF-16LE eol=CRLF
```

> **Note**
> Reencoding might slow down certain Git operations.

Defining attribute macros

In the *Identifying binary files and end-of-line conversions* section, we learned how to mark binary files with the `binary` attribute. The `binary` attribute is the **attribute macro**, expanding to `-diff -merge -text` (unsetting three file attributes). It would be nice to define such macros for arbitrary combinations of attributes. There can be more than one pattern matching a given type of file, but one `.gitattributes` line can contain only one file pattern. If we want to have the same attributes for different types of files, attribute macros allow avoiding duplication.

Git allows us to define such macros, but only in top-level `.gitattributes` files, namely `core.attributesFile`, `.git/info/attributes`, or `.gitattributes` in the main (top-level) directory of a project. The built-in `binary` macro could have been defined as follows:

```
[attr]binary -diff -merge -text
```

You can also define your own attributes. In this case, you can use the `git check-attr` command to programmatically check which attributes are set for a given file, or what the value is of an attribute for a set of files.

Fixing mistakes with the reset command

At any stage during development, you might want to **undo** an operation, fix mistakes, or abandon your current work. There is no `git undo` command in core Git, and neither is there support for the universal `--undo` option in Git commands, though many commands have an `--abort` option to abandon current **work in progress (WIP)**. One of the reasons why there is no such command or option yet is the ambiguity on what needs to be undone (especially for multi-step operations).

Many mistakes can be fixed with the help of the `git reset` command. It can be used for various purposes and in various ways; understanding how this command works will help you in using it in any situation, which is not limited to the provided example usage.

Note that this section only covers the full-tree mode of `git reset`; the description of what `git reset -- <file>` does, which is an alternative to using the more modern `git restore <file>` command, has been left for the *Managing worktree and staging area* section at the end of this chapter.

Rewinding the branch head, softly

The `git reset` command in its full-tree mode affects the current branch head, and can also affect the index (the staging area) and the working directory. This reset does not change which branch is current, as opposed to `git checkout` or `git switch`.

To reset only the current branch head and not touch the index or the working tree, you can use `git reset --soft [<revision>]` (if a revision is not given, it defaults to HEAD):

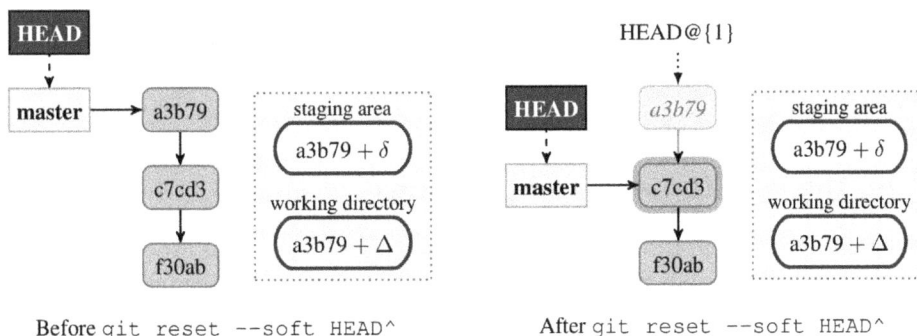

Figure 3.3 – Before and after a soft reset

Effectively, we are just changing the pointer of the current branch (`master` in the example shown in *Figure 3.3*) to point to a given revision (`HEAD^` – the previous commit in the example). Neither the staging area nor the working directory is affected. This leaves all your changed files (and all files that differ between the old and new revision pointed by branch) in the **Changes to be committed** state, as `git status` would put it.

Removing or amending a commit

The way the command works means that a **soft reset** can be used to undo the act of creating a commit. This can be used to **amend a commit**, though it is far easier to simply use the `--amend` option of `git commit`.

Let's take a look at the   following command:

```
$ git commit --amend [<options>]
```

This is equivalent to the following:

```
$ git reset --soft HEAD^
$ git commit --reedit-message=ORIG_HEAD [<options>]
```

The `git commit --amend` command also works for merge commits as opposed to using a soft reset. When amending a commit, if you want to just fix the commit message, there will be no additional options. If you want to include a fix from the working directory without changing the commit message, you can add `--all --no-edit`. If you want to fix the authorship information after correcting the Git configuration, use `--reset-author --no-edit`.

You learned how amending a commit changes the graph of revisions in *Chapter 2*, *Developing With Git*, in the *Amending a commit* section.

Squashing commits with reset

You are not limited to rewinding the branch head to just the previous commit. Using a soft reset, you can squash a few earlier commits (for example, commit and bugfix, or introducing new functionality and using it), making one commit out of two (or more); alternatively, you can instead use the `squash` instruction of **interactive rebase**, as described in *Chapter 10*, *Keeping History Clean*. With the latter, you can squash any series of commits, also in the middle of the history, not just the most recent ones. You can also use git `merge --squash` for this.

Resetting the branch head and the index

The default mode of reset command – the so-called **mixed reset** (because it is between the soft and hard forms) – changes the current branch head so that it points to a given revision, and also resets the index, putting the contents of that revision into the staging area. This mode is shown in *Figure 3.4*:

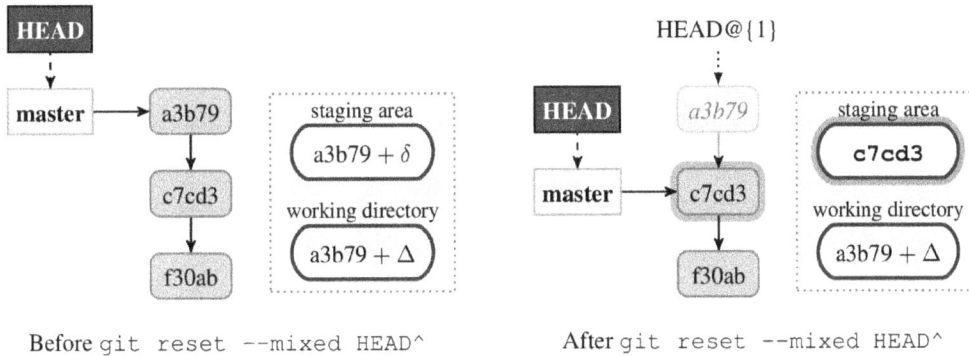

Figure 3.4 – Before and after a mixed reset

This leaves all your changed files (and all files that differ between the old and new revision pointed by branch) in the **Changes not staged for commit** state, as `git status` would put it. The `git reset --mixed` command will also report what has not been updated using the short status format:

```
$ git reset HEAD^
Unstaged changes after reset:
M       README.md
```

This version of the `reset` command can be used, for example, to undo all additions of new files. This can be done by running `git reset`, if you didn't stage any changes (or that you can put up with losing them). If you want to un-add a particular file, use `git rm --cached <file>`.

Splitting a commit in two with reset

You can use a mixed reset to split a commit in two. First, run `git reset HEAD^` to reset the branch head and the index to the previous revision. Then, interactively add changes that you want to have in the first commit, and then create this first commit from the index (`git add -i` and `git commit`). A second commit can then be created from the working directory state (`git commit -a`).

If it is easier to interactively remove changes, that's also an option. Use `git reset --soft HEAD^`, interactively un-stage changes with an interactive per-file reset, create the first commit from the constructed state in the index, and create the second commit from the working directory.

Here, again, like for squashing commits, you can use the interactive rebase to split commits further in the history. The rebase operation will switch to the appropriate commit, at which point the actual splitting can be done, as described here.

Saving and restoring state with the WIP commit

Suppose you are interrupted by an urgent bugfix request while you are in the middle of work on the development branch. You don't want to lose your changes, but the worktree is a bit of a mess, and you are unable to finish the commit in time. One possible solution is to save the current state of the working area by creating a temporary commit:

```
$ git commit -a -m 'snapshot WIP (Work In Progress)'
```

Then, you can handle the interruption, switching to the maintenance branch and creating a commit to fix the issue. At this point, you need to go back to the previous branch (by using checkout), remove the WIP commit from the history (using a soft reset), and go back to the un-staged starting state (with a mixed reset), as follows:

```
$ git switch -
$ git reset --soft HEAD^
$ git reset
```

Usually, though it is much easier to just use `git stash` instead to handle interruptions, see the *Stashing away your changes* section in this chapter. On the other hand, such temporary commits can be shared with other developers, as opposed to stash (because stash stack is based on purely local data – the reflog).

Discarding changes and rewinding the branch

Sometimes, your files will get in such a mess that you want to discard all changes and return the working directory and the staging area (the index) to the last committed state to the last good version. In other cases, you might want to rewind the state of the repository to an earlier version. In such instances, a **hard reset** is what you need; it will change the current branch head while resetting the index and the working tree. Any changes to any tracked files will be discarded:

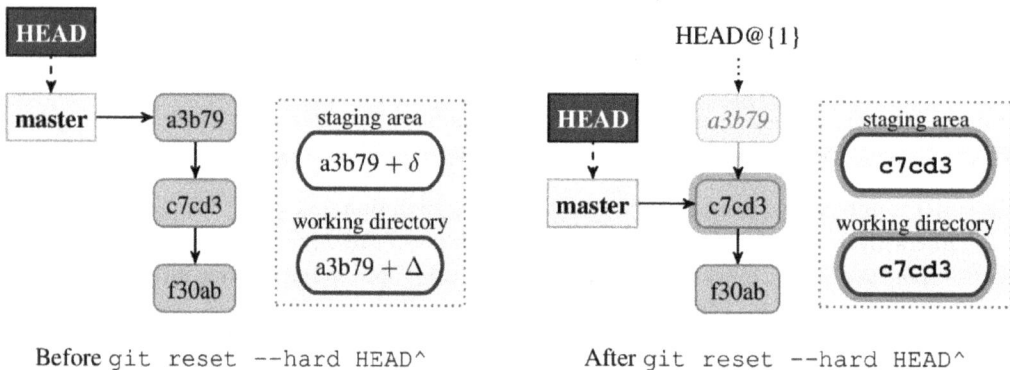

Figure 3.5 – Before and after a hard reset

This command can be used to undo a commit as if it had never happened, by removing it. Running `git reset --hard HEAD^` will effectively discard the last commit (though it will be available for a limited time via reflog) unless this commit can be reached from some other branch.

Another common usage is to discard changes to the working directory with `git reset --hard`, which resets to the last committed state.

> **Important note**
> It is very important to remember that a hard reset would irrecoverably remove all changes from the staging area and working directory. You cannot undo this part of the operation! Changes are lost forever!

Moving commits to a feature branch

Say that you were working on something on the `master` branch, and you have already created a sequence of commits. You have realized that the feature you are working on is more involved, and you want to continue polishing it on a separate topic branch, as described in *Chapter 8, Advanced Branching Techniques*. You want to move all those commits that are in `master` (let's say, the last three revisions) to the aforementioned feature branch.

You need to create the feature branch, save uncommitted changes (if any), rewind the `master` branch while removing those topical commits from it, and switch to the feature branch to continue working (or you can use rebase instead):

```
$ git branch feature/topic
$ git stash
No local changes to save
$ git reset --hard HEAD~3
HEAD is now at f82887f before
$ git switch feature/topic
Switched to branch 'feature/topic'
```

Of course, if there were local changes to save (there were none in the preceding example), this preceding series of commands would have to be followed by `git stash pop`.

Undoing a merge or a pull

Hard resets can also be used to abort a failed merge. You can use `git reset --hard HEAD` (here, HEAD is the default value for revision and can be omitted), for example, if you decide that you don't want to resolve the merge conflict at this time (though with modern Git you can use `git merge --abort` instead).

You can also remove a successful fast-forward pull or undo a rebase (and many other operations while moving the branch head) with `git reset --hard ORIG_HEAD`. (Here, you can use HEAD@{1} instead of ORIG_HEAD.)

Safer reset – keeping your changes

A hard reset will discard your local changes, similar to how `git switch --discard-changes` or `git checkout --force` would. Sometimes, you might want to rewind the current branch while keeping the local changes: that's what `git reset --keep` is for.

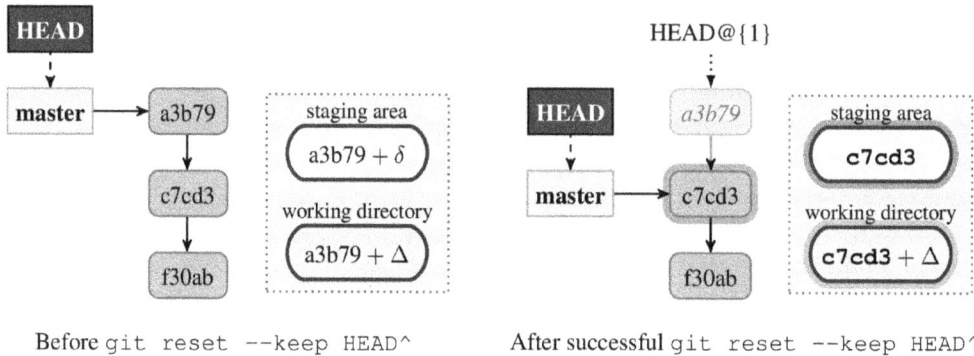

Figure 3.6 – Before and after a successful git reset --keep HEAD^ command

This mode resets the staging area (index entries) but retains the unstaged (local) changes that are currently in the working directory; see *Figure 3.6*. If it is not possible, the reset operation is aborted:

```
$ git reset --keep HEAD^
error: Entry 'README' not uptodate. Cannot merge.
fatal: Could not reset index file to revision 'HEAD^'.
```

This means that local changes in the worktree are preserved and moved to the new commit, in a similar way to how `git checkout <branch>` works with uncommitted changes. The successful case is a bit like stashing changes away, hard resetting, and then unstashing (but with a single atomic command).

> **How does safe reset work?**
>
> The way `git reset --keep <revision>` works is by updating the version (in the working directory) of only those files that are different between the revision we rewind to and HEAD. The reset is aborted if there is any file that is different between HEAD and `<revision>` (and thus would need to be updated) and has local uncommitted changes.

Rebasing current changes to an earlier revision

Suppose that you are working on something but you realize that what you have in your working directory should be in another branch, unrelated to a previous commit. For example, you might have started to work on a bug while on the `master` branch, and only then realized that it also affects the maintenance branch, `maint`.

This means that the fix should be put earlier in a branch, starting from the common ancestor of those branches (or a place where the bug was introduced). This would make it possible to merge the same fix both into master and maint, as described in *Chapter 15, Git Best Practices*:

```
$ edit
$ git checkout -b bugfix-127
$ git reset --keep start
```

An alternate solution would be to simply use git stash to move changes:

```
$ edit
$ git stash
$ git switch -c bugfix-127 start
$ git stash pop
```

Stashing away your changes

Often, when you've been working on a project, and things are in a messy state not suitable for a permanent commit, you want to temporarily save the current state and go to work on something else. The answer to this problem is the git stash command.

Stashing takes the dirty state of your working area – that is, your modified *tracked* files in your worktree and the state of the staging area – saves this state, and resets both the working directory and the index to the last committed version (to match the HEAD commit), effectively running git reset --hard HEAD. You can then reapply the stashed changes at any time.

You can also stash *untracked* files with the --include-untracked option.

Stashes are saved on a stack: by default, you apply the last stashed changes (stash@{0}), though you can list stashed changes (with git stash list) and explicitly select any of the stashes.

Using git stash

If you don't expect the interruption to last long, you can simply **stash away** your changes, handle the interruption, and then unstash them:

```
$ git stash
$ # ... handle interruption ...
$ git stash pop
```

By default, git stash pop will apply the last stashed changes and delete the stash if applied successfully. To see what stashes you have stored, you can use git stash list:

```
$ git stash list
stash@{0}: WIP on master: 049d078 atoi() is deprecated
stash@{1}: WIP on master: c264051 Add error checking
```

You can use any of the older stashes by specifying the stash name as an argument, or simply its number. For example, you can run `git stash apply stash@{1}` or `git stash apply 1` to apply it, and you can drop it (remove it from the list of stashes) with `git stash drop stash@{1}` or `git stash drop 1`; the `git stash pop` command is just a shortcut for apply + drop.

The default description that Git gives to stashed changes (namely **WIP on <branch>**) is useful for remembering where you were when stashing the changes (it specifies the branch and commit) but doesn't help you remember what you were working on, and what has been stashed away. However, you can examine the changes that were recorded in the stash as a diff with `git stash show -p`. But if you expect that the interruption might be more involved, you should save the current state to a stash while describing what you were working on:

```
$ git stash push -m 'Add <count>'
Saved working directory and index state On master: Add <count>
HEAD is now at 049d078 atoi() is deprecated
```

Git would then use the provided message to describe stashed changes when listing stashes:

```
$ git stash list
stash@{0}: On master: Add <count>
stash@{1}: WIP on master: c264051 Add error checking
```

Sometimes, the branch you were working on when you ran `git stash save` has changed enough that `git stash pop` fails because there are too many new revisions past the commit you were on when stashing the changes. If you want to create a regular commit out of the stashed changes, or just test stashed changes, you can use `git stash branch <branch name>`. This will create a new branch at the revision you were at when saving the changes, switch to this branch, reapply your work there, and drop stashed changes.

Stash and the staging area

By default, stashing resets both the working directory and the staging area to the HEAD version. You can make `git stash` keep the state of the index and reset the working area to the staged state with the `--keep-index` option:

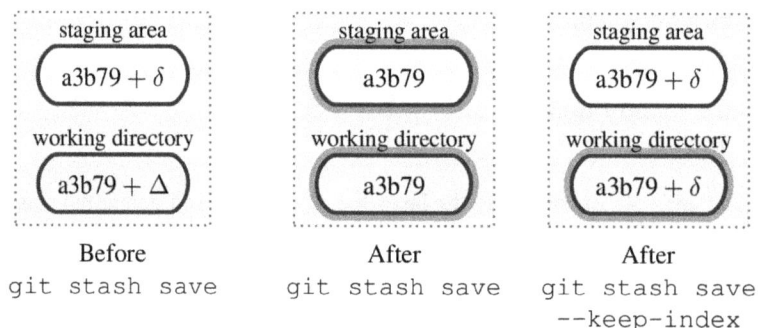

Figure 3.7 – The difference between git stash with and without --keep-index

This is very useful if you used the staging area to untangle changes in the working directory, as described in the *Selective commit* section in *Chapter 2, Developing with Git*, or if you want to split the commit in two, as described in the *Splitting a commit with reset* section in this chapter. In both cases, you would want to test each change before committing.

The workflow would look like this:

```
$ git add --interactive
$ git stash --keep-index
$ make test
$ git commit -m 'First part'
$ git stash pop
```

You can also use `git stash --patch` to select how the working area should look after stashing away the changes.

When restoring stashed changes, Git will ordinarily try to apply only saved worktree changes, adding them to the current state of the working directory (which must match the staging area). If there are conflicts while applying the state, they are stored in the index as usual – Git won't drop the stash if there are conflicts.

You can also try to restore the saved state of the staging area with the `--index` option; this will fail if there are conflicts when you're applying working tree changes (because there is no place to store conflicts since the staging area is busy).

Stash internals

Perhaps you applied stashed changes, did some work, and then for some reason want to un-apply those changes that originally came from the stash. Maybe you mistakenly dropped the stash or cleared all stashes (which you can do with `git stash clear`) and would like to recover them. Or perhaps you want to see how the file looked when you stashed away changes. To do any of this, you will need to know what Git does when creating a stash entry.

To stash away your changes, Git creates two automatic commits: one for the index (staging area) and one for the working directory. With `git stash --include-untracked`, Git creates an additional third automatic commit for untracked files.

The commit containing the work in progress (WIP) in the working directory (the state of files tracked from there) has the commit with the contents of the staging area (the index) as its second parent. This WIP containing commit is stored in a special ref: `refs/stash`. Both the WIP (stash) and index commits have the revision you were on when saving changes as their first parent.

We can see this by running `git log --graph` or `gitk --all`:

```
$ git stash save --quiet 'Add <count>'
$ git show-ref --abbrev
```

```
765b095 refs/heads/master
81ef667 refs/stash
$ gitk --all
```

This can be seen in the following figure:

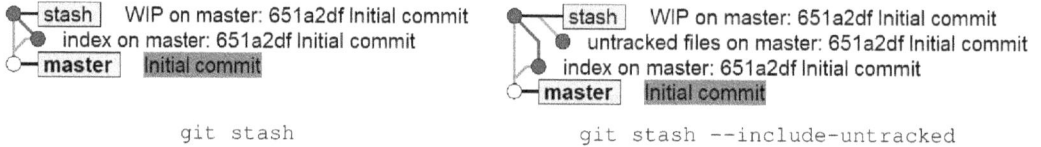

git stash   git stash --include-untracked

Figure 3.8 – The structure of the stash without and with untracked file information. Graphs were generated with gitk --all on a newly created repository with a single commit and a stash

We had to use git show-ref here (we could have used git for-each-ref instead) because git branch -a only shows branches, not arbitrary refs.

When saving untracked changes, with git stash --include-untracked, the situation is similar. *Figure 3.8* shows that the untracked file commit is the third parent of the WIP commit and that it doesn't have any parents. It only consists of untracked files, which you can check with git ls-tree -r stash@{<n>}^3.

Well, that's how stashing works, but how does Git maintain the stack of stashes? You may have noticed that the git stash list output and the stash@{<n>} notation therein looks like reflog; Git finds older stashes in the reflog for the refs/stash reference:

```
$ git reflog stash --no-decorate
81ef667 stash@{0}: On master: Add <count>
bb76632 stash@{1}: WIP on master: Added .gitignore
```

This is why you cannot share the stack of stashes: reflogs are local to the repository and are not and cannot be synchronized when pushing or fetching.

Un-applying a stash

Let's take the first example from the beginning of this section: un-applying changes from the earlier git stash apply. One possible solution to achieve the required effect is to retrieve the patch associated with working directory changes from a stash, and apply it in reverse:

```
$ git stash show -p stash@{0} | git apply -R -
```

Note the -p option that was applied to the git stash show command – it forces patch output instead of a summary of changes. We could use git show -m stash@{0} (the -m option is necessary because a WIP commit representing the stash is a merge commit), or even simply git diff stash@{0}^-1, in place of git stash show -p.

Recovering stashes that were dropped erroneously

Let's try the second example: recovering stashes that were accidentally dropped or cleared. If they are still in your repository and were not removed during the repository maintenance phase, you can search all commit objects that are unreachable from other refs and look like stashes (that is, they are merged commits and have a commit message using a strict pattern).

A simplified solution might look like this:

```
$ git fsck --unreachable |
grep "unreachable commit " | cut -d" " -f3 |
git log --stdin --merges --no-walk --grep="WIP on "
```

The first line of this pipeline finds all unreachable (lost) objects, the second one filters out everything but commits and extracts their SHA-1 identifiers, and the third line filters out even more, showing only merge commits with a commit message containing the `"WIP on "` string.

This solution would not, however, find stashes with a custom message (those created with `git stash save "message"`); you would need to add another `--grep`.

Managing worktrees and the staging area

In *Chapter 2, Developing with Git*, we learned that, besides the *working directory* (*worktree*) where you work on changes and the local repository where you store committed changes as revisions, there is also a third section between them: the *staging area*, sometimes called the *index*.

In the same chapter, we learned how to examine the status of the working directory, as well as how to view the differences. We saw how to create a new commit out of the working directory or out of the staging area.

Now, it is time to learn how to examine and modify the state of individual files.

Examining files and directories

It is easy to examine the contents of the working directory: you can just use the standard tools for viewing files (for example, an editor or pager) and examining directories (for example, a file manager or the `dir` command). But how do we view the staged contents of a file or the last committed version?

One possible solution is to use the `git show` command with the appropriate selector. *Chapter 4, Exploring Project History*, will introduce and explain the `<revision>:<pathname>` syntax to examine the contents of a file at a given revision. A similar syntax can be used to retrieve the staged contents, namely `:<pathname>` (or `:<stage>:<pathname>` if there is a merge conflict involving the given file; `:<pathname>` in itself is equivalent to `:0:<pathname>`).

Let's assume that we are in the `src/` subdirectory and want to see the contents of the `rand.c` file there as it's in the working directory, in the staging area (using the absolute and relative path), and in the last commit (also using the absolute and the relative path):

```
src $ less -FRX rand.c
src $ git show :src/rand.c
src $ git show :./rand.c
src $ git show HEAD:src/rand.c
src $ git show HEAD:./rand.c
```

To see the list of files that are staged in the index, there is the `git ls-files` command. By default, it operates on the staging area contents, but it can also be used to examine the working directory. The latter feature can, as we have seen in this chapter, be used to list ignored files. This command lists all files in the specified directory. Alternatively, in the current directory, you can use `:/` to denote the top-level directory of a project. The recursive behavior is caused by the fact that the index is a flat list of files, similar to `MANIFEST` files.

Without using the `--full-name` option, it would show filenames relative to the current directory (or the one specified as a parameter). In all examples, it is assumed that we are in the `src/` subdirectory, as seen in the command prompt:

```
src $ git ls-files
rand.c
src $ git ls-files --full-name :/
COPYRIGHT
Makefile
README
src/rand.c
```

What about committed changes? How can we examine which files were in a given revision? This is where `git ls-tree` comes to the rescue (note that it is a plumbing command and does not default to the HEAD revision):

```
src $ git ls-tree --name-only HEAD
rand.c
src $ git ls-tree --abbrev --full-tree -r -t HEAD
100644 blob 862aafd    COPYRIGHT
100644 blob 25c3d1b    Makefile
100644 blob bdf2c76    README
040000 tree 7e44d2e    src
100644 blob b2c087f    src/rand.c
```

Note that `git ls-tree` is not recursive by default; you need to use the `-r` option.

Searching file contents

Let's assume that you were reviewing code in the project and noticed an erroneous doubled semicolon, ; ;, in the C source code. Or perhaps you were editing the file and noticed a bug nearby. You fixed it, but you're wondering, "*How many of those mistakes are there?*" You would like to create a commit to fix such errors.

Or perhaps you want to search the version that was scheduled for the next commit – that is the contents of the staging area. Perhaps you want to examine how it looks in the next branch.

With Git, you can use the git grep command:

```
$ git grep -e ';;'
```

By default, this command will search tracked files in the working directory, from the current directory downwards, recursively. Note that when running the example command, we will get many false positives from shell scripts, for example. So, let's limit the search space to only C source files:

```
$ git grep -e ';;' -- '*.c'
```

The quotes around *.c are necessary for Git to do the glob pattern expansion (path limiting) instead of git grep getting the list of files expanded by the shell. We still have many false matches from the forever loop C idiom:

```
for (;;) {
```

With git grep, you can construct complex conditions, excluding false positives. Say that we want to search the whole project, not only the current directory, and avoid false positives:

```
$ git grep -e ';;' --and --not 'for *(.*;;' -- '**/*.c'
```

To search the staging area, use git grep --cached or the equivalent – and perhaps easier to remember – git grep --staged. To search the next branch, use git grep next --; this construction can be used to search any version.

Un-tracking, un-staging, and un-modifying files

If you want to undo some file-level operation (if, for example, you have changed your mind about tracking files or staging changes), then look no further than git status hints (add --ignored to get advice about ignored files):

```
$ git status --ignored
On branch master
Changes to be committed:
  (use "git restore --staged <file>..." to unstage)
```

```
Changes not staged for commit:
  (use "git add <file>..." to update what will be committed)
  (use "git restore <file>..." to discard changes in working
directory)

Untracked files:
  (use "git add <file>..." to include in what will be committed)

Ignored files:
  (use "git add -f <file>..." to include in what will be committed)
```

You need to remember that only the contents of the working directory and the staging area can be changed. Committed changes are immutable (though you can *rewind* the history or replace it).

If you want to undo adding a previously untracked file to the index – or remove a formerly tracked file from the staging area so that it will be deleted (not present) in the next commit while keeping it in the working directory – use `git rm --cached <file>`.

> **The difference between the --cached (--staged) and --index options**
>
> Many Git commands, including `git diff`, `git grep`, and `git rm`, support the `--cached` option (or its alias, `--staged`). Others, such as `git stash`, have the `--index` option (the index is an alternate name for the staging area). These are *not* synonyms (as we will later see with `git apply` command, which supports both).
>
> The `--cached` option is used to ask the command that usually works on files in the working directory to *only* work on the staged contents *instead*. For example, `git grep --cached` will search the staging area instead of the working directory, and `git rm --cached` will only remove a file from the index, leaving it in the worktree.
>
> The `--index` option is used to ask the command that usually works on files in the working directory to *also* affect the index, *additionally*. For example, `git stash apply --index` not only restores stashed working directory changes but also restores the index.

If you asked Git to record the state of some file in the staging area but changed your mind, you can reset the staged contents of the file to the committed version with `git restore --staged <file>` (`--source=HEAD` is the default) or `git reset HEAD -- <file>`.

If you edited a file incorrectly to the point that the working directory version is a mess and you want to restore it to the version from the index, use `git restore <file>` (`--worktree` is the default if `--staged` is not given) or `git checkout -- <file>`. If you staged some of this mess and would like to reset both the worktree and the staging area to the last committed version, use `git restore --worktree --staged <file>` or `git checkout HEAD -- <file>` instead.

> **Important note**
>
> These commands do *not undo operations*; they restore the previous state based on a backup that is the worktree, the index, or the committed version. For example, if you staged some changes, modified a file, and then added modifications to the staging area, you can reset the index to the committed version, but not to the state after the first and before the second `git add`.

Resetting a file to the old version

You can use any revision when restoring a file, with a per-file reset and per-file checkout. For example, to replace the current worktree version of the `src/rand.c` file with the one from the previous commit, you can use `git restore -s HEAD^ src/rand.c` or `git checkout HEAD^ -- src/rand.c` (or redirect the output of `git show HEAD^:src/rand.c` to a file). To put the version from the next branch into the staging area, run `git restore -s next src/rand.c` or `git reset next -- src/rand.c`.

Note that `git add <file>`, `git restore <file>`, `git reset <file>`, and `git checkout <file>` all enter interactive mode for a given file when invoked with the `--patch` option. This can be used to hand-craft a staged or worktree version of a file by selecting which changes should be applied (or un-applied).

> **Tip**
>
> When using Git from the command line, you might need to put a double dash, `--`, after other options and before the filename if, for example, you have a file with the same name as a branch.

Cleaning the working area

Untracked files and directories may pile up in your working directory. They can be leftovers from merges or be temporary files, proof of concept work, or perhaps mistakenly put there. Whatever the case, often, there is no pattern to them, and you don't need and don't want to make Git ignore them (see the *Ignoring files* section of this chapter); you just want to remove them. You can use the `git clean` command for that.

Because untracked files do not have a backup in the repository, and you cannot undo their removal (unless the operating system or the filesystem supports undo or trashcan), it's advisable to first check which files *can be removed* with `--dry-run/-n`. By default, actual removal requires the `--force/-f` option:

```
$ git clean --dry-run
Would remove patch-1.diff
```

Git will clean all untracked files recursively, starting from the current directory. You can select which paths are affected by listing them as an argument; you can also exclude additional types of files with the --exclude=<pattern> option. You can also interactively select which untracked files to delete with the --interactive option:

```
$ git clean --interactive
Would remove the following items:
  src/rand.c~
  screenlog.0
*** Commands ***
    1: clean          2: filter by pattern    3: select by numbers
    4: ask each       5: quit                 6: help
What now>
```

The clean command also allows us to only remove ignored files, for example, to remove build products but keep manually tracked files, with the -X option. However, usually, it is better to leave removing build byproducts to the build system, so that the project files can be cleaned without having to clone the repository.

You can also use git clean -x in conjunction with git reset --hard to create a pristine working directory to test a clean build by removing both ignored and not-ignored untracked files and resetting tracked files to the committed version.

Multiple working directories

For a long time, Git allowed you to specify where to find the administrative area of the repository (the .git directory). This can be done by appending the --git-dir=<path> option to the git command (that is, the git --git-dir=<path> <command> construct), or by setting the GIT_DIR environment variable. This feature makes it possible to work from the **detached working directory**.

With modern Git, you have a better solution to creating a new linked work tree than manual configuration: git worktree add <path> <branch>. This feature allows us to have more than one branch checked out. For convenience, if you omit the <branch> argument, then the new branch will be created based on the name of the created worktree.

This mechanism can be used instead of git stash if you need to switch to a different branch, but your current working directory, and possibly also the staging area, is in a state of high disarray. Instead of disturbing it, you can create a temporary linked working tree to make a fix and remove it when you're done. For example, you might need to do this to urgently fix a security bug in a separate branch.

Each detached worktree should be associated with and have checked out different branches or be on the anonymous branch (detached HEAD) to avoid problems. You can override this safety with the --force option.

You can remove any detached worktree with `git worktree remove` or by removing its directory and allowing it to be pruned. If a working tree is on a portable device or network disk, which may not always be available, we can `lock` the worktree so that it can't be pruned (and `unlock` if it is no longer needed).

To examine details about each working directory, such as the currently checked-out branch, and see if it is locked, you can use the `git worktree list` command.

Summary

In this chapter, we learned how to better manage the contents of the working directory and the staging area in preparation for creating a new commit.

We now know how to undo the last commit, how to drop changes to the working area, how to retroactively change the branch we are working on, and other uses of the `git reset` command. We also understand the three (and a half) forms of reset.

We also learned how to examine and search the contents of the working directory, the staging area, and committed changes. We now know how to use Git to copy the file version from the worktree, the index, or the `HEAD` commit into the worktree or the index. We can use Git to clean (remove) untracked files.

This chapter explained how to configure how files are handled in the working directory and how to make Git ignore files (by making them intentionally untracked) and why. It described how to handle the differences between line-ending formats between operating systems. It also explained how to enable (and write) keyword expansion, how to configure how binary files are handled, and how to enhance `diff` and `merge` specific classes of files.

Finally, we learned to stash away changes to handle interruptions and to make it possible to test interactively prepared commits, before creating a commit. This chapter explained how Git manages stashes, enabling us to go beyond built-in operations.

This chapter, together with *Chapter 2, Developing with Git*, taught you how to contribute to a project.

The following chapters will teach you how to collaborate with other people, how to send what you contributed, and how to merge changes from other developers. We will start with two chapters explaining how to explore ad search project history with *Chapter 4, Exploring Project History* and *Chapter 5, Searching Through the Repository*.

Questions

Answer the following questions to test your knowledge of this chapter:

1. How can you avoid having a large number of build artifacts appear in the `git status` output?
2. Let's assume that you use a custom **domain-specific language** (**DSL**) or a programming language without built-in support in Git, such as Julia. How can you configure Git so that it provides better support for this language?

3. How can you squash the two most recent commits while making one commit out of them?

4. How can you split the most recent commit into two commits?

5. What should you do if an urgent change is needed (for example, because of a security bug) but the working area is in a messy state and you don't want to lose your work?

6. How can you search through an old revision of the project – for example, a version tagged v0.1 – without checking out that revision?

Answers

Here are the answers to the questions given above:

1. Add patterns matching the pathnames of those build artifacts to a .gitignore file.

2. Define a custom diff driver and provide the regular expression pattern matching the main "sections" of code with xfuncname. Also, add an appropriate regular expression defining words in that programming language with wordRegex, and perhaps also define whitespace problems with the whitespace attribute.

3. Use git reset --soft HEAD~2 to rewind the branch and create a joined commit with git commit, or use interactive rebase.

4. Perform a soft reset, git reset --soft HEAD^, construct the first commit with interactive add, test the code with git stash --keep-index, pop the stash if the tests pass, and create the first commit with git commit and the second with git commit -a; there are other solutions.

5. Use git stash to stash away current changes, create a WIP commit, or create a new detached working area for the urgent work with git worktree add.

6. To search file contents from a revision tagged as v0.1, you can use git grep -e <pattern> v0.1.

Further reading

To learn more about the topics that were covered in this chapter, take a look at the following resources:

- Scott Chacon and Ben Straub, *Pro Git, 2.2 Git Basics – Recording Changes to the Repository,* the *Ignoring files* section: https://git-scm.com/book/en/v2/Git-Basics-Recording-Changes-to-the-Repository#_ignoring

- Scott Chacon and Ben Straub, *Pro Git, 7.3 Git Tools – Stashing and Cleaning*: https://git-scm.com/book/en/v2/Git-Tools-Stashing-and-Cleaning

- Scott Chacon and Ben Straub, *Pro Git, 8.2 Customizing Git – Git Attributes*: https://git-scm.com/book/en/v2/Customizing-Git-Git-Attributes

- *gitattributes manpage - Defining attributes per path*: `https://www.git-scm.com/docs/gitattributes`

- *gitignore manpage - Specifies intentionally untracked files to ignore*: `https://www.git-scm.com/docs/gitignore`

- Pragati Verma, *A Guide to Git Stash* (2021): `https://dev.to/pragativerma18/a-guide-to-git-stash-2h5d`

- Andrew Knight, *Ignoring Files with Git* (2018): `https://automationpanda.com/2018/09/19/ignoring-files-with-git/`

- Dragos Barosan, *New in Git: switch and restore* (2021): `https://www.banterly.net/2021/07/31/new-in-git-switch-and-restore/`

4

Exploring Project History

One of the important parts of mastering a **version control system** (**VCS**) is exploring project history and making use of the fact that with the VCS, we have an archive of every version that has ever existed. For example, you might want to examine what other developers did or remind yourself what you are about to publish.

In this chapter, you will learn how to select and view a revision or a range of revisions, as well as how to refer to them. The following chapter will continue this topic and explain how to find revisions using different criteria, as well as how to search through selected revisions; it will also describe how to search through project content.

This chapter will also introduce the concept of **Directed Acyclic Graph** (**DAG**) of revisions and explain how this concept relates to the idea of the project history, as well as the ideas of branches, tags, and the current branch in Git.

Here is the list of topics we will cover in this chapter:

- DAG of revisions as a way of representing history
- Different ways of revision selection
- Selecting starting branches and tags
- Using data from reflog to select revisions
- Double-dot (A..B) and triple-dot notation (A...B) for revision range selection
- Advanced revision range selection

The purpose of this chapter is to teach you how to select relevant parts of project history. The next chapter will explain how to investigate this further by searching through what you've selected.

DAGs

What makes VCSs different from backup applications is *the ability to represent more than linear history*. This is necessary both to support the simultaneous parallel development by different developers (each developer in their own clone of the repository) and to allow independent parallel lines of development – branches. For example, with a VCS, you might want to keep the ongoing development and the work on bug fixes for the stable version isolated. You can do this by using individual branches for those separate lines of development. So, the VCS needs to be able to model such a non-linear way of development and needs to have some structure to represent it.

The structure that Git uses (on the abstract level) to represent the possibly non-linear history of a project is called a **Directed Acyclic Graph (DAG)**.

The following diagram (*Figure 4.1*) shows an example of a DAG, drawn in two different ways. The same graph is represented on both sides of the figure: using the free-form layout on the left and the left-to-right layout on the right.

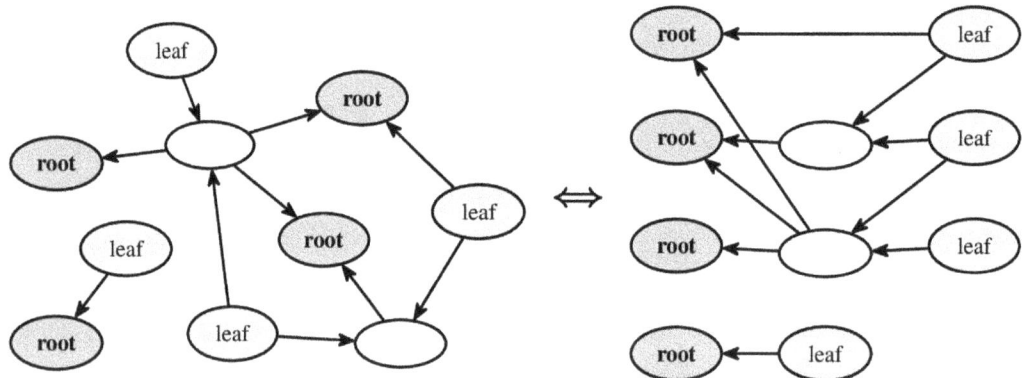

Figure 4.1 – Generic example of a DAG, with the same graph drawn with different layouts

A **directed graph** is a data structure from computer science and mathematics that's composed of *nodes* (vertices) connected with *directed edges* (arrows). A directed graph is *acyclic* if it doesn't contain any cycles, which means that there is no way to start at some node and follow a sequence of directed edges to end up at the starting node again.

Understanding this topic, in my opinion, helps with mastering the art of exploring, searching, and shaping the project history. You might want to read it more times to internalize this knowledge. It is not, however, required to be able to use Git successfully, so this section can be skipped on the first pass.

In a specific realization of a graph, each node represents some object (or a piece of data), and each edge from one node to another represents some kind of relationship between the objects it connects (or between data represented by nodes).

The DAG of revisions in **distributed version control systems** (**DVCSs**) uses the following representation:

- **Nodes**: In DVCSs, each node represents one **revision** (one version) of a project (of the entire tree). Those objects are called **commits**.

- **Directed edges**: In DVCSs, each edge represents the *this revision is based on that revision* relationship between two revisions. The arrow goes from a later **child** revision to an earlier **parent** revision it was based on – that is, the revision it was created from. This is *the reverse* of the way most people like to think of *the arrow of time* – that is, the arrow pointing from an earlier commit to a later one.

Because directed edges represent the *is based on* causal relationship between revisions, the arrows in the DAG of revisions cannot form a cycle. Usually, the DAG of revisions is laid out left-to-right (root nodes on the left, leaves on the right) or bottom-to-top (most recent revisions on the top). The figures in this book and ASCII-art examples in the Git documentation use the left-to-right convention, while the Git command line uses bottom-to-top, which is the most recent first convention.

There are two special types of nodes in any DAG (see *Figure 4.1*):

- **Root nodes** (or **roots**): These are nodes (revisions) that have no parents (have no outgoing edges). There is at least one root node in the DAG of revisions, which represents the initial (starting) version of a project.

- **Leaf nodes** (or **leaves**): These are nodes that have no children (no incoming edges); there is at least one such node. They represent the most recent versions of the project, not having any work based on them. Usually, each leaf in the DAG of revisions has a branch head pointing to it.

> Important note
>
> There can be more than one root node in Git's DAG of revisions. Additional root nodes can be created when you're joining two formerly originally independent projects together; each joined project brings its own root node. As this is a very rare occasion, with modern Git, you need to pass the `--allow-unrelated-histories` option to the `git merge` or `git pull` command to do so, to help avoid mistakes.
>
> Another source of root nodes is **orphan branches** – that is, disconnected branches with no history in common. They are, for example, used by GitHub to manage a project's web pages together in one repository with the code (in the `gh-pages` branch), and by the Git project itself to store pre-generated documentation (the `man` and `html` branches) and related projects (the `todo` branch). To create such a branch, you need to use the `--orphan` option in `git checkout` or `git switch`.

The fact that the DAG can have more than one leaf node means that there is no inherent notion of the latest version, as it was in the linear history paradigm.

Whole-tree commits

In DVCSs, each node of the DAG of revisions (DVCS's model of history) represents a version of the project as a whole single entity: a **snapshot** of the whole directory tree contents of a project.

This means that by default, each developer will get the history of all the files in their clone of the repository. Where needed, they can choose to get only a part of the history (shallow clone and/or cloning only selected branches), they can checkout only selected files (sparse checkout), or they can use the partial clone feature (with, for example, different versions of files contents loaded on demand). Those special cases, and more, will be described in *Chapter 12, Managing Large Repositories*.

Branches and tags

A **branch operation** is what you use when you want your development process to fork into two different directions, to create another line of development. For example, you might want to create a separate branch, called maintenance, to help in managing bug fixes to the released stable version of a project, while isolating this activity from the rest of the development.

A **tag operation** is a way to associate a meaningful symbolic name with the specific revision in the repository. For example, let's assume that you're preparing to release version 1.3 of your project. To do this, you're preparing a *release candidate* version to send to beta testers. Let's also assume that this is your third such attempt. You might want to create a v1.3-rc3 tag, among others, to be able to go back to this specific version, check the validity of bug reports from your testers, and find the sources of reported bugs.

Both branches and tags, sometimes called **references** (**refs**) when used together, have the same meaning and almost the same representation within the DAG of revisions. They are external references (pointers) to the graph of revisions, as shown in *Figure 4.2*:

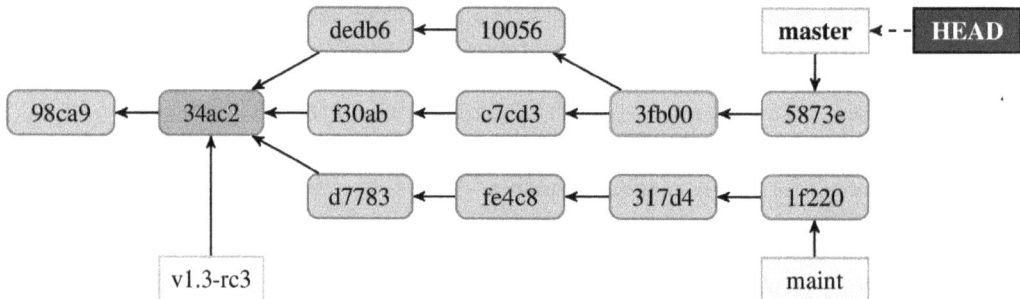

Figure 4.2 – Example DAG of revisions in a VCS with two branches, a single tag, one fork point, and a single merge commit

A **tag** is a symbolic name for a given revision – for example, v1.3-rc3 in *Figure 4.2*. It always points to the same object; it doesn't change. The idea behind having tags is to be able to refer to the given revision with a symbolic name and to have this symbolic name mean the same for every developer. Checking out or viewing the given tag should provide the same results for everyone.

A **branch** is a symbolic name for the line of development. The most recent commit (leaf revision) on such a line of development is referred to as the *top* or *tip of the branch*, **branch head**, or just a branch. Creating a new commit (on a current branch) will generate a new node in the DAG and advance the appropriate branch reference. (*Figure 4.2* shows two branch heads: `maint` and `master`.)

The branch in the DAG is, as a line of development, the subgraph of the DAG composed of those revisions that can be **reached** from the tip of the branch (from the branch head) – in other words, of those revisions that you can walk to by following the parent edges starting from the branch head.

Git needs to know which branch tip to advance when creating a new commit. It needs to know which branch is the current one and is **checked out** into the working directory. Git uses the **HEAD** pointer for this, as shown in *Figure 4.2*. Usually, this points to one of the branches, which, in turn, points to some node in the DAG of revisions. However, this isn't always the case – see *Chapter 2, Developing with Git*, for an explanation of the **detached HEAD** situation, where HEAD points directly to a node in the DAG.

Full names of references (branches and tags)

Originally, Git stored branches and tags in files inside the `.git` administrative area, in the `.git/refs/heads/` and `.git/refs/tags/` directories, respectively. Modern Git can store information about tags and branches inside the `.git/packed-refs` file to avoid handling a very large number of small files. Nevertheless, active references use the original *loose* format – one file per reference.

The HEAD pointer (denoting the current branch) is stored in `.git/HEAD`. It is usually a symbolic reference – for example, **ref: refs/heads/master**.

The `master` branch is stored in `.git/refs/heads/master` and has `refs/heads/master` as its full name (in other words, branches reside in the `refs/heads/` namespace). The tip of the branch is referred to as the **head** of a branch, hence the name of this namespace. In *loose* format, the file contents are an SHA-1 identifier of the most current revision on the branch (the **branch tip**), in plain text as hexadecimal digits. It is sometimes required to use the full name if there is ambiguity among refs.

The remote-tracking branch, `origin/master`, which remembers the last seen position of the `master` branch in the remote repository, `origin`, is stored in `.git/refs/remotes/origin/master` and has `refs/remotes/origin/master` as its full name. The concept of **remotes** will be explained in *Chapter 6, Collaborative Development with Git*, while **remote-tracking branches** will be covered in *Chapter 8, Advanced Branching Techniques*.

The `v1.3-rc3` tag has `refs/tags/v1.3-rc3` as its full name (tags reside in the `refs/tags/` namespace). To be more precise, in the case of **annotated** and **signed tags**, this file stores a reference to the **tag object**, which, in turn, points to the node in the DAG, and not directly to a commit. This is the only type of ref that can point to any type of object; branches and remote-tracking branches always point to a commit.

These full names (fully qualified names) can be seen when using commands intended for scripts (so-called **plumbing commands**), such as `git show-ref`:

```
$ git show-ref
98cbfdf5c1be9a4f6c0f7e3b97608b39274463df refs/heads/master
d81ce7b6aeedb51aa2d5e18d110333aea080fdd4 refs/stash
```

Branch points

When you create a new branch starting at a given version, the lines of development usually diverge. The act of creating a divergent branch is denoted in the DAG by a commit that has more than one child – that is, a node pointed to by more than one arrow.

> **Important note**
>
> Git does not track information about creating (forking) a branch and does not mark branch points in any way that they're preserved across clones and pushes. There is information about this event in the **reflog** (**branch: Created from HEAD**), but this is local to the repository where branching occurred and is temporary. However, if you know that the B branch started from the A branch, you can find the branching point with `git merge-base A B`. In modern Git, you can use the `--fork-point` option to make this command use the information from the reflog, when available.

In *Figure 4.2*, the `34ac2` commit is a branching point, or a **fork point**, for the `master` and `maint` branches.

Merge commits

Typically, when you've used branches to enable independent parallel development, you will want to join them later. For example, you would want bug fixes that have been applied to the stable (maintenance) branch so that they're included in the main line of development as well (if they're applicable and weren't fixed accidentally during the main line development).

You would also want to merge changes created in parallel by different developers working simultaneously on the same project, each using their own clone of the repository, and creating their own lines of commits.

Such a **merge operation** will create a new revision, joining two lines of development. The result of this operation will be based on more than one commit. The node in the DAG representing said revision will have more than one parent and more than one outgoing edge. Such an object is called a **merge commit**.

You can see a merge commit, `3fb00`, in *Figure 4.2*.

Single revision selection

During development, often, you'll want to select a single revision in the history of the project so that you can examine it or compare it with the current version. The ability to select a revision is also the basis for selecting a revision range – for example, selecting a subsection of history to examine.

Many Git commands take revision parameters as arguments, which are typically denoted by `<rev>` in the Git reference documentation. Git allows you to specify a commit or a range of commits in several ways. This will be described in this and the next section.

HEAD – the implicit revision

Most, but not all, Git commands that require the revision parameter default to using HEAD. For example, `git log` and `git log HEAD` will show the same information. You can also use @ alone as a shortcut for HEAD.

Here, HEAD denotes the **current branch**, or in other words, the commit that was checked out into the working directory and forms a base of current work (a current revision).

There are a few other references that are similar to HEAD:

- `FETCH_HEAD`: This records the information about the remote branches that were fetched from a remote repository with your last `git fetch` or `git pull` invocation. It is very useful for one-off fetching, with a repository to fetch from given by a URL (`git fetch <URL>`), unlike when we're fetching from a named repository such as `origin`, where we can use a remote tracking branch instead, such as `origin/master`. Moreover, with named repositories, we can use the reflog for the remote-tracking branch – for example, `origin/master@{1}` – to get the position before the fetch. Note that FETCH_HEAD is overwritten by each fetch from any repository.

- `ORIG_HEAD`: This records the previous position of the current branch. This reference is created by commands that move the current branch in a drastic way (creating a new commit doesn't set ORIG_HEAD) to record the position of HEAD before the operation. This is very useful if you want to undo or abort such an operation. However, nowadays, the same can be done using reflogs, which store additional information that can be examined in their use; see the *Reflogging shortnames* section for more details.

You can also stumble upon the short-lived temporary references that are used during specific operations:

- During a merge, before creating a merge commit, `MERGE_HEAD` records the commit(s) that you're merging into your branch. It vanishes after creating a merge commit

- During a cherry-pick, before creating a commit that copies picked changes into another branch, `CHERRY_PICK_HEAD` records the commit that you've selected for cherry-picking

Branch and tag references

The most straightforward and commonly used way to specify a revision is to use symbolic names: branches, naming the line of development, pointing to the tip of said line; and tags, naming the specific revision. This way of specifying revisions can be used to view the history of a line of development, examine the most current revision (current work) on a given branch, or compare a branch or a tag with the current work.

You can use any of the *refs* (external references to the DAG of revisions) to select a commit. You can use a branch name, a tag name, and a remote-tracking branch name in any Git command that requires a revision as a parameter.

Usually, it is enough to give the *short* name of a branch or tag, such as `git log master`, to view the history of a `master` branch, or `git log v1.3-rc3` to see how version `v1.3-rc1` came about. It can, however, happen that there are different types of refs with the same name, such as both the branch and tag being named `dev` (though it is recommended to avoid such situations). Alternatively, you could have created (usually by accident) the local `origin/master` branch when there was a remote-tracking branch with a short named `origin/master`, tracking where the `master` branch was in the remote repository named `origin`.

In such a situation, when the ref name is ambiguous, it is disambiguated by taking the first match in the following rules (this is a shortened and simplified version – for the full list, see the **gitrevisions(7)** manpage):

1. The top-level symbolic name – for example, HEAD.

2. Otherwise, the name of the tag (the `refs/tags/` namespace).

3. Otherwise, the name of the local branch (the `refs/heads/` namespace).

4. Otherwise, the name of the remote-tracking branch (the `refs/remotes/` namespace).

5. Otherwise, the name of the remote if a default branch exists for it; the revision is said to be the default branch (for example, `refs/remotes/origin/HEAD` for `origin` as a parameter).

The --branches, --tags, and similar options

If you want to see the whole graph of revisions, you need a way to specify all the refs – that is, branches, remote-tracking branches, and tags. That's what the `--all` option to the `git log` command is for. With this option, Git pretends as if all the refs in the `refs/` namespace, along with HEAD, were listed as starting points for revision traversal (for viewing the history of a project).

If you want to limit yourself to branches, remote-tracking branches, or tags, you can use the `--branches`, `--remotes`, or `--tags` option, respectively. All of those options take an optional `<pattern>` parameter, which limits respective refs to ones matching the given shell glob. If the pattern lacks glob wildcards (that is, `*`, `?`, or `[`), then `/*` at the end of the pattern is implied. For example, to pretend as if all topic branches (with hierarchical names that begin with author initials) and all remote-tracking branches for the `origin` remote were listed on the command line, you can use the following command:

```
$ git log --branches=??/* --remotes=origin
```

The `--all` option with the `<pattern>` parameter is named `--glob=<pattern>`.

Glob patterns

In computer science, **glob patterns** are used to match strings using a specific set of wildcard characters. This is the syntax that's used by UNIX shells and is described on the **glob(7)** manpage. It is simpler but less expressive than **regular expressions**.

The most common glob wildcards are `*`, `?`, and `[...]`. The `*` wildcard character matches any number of characters including none, `?` matches a single character, and `[abc]` matches one character from the one listed inside brackets. You can simplify the list of characters using the character range – for example, `[a-z]`.

Pattern matching can be enhanced with the help of the `--exclude=<pattern>` option, which affects `--all`, `--branches`, `--tags`, `--remotes`, and `--glob`, excluding refs that the next such option would otherwise consider. This option can be given multiple times, accumulating exclusion patterns. For example, to include all topic branches but exclude your own topic branches (which have names starting with `jn/`), you can use the following command:

```
$ git log --exclude=jn/* --branches=??/*
```

SHA-1 and the shortened SHA-1 identifier

In Git, each revision is given a unique identifier (object name), which is a **SHA-1 hash function**, based on the contents of the revision (though the exact hash function will change from SHA-1 to SHA-256 in the future). You can select a commit by using its SHA-1 identifier as a 40-character long hexadecimal number (120 bits). Git shows full SHA-1 identifiers in many places. For example, you can find them in the full `git log` output:

```
$ git log
commit 50f84e34a1b0bb893327043cb0c491e02ced9ff5
Author: Junio C Hamano <gitster@pobox.com>
Date:   Mon Jun 9 11:39:43 2014 -0700

    Update draft release notes to 2.1

    Signed-off-by: Junio C Hamano <gitster@pobox.com>

commit 07768e03b5a5efc9d768d6afc6246d2ec345cace
Merge: 251cb96 eb07774
```

```
Author: Junio C Hamano <gitster@pobox.com>
Date:    Mon Jun 9 11:30:12 2014 -0700

    Merge branch 'jc/shortlog-ref-exclude'
```

It isn't necessary to give the full 40 characters of the SHA-1 identifier. Git is smart enough to figure out what you meant if you provide it with the first few characters of the SHA-1 revision identifier, so long as the partial SHA-1 is at least 4 characters long. To be able to use a shortened SHA-1 to select revision, it must be long enough to be unambiguous – that is, there must be one and only one commit object where the SHA-1 identifier begins with given characters.

For example, both `dae86e1950b1277e545cee180551750029cfe735` and `dae86e` name the same commit object, assuming, of course, that that there is no other object in your repository whose object name starts with `dae86e`. If there is any ambiguity, Git will tell us about all the choices, like so:

```
error: short object ID dae86e is ambiguous
hint: The candidates are:
hint:    dae86e19 commit 2021-03-17 - README: Add CI badges
hint:    dae86e1f tree
hint:    dae86ebf blob
fatal: ambiguous argument 'dae86e': unknown revision or path not in
the working tree.
Use '--' to separate paths from revisions, like this:
'git <command> [<revision>...] -- [<file>...]'
```

In many places, Git shows unambiguously shortened SHA-1 identifiers in its command output. For example, in the preceding example of the `git log` output, we can see the shortened SHA-1 identifiers in the `Merge:` line.

You can also request that Git use the shortened SHA-1 in place of the full SHA-1 revision identifiers with the `--abbrev-commit` option. By default, Git will use at least 7 characters for the shortened SHA-1; you can change this with the optional parameter – for example, `--abbrev-commit=12`.

Note that Git will use as many characters as is required for the shortened SHA-1 to be unique at the time the command was issued. The parameter to `--abbrev-commit` (and the similar `--abbrev` option) is the minimal length of the abbreviation.

A short note about shortened SHA-1

Generally, 8 to 10 characters is more than enough for the shortened SHA-1 (for the SHA-1 prefix) to be unique within a project. One of the largest Git projects, the Linux kernel, is beginning to need 12 characters out of the possible 40 to stay unique. While a hash collision, which means having two revisions (two objects) that have the same full SHA-1 identifier, is extremely unlikely (with $1/2^{80} \approx 1/1.2 \times 10^{24}$ probability), the formerly unique shortened SHA-1 identifier may stop being unique due to repository growth.

The SHA-1 and the shortened SHA-1 are often copied from the command output and pasted as revision parameters in another command. They can also be used to communicate between developers in case of doubt or ambiguity as SHA-1 identifiers are the same in any clone of the repository. *Figure 4.2* uses a five-character shortened SHA-1 to identify revisions in the DAG.

Ancestry references

The other main way to specify a revision is via its **ancestry**. You can specify a commit by starting from some child of it (for example, you can start from the current commit – that is, HEAD, a branch head, or a tag), and then follow through parent relationships to the commit in question. There is a special suffix syntax to specify such ancestry paths.

If you place ^ at the end of a revision name, Git resolves it to mean a (first) parent of that revision. For example, HEAD^ means the parent of HEAD – that is, the previous commit.

This is a shortcut syntax. For merge commits, which have more than one parent, you might want to follow any of the parents. To select a parent, put its number after the ^ character: using the ^<n> suffix means the *n-th* parent of a revision. We can see that ^ is a short version of ^1.

As a special case, ^0 means the commit itself; it is only important when a command behaves differently when you're using the branch name as a parameter and when you're using other revision specifiers. It can be also used to get the commit that an annotated (or a signed) tag points to; compare git show v0.9 and git show v0.9^0. Note that you can do the latter operation with <tag>^{commit}; in most cases, it is what <tag>^{} would do (follow this until you find an object that isn't a tag).

This suffix syntax is composable. You can use HEAD^^ to mean the grandparent of HEAD and the parent of HEAD^. There is another shortcut syntax for specifying a chain of first parents. Instead of writing *n* times the ^ suffix – that is, ^^...^ or ^1^1...^1 – you can simply use ~<n>. As a special case, ~ is equivalent to ~1, so HEAD~ and HEAD^ are equivalent. In a similar vein, HEAD~2 means the first parent of the first parent or the grandparent and is equivalent to HEAD^^.

You can also combine everything. For example, you can get the second parent of a great-grandparent of HEAD (assuming it was a merge commit) by using HEAD~3^2, and so on. You can use git name-rev or git describe --contains to find out how a revision is related to local refs, like so:

```
$ git log | git name-rev --stdin
commit 82006acd359717624fb33a7ae554cba6be717911 (master)
Merge: 20cfc7c 3a59408
Author: Bob Hacker <bob@company.com>
Date:   Sun May 30 00:58:23 2021 +0200

    Merge branch 'master' of https://git.company.com/random

commit 20cfc7c25ff82e36d6e72b6a31f5839331f270e7 (master~1)
Author: Bob Hacker <bob@company.com>
```

```
Date:    Sun May 30 00:44:59 2021 +0200

    Added COPYRIGHT
[...]
```

As you can see, with `git name-rev --stdin` used as a filter for `git log`, after each SHA-1 identifier, you get its ancestry reference in parentheses – for example, (**master~1**).

Reverse ancestry references – git-describe output

The ancestry reference describes how a historic version relates to the current branches and tags. It depends on the position of the starting revision. For example, HEAD^ would usually mean a completely different commit next month.

Sometimes, we want to describe how the current version relates to the prior named version. For example, we might want to have a human-readable name of the current version to store in the generated binary application. We want this name to refer to the same revision for everybody. This is the task of `git describe`.

Here, `git describe` finds the most recent tag that can be reached from a given revision (by default, from HEAD) and uses it to describe that version. If the found tag points to the given commit, then (by default) only the tag is shown. Otherwise, `git describe` suffixes the tag name with the number of additional commits on top of the tagged object and uses the abbreviated SHA-1 identifier of the given revision. For example, `v1.0.4-14-g2414721` means that the given commit was based on the named (tagged) version `v1.0.4`, which was 14 commits ago, and that it has `2414721` as a shortened SHA-1. Without the SHA-1 abbreviation, the notation would be ambiguous; in the presence of non-linear history, there can be many revisions that are 14 commits away from the given tag.

Git understands this output format as a revision specifier.

Reflogging shortnames

To help you recover from some types of mistakes, and to be able to undo changes (to go back to the state before the change), Git keeps a **reflog** – a *temporary* log of where your HEAD and branch references have been for the last few months, and how they got there, as described in *Chapter 2, Developing with Git*. The default is to keep reflog entries up to 90 days; 30 days for revisions that can only be reached through reflog (for example, amended commits). This can be configured on a ref-by-ref basis; see *Chapter 13, Customizing and Extending Git*.

You can examine and manipulate your reflog with the `git reflog` command and its subcommands. You can also display history with `git log -g` (or `git log --walk-reflog`):

```
$ git reflog
ba5807e HEAD@{0}: pull: Merge made by the 'recursive' strategy.
3b16f17 HEAD@{1}: reset: moving to HEAD@{2}
```

```
2b953b4 HEAD@{2}: reset: moving to HEAD^
69e0d3d HEAD@{3}: reset: moving to HEAD^^
3b16f17 HEAD@{4}: commit: random.c was too long to type
```

Every time HEAD and your branch head are updated for any reason, Git stores that information for you in this local temporary log of ref history. The data from the reflog can be used to specify references (and therefore to specify revisions):

- To specify the nth prior value of HEAD in your local repository, you can use HEAD@{n} notation that you can see in the git reflog output. It's the same with the nth prior value of the given branch – for example, master@{n}. The special syntax, @{n}, means the nth prior value of the current branch, which can be different from HEAD@{n}.

- You can also use this syntax to see where a branch was some specific amount of time ago. For instance, to denote where your master branch was yesterday in your local repository, you can use master@{yesterday}.

- You can use the @{-n} syntax to refer to the nth branch that was checked out (used) before the current one. In some places, you can simply use – (dash) in place of @{-1}. For example, git checkout – or git switch – will switch to the previous branch.

Upstreaming remote-tracking branches

The local repository that you use to work doesn't usually live in isolation. It interacts with other repositories, usually at least with the origin repository it was cloned from (unless it was started from scratch with git init).

> **Note**
> The name of the default remote can be set using clone.defaultRemoteName.

For these remote repositories with which you interact often, Git will track where their branches were at the time of the last contact.

To follow the movement of branches in the remote repository, Git uses **remote-tracking branches**. You cannot create new commits on remote-tracking branches as they would be overwritten on the next contact with the remote. If you want to create your own work based on work on some branch in the remote repository, you need to create a local branch based on the respective remote-tracking branch. Git can do that automatically: when there is no local branch with the same name as the remote-tracking branch, some-branch, then the git checkout <some-branch> command will create a local branch based on this remote-tracking branch for you.

For example, when working on a line of development that is to be ultimately published to the next branch in the origin repository, which is tracked by the origin/next remote-tracking branch, you would create a local next branch. We say that origin/next is upstream of the next branch, and we can refer to it as next@{upstream}.

The @{upstream} suffix (short form <refname>@{u}), which can only be applied to a local branch name, selects the branch that the ref is set to build on top of. A missing ref defaults to the current branch – that is, @{u} is the upstream for the current branch.

There is also [<branch>]@{push}, which is useful for triangular workflows, where the repository you push your changes to is different from the repository you get updates from.

Selecting revisions via a commit message

You can specify the revision by matching its commit message with a regular expression. The :/<pattern> notation (for example, :/^Bugfix) specifies the youngest matching commit that can be reached from any ref, while <rev>^{/<pattern>} (for example, next^{/fix bug}) specifies the youngest matching commit that can be reached from <rev>:

```
$ git log 'origin/pu^{/^Merge branch .rs/ref-transactions}'
```

This revision specifier gives similar results to the --grep=<pattern> option to git log, but it's composable. This means that it can be combined with other components, such as ancestry references. On the other hand, it only returns the first (youngest) matching revision, while the --grep option returns all matching revisions.

Selecting the revision range

Now that you can specify individual revisions in multiple ways, let's learn how to specify ranges of revisions, a subset of the DAG we want to examine. **Revision ranges** are particularly useful for viewing selected parts of the history of a project.

For example, you can use range specifications to answer questions such as, "What work is on this branch that I haven't yet merged into my main branch?", "What works on my main branch I haven't yet published?", or simply "What was done on this branch since its creation?"

Single revision as a revision range

History traversing commands such as git log operate on a set of commits, walking down a chain of revisions from child to parent. These kinds of commands, given a single revision as an argument (as described in the *Single revision selection* section of this chapter), will show the set of commits that can be reached from that revision, following the commit ancestry chain, all the way down to root commits. Thanks to Git using pager by default, Git will stop after one full page – that is, one full screen of commits.

For example, `git log master` would show all commits that can be found from the tip of a `master` branch (all revisions that are or were based on the current work on the said branch), which means that it would show the whole `master` branch – that is, the whole line of development.

Double-dot notation

The most common range specification is the double-dot syntax, `A..B`. For a linear history, it means all revisions between A and B, or to be more exact all the commits that are in B but not in A, as shown in *Figure 4.3*. For example, the `HEAD~4..HEAD` range means four commits: `HEAD`, `HEAD^`, `HEAD^^`, and `HEAD^^^`. In other words, it means `HEAD~0`, `HEAD~1`, `HEAD~2`, and `HEAD~3`, assuming that there is no merge commit between the current branch and its fourth ancestor:

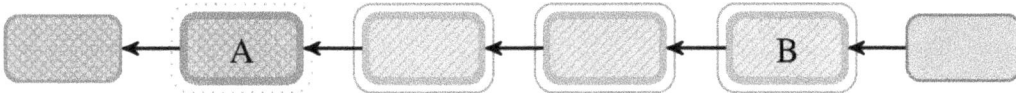

Figure 4.3 – Double-dot notation A..B for linear history. The selected
revision range is marked with a thin halo (with an outline)

> **Tip**
> If you want to include a starting commit (in the general case, boundary commits), which Git considers uninteresting by default, you can use the `--boundary` option with `git log`.

The situation is more complicated for a history that is not a straight line. One such case is when A is not the ancestor of B (there is no path in the DAG of revisions leading from B to A), but both have a common ancestor, as shown in *Figure 4.4*:

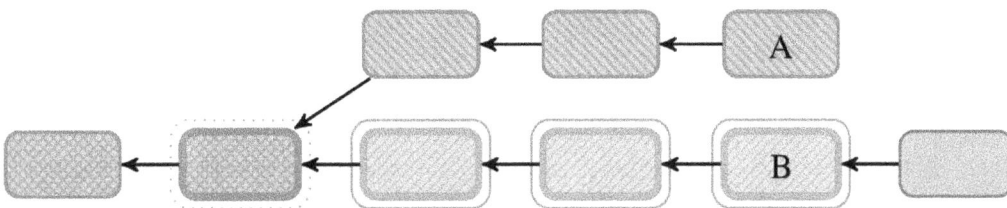

Figure 4.4 – Double-dot notation A..B for non-linear history, where revision A is not an
ancestor of revision B, showing the case with a divergent history (with a fork point)

Another situation with non-linear history is when the path from B to A is not a simple line – that is, when there are merge commits between A and B, as shown in *Figure 4.5*. In the view of nonlinear history, the double-dot notation, `A..B`, or *between A and B*, is defined as those commits that can be reached from A while being not reachable from B:

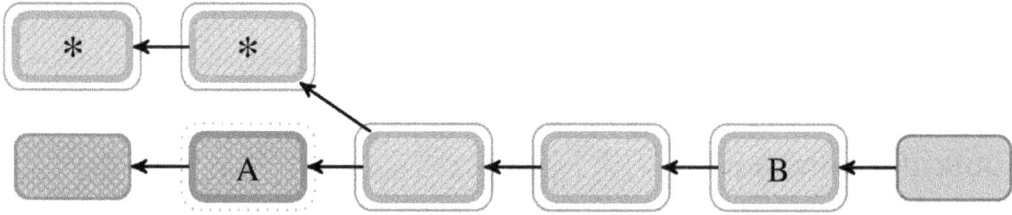

Figure 4.5 – Double-dot notation A..B for a non-linear history, with merge commit between A and B. To exclude commits marked with *, use the --strict-ancestor option

For Git, A..B means a range of all commits that can be reached from one revision (B) but can't be reached from another revision (A) while following the ancestry chain. In the case of divergent A and B, as shown in *Figure 4.4*, this is simply all commits in B from the branch point of A.

For example, say your master and experiment branches diverge. You want to see what's in your experiment branch that hasn't been merged into your master branch yet. You can ask Git to show you a log of just those commits with master..experiment.

If, on the other hand, you want to see the opposite – all the commits in master that aren't in experiment – you can reverse the branch names. The experiment..master notation shows you everything in master that can't be reached from experiment.

Another example is that origin/master..HEAD shows what you're about to push to a remote repository (commits in your current branch that are not yet present in the master branch in origin), while HEAD..origin/master can show what you have fetched but not yet merged in.

> **Tip**
> You can also leave off one side of the syntax to have Git assume HEAD: origin/master..
> is origin/master..HEAD and ..origin/master is HEAD..origin/master; Git
> substitutes HEAD if one side is missing.

Git uses double-dot notation in many places, such as in the output of git fetch and git push for ordinary fast-forward cases. Here, you can just copy and paste a fragment of output as parameters to git log. In this case, the beginning of the range is the ancestor of the end of the range – that is, the range is linear:

```
$ git push
To https://git.company.com/random
   8c4ceca..493e222  master -> master
```

Creating the range by including and excluding revisions

The double-dot A..B syntax is very useful and quite intuitive, but it is a shorthand notation. Usually, it's enough, but sometimes, you might want more than it provides. Perhaps you want to specify more than two branches to indicate your revision, such as seeing what commits are present in any of several branches that aren't in the branch you're currently on. Perhaps you want to see only those changes on the master branch that aren't in any of the other long-lived branches.

Git allows you to exclude the commits that can be reached from a given revision by *prefixing* said revision with ^. For example, to view all revisions that are on maint or master, but are not in next, you can use git log maint master ^next. This means that the A..B notation is just a shorthand for B ^A.

Instead of having to use the ^ character as a prefix for each of the revisions we want to exclude, Git allows us to use the --not option, which *negates* all the following revisions. For example, B ^A ^C might be written as B --not A C. This is useful, for example, when we're generating excluded revisions programmatically.

Thus, these three commands are equivalent:

```
$ git log A..B
$ git log B ^A
$ git log B --not A
```

The revision range for a single revision

There is another useful shortcut, A^!, that is a range composed of a single commit. For non-merge commits, it is simply A^..A.

For merge commits, A^! excludes all the parents. With the help of yet another special notation, namely A^@, denoting all the parents of A (that is, A^1, A^2,... A^n), we can say that A^! is a shortcut for A --not A^@.

Triple-dot notation

The last major syntax for specifying revision ranges is the triple-dot syntax, A...B. It selects all the commits that can be reached by either of two references, but not by both of them; see *Figure 4.6*. In mathematics, this notation is called the **symmetric difference of A and B**:

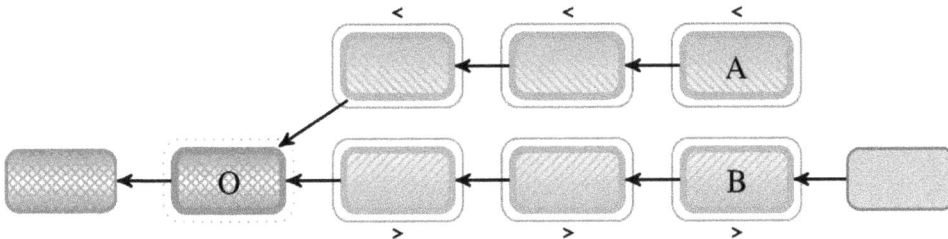

Figure 4.6 – A triple-dot notation, A...B, for a non-linear history, where the selected range is shown with a thin outline, and O is the boundary commit – the merge base of A and B

It is a shortcut notation for A B --not $(git merge-base --all A B), where $(...) denotes shell **command substitution** (using POSIX shell syntax). Here, it means that the shell will run the git merge-base command to find out all the best common ancestors (all merge bases), and then paste its output on the command line so that it can be negated.

A common switch to the git log command to use with the triple dot notation is --left-right. This option makes Git show which side of the range each commit is in by prefixing commits from the left-hand side (A in A...B) with <, and those from the right (B in A...B) with >, as shown in *Figure 4.6* and the following example. This helps make the data more useful:

```
$ git log --oneline --left-right 37ec5ed...8cd8cf8
>8cd8cf8 Merge branch 'fc/remote-helper-refmap' into next
>efcd02e Merge branch 'rs/more-starts-with' into next
>831aa30 Merge branch 'jm/api-strbuf-doc' into next
>1aeca19 Merge branch 'jc/count-parsing' into next
<1a7e8e8 Revert "replace: add --graft option"
<7a30690 t9001: avoid non-portable '\n' with sed
>5cc3268 fetch doc: remove "short-cut" section
```

> **Important note**
>
> If the --left-right option is combined with --boundary, these normally uninteresting boundary commits are prefixed with -.
>
> In the case of using the triple-dot A...B revision range, these boundary commits are git merge-base --all A B.

Git uses triple-dot notation in git fetch and git push output when there is a **forced update**, in cases where the old version (left-hand side) and the updated version (right-hand side) diverged, and the new version was forced to overwrite the old version:

```
$ git fetch
From git://git.kernel.org/pub/scm/git/git
 + 37ec5ed...8cd8cf8 next     -> origin/next  (forced update)
 + 9478935...16067c9 pu       -> origin/pu  (forced update)
   d0b0081..1f58507  todo     -> origin/todo
```

> **Using the revision range notation in diff**
>
> To make it easier to copy and paste between the `log` and `diff` commands, Git allows us to use the *revision range* double-dot notation, `A..B`, and triple-dot, `A...B`, as a *set of revisions (endpoints)* in the `git diff` command.
>
> For Git, using `git diff A..B` is the same as `git diff A B`, which means the difference between revision A and revision B. If the revision on either side of the double dot is omitted, it will have the same effect as using HEAD instead. For example, `git diff A..` is equivalent to `git diff A HEAD`.
>
> The `git diff A...B` notation is intended to show incoming changes on branch B. Incoming changes mean revisions up to B, starting at a common ancestor – that is, a merge base of both A and B. Thus, writing `git diff A...B` is equivalent to `git diff $(git merge-base A B) B`; note that `git merge-base` is without `--all` here. The result of this convention makes it so that a copy and paste of the `git fetch` output (whether with double-dot or triple-dot) as an argument to `git diff` will always show fetched changes. Note, however, that it doesn't include changes that were made on A since divergence at all!
>
> With modern Git, you can use the less cryptic `git diff --merge-base A B` instead of using triple-dot notation – that is, `git diff A...B`.
>
> Additionally, this feature makes it possible to use `git diff A^!` to view how revision A differs from its parent (it's the shortcut for `git diff A^ A`).

Summary

This chapter covered the various ways you can explore project history: to find relevant revisions, select revisions to display, and for further analysis.

We started by describing the conceptual model of project history: the DAG of revisions. Understanding this concept is very important because many selection tools refer directly or indirectly to the DAG.

Then, you learned how to select a single revision and range of revisions, as well as how the concept of revision range works for a non-linear history. We can use this knowledge to see what changes were made on a branch since its divergence from the base branch, and vice versa; we can also examine what happened to both branches since their divergence.

Selecting revisions is an important first step in searching through project history. This will be described in the next chapter.

Questions

Answer the following questions to test your knowledge of this chapter:

1. How would you list all revisions that are present upstream for the current branch but are not present in the current branch (are not integrated)?

2. How would you list all revisions that you would send using `git push`, allowing for a triangular workflow (remote to push to is different from remote to pull from)?

3. How can you find all divergent changes in two branches, A and B, starting from a fork point, and show which changeset is present on which branch?

4. How can you list all commits that were made on any remote-tracking branch whose name starts with `fix-`, from any remote repository?

5. What is the simplest way of switching to the previous branch, and how does it work?

Answers

Here are the answers to this chapter's questions:

1. Combine the double-dot notation with the notation for the upstream branch: `git log ..@{upstream}`.

2. Use `git log @{push}..HEAD`, combining double-dot notation with the "where to push to" notation. Note that for simple workflows, `@{push}` is the same as `@{upstream}`.

3. Use the triple-dot notation and the appropriate `git log` option: `git log --left-right A...B`.

4. Use the `--remotes[=<pattern>]` option with the appropriate glob pattern: `git log --remotes=*/fix-*`.

5. Use `git checkout -` or `git switch -`. In those commands, `-` means `@{-1}`, which uses the reflog to find the previous value of the current branch.

Further reading

To learn more about the topics that were covered in this chapter, take a look at the following resources:

- gitrevisions(7) – specifying revisions and ranges for Git: `https://git-scm.com/docs/gitrevisions`

- Scott Chacon, Ben Straub: *Pro Git*, 2nd Edition (2014), Apress *Chapter 2.3: Git Basics - Viewing the Commit History*: `https://git-scm.com/book/en/v2/Git-Basics-Viewing-the-Commit-History`

- *glob(7)* – globbing pathnames (shell wildcard patterns): `https://man7.org/linux/man-pages/man7/glob.7.html`

- Jan Goyvaerts: *Regular Expressions Tutorial: Learn How to Use and Get The Most out of Regular Expressions*: `https://www.regular-expressions.info/tutorial.html`

5

Searching Through the Repository

After selecting the parts of the project history that you want to search, the next task is to extract the information you want from selected commits. You can limit your search according to the revision metadata, such as the author of the commit, the date that the change was created, or the contents of the commit message. You can look at the changes themselves, or you may be interested in how a given file or subsystem evolved. With access to the project history, you can find who wrote a given section of the code or which commit introduced a regression (first buggy commit).

Another important skill is to format Git output so that it is easy to find the information you want. This task is made possible by various predefined pretty `git log` output formats and the ability to define and compose one's own output format.

Here is the list of topics we will cover in this chapter:

- Limiting the history and history simplification
- Searching the history with the pickaxe tool and diff search
- Finding bugs with `git bisect`
- Line-wise history of file contents with `git blame` and rename detection
- Selecting and formatting output (the pretty formats)
- Summarizing contribution with `git shortlog`
- Specifying a canonical author name and email with `.mailmap`
- Viewing specific revisions, files at revision, and diff output options

The purpose of this chapter is to show how to extract information from the project history.

Searching the history

A huge number and variety of useful options for the git log command are revising limiting options — that is, options that let you show only a subset of commits. This complements selecting commits to view by passing the appropriate revision range and allows us to search the history for specific versions, utilizing information other than the shape of the graph of revisions.

Limiting the number of revisions

The most basic way of limiting git log output is to show only the specified number of the most recent commits. This is done using the -<n> option (where n is any integer); this can also be written as -n <n>, or in long form as --max-count=<n>. For example, git log -2 would show the two last (most recent) commits in the current line of development, starting from the implicit HEAD revision.

You can skip the first few commits shown with --skip=<n>.

Matching revision metadata

History limiting options can be divided into those that check the information stored in the commit object itself (the revision metadata) and those that filter commits based on the changeset (based on changes from the parent commit or commits).

Time-limiting options

If you are interested in commits created within some specific date range, you can use a number of options such as --since and --until, or --before and --after. For example, the following command returns the list of commits made in the last two weeks:

```
$ git log --since=2.weeks
```

These options work with various *date* formats. You can specify a specific date such as *2008-04-21*, or a relative date such as *2 years 3 months 3 days ago*; you can also use a dot in place of a space.

When using a specific date, you must remember that if the date does not include a time zone, it will be interpreted in the local time zone. It is important because, in such a situation, Git will not yield identical results when run by colleagues who may be situated in other time zones around the world. For example, --since="2014-04-29 12:00:00" would catch six hours' worth more commits when issued in Birmingham, England, UK (where it means 2014-04-29Z11:00:00 universal time) than when issued in Birmingham, Alabama, USA (where it means 2014-04-29Z17:00:00). To get the same results, you need to include the time zone in the time limit — for example, --after="2013-04-29T17:07:22+0200".

Note that commits in Git are described not by a single date, but by two possibly different dates: the author date and the committer date. Time-limiting options described here examine the **committer date**, which means the date and time when the revision object was created. This might be different from the **author date**, which means the date and time when a changeset was created (when the change was made).

The date of authorship can be different from the date of committership in a few cases:

- One case is when the commit was created in one repository, converted to email, and then applied by another person in another repository.

- Another way to have those two dates differ is to have the commit recreated while rebasing; by default, this keeps the author date and gets a new committer date (see *Chapter 9, Merging Changes Together*, the *Rebasing a branch* section, and *Chapter 10, Keeping History Clean*, the *An interactive rebase* section).

Matching commit contents

If you want to filter your commit history to only show those done by a specific author or committer, you can use the --author or --committer options, respectively. For example, let's say you're looking for all the commits in the Git source code authored by Linus. You could use something like git log --author=Linus. The search is, by default, case-sensitive, and uses **regular expressions**. Git will search both the name and the email address of the commit author; to match the first name only, use --author=^Linus. Using ^ here means that the authorship information should start with **Linus**.

The --grep option lets you search commit messages (which should contain descriptions of the changes). Let's say that you want to find all the security bug fixes that mention **Common Vulnerabilities and Exposures (CVE)** identifiers in the commit message. You could generate a list of such commits with git log --grep=CVE.

If you specify both --author and --grep options, or more than one --author or --grep option, Git will show commits that match either query. In other words, Git would logically OR all the commit matching options. If you want to find commits that match all the queries, with matching options logically AND-ed, you need to use the --all-match option.

There is also a set of options to modify the meaning of matching patterns, similar to the ones used by the grep program. To make the search case-insensitive, use the -i / --regexp-ignore-case option. If you want to match simply a substring, you can use -F / --fixed-strings (you might want to do this to avoid having to escape regular expression metacharacters such as . and ?). To write more powerful search terms, you can use --extended-regexp or --perl-regexp (you can use the latter only if Git was compiled and linked with the **Perl Compatible Regular Expressions (PCRE)** library). To find non-matching commits, use --invert-grep.

When walking reflogs with `git log -g` (see the *Reflog shortnames* section), you can use the `--grep-reflog=<regexp>` option to show only positions with the matching reflog entry. For example, to show all operations on HEAD that were not a simple commit operation, you can use the following:

```
$ git log -g --invert-grep --grep-reflog="^commit:"
```

Commit parents

Git, by default, will follow all the parents of each merge commit when walking down the ancestry chain. To make it follow only the first parent, you can use the aptly named `--first-parent` option. This would show you the main line of history (sometimes called the trunk), assuming that you follow the specific practices with respect to merging changes; you will learn more about this in *Chapter 8, Advanced Branching Techniques*, and *Chapter 9, Merging Changes Together*.

Consider the following command (this example uses the very nice `--graph` option that makes an ASCII-art diagram of the history):

```
$ git log -5 --graph --oneline
* 50f84e3 Update draft release notes to 2.1
*   07768e0 Merge branch 'jc/shortlog-ref-exclude'
|\
| * eb07774 shortlog: allow --exclude=<glob> to be passed
* |   251cb96 Merge branch 'mn/sideband-no-ansi'
|\ \
| * | 38de156 sideband.c: do not use ANSI control sequence
```

Compare it with this:

```
$ git log -5 --graph --oneline --first-parent
* 50f84e3 Update draft release notes to 2.1
* 07768e0 Merge branch 'jc/shortlog-ref-exclude'
* 251cb96 Merge branch 'mn/sideband-no-ansi'
* d37e8c5 Merge branch 'rs/mailinfo-header-cmp'
* 53b4d83 Merge branch 'pb/trim-trailing-spaces'
```

You can filter the list to show only the merge commits or only the non-merge commits with the `--merges` and `--no-merges` options, respectively. These options can be considered simply shortcuts for more generic options: `--min-parents=<number>` (`--merges` is `--min-parents=2`) and `--max-parents=<number>` (`--no-merges` is `--max-parents=1`).

Let's say that you want to find the starting point(s) of your project. You can do this with the help of
--max-parents=0, which would give all the root commits:

```
$ git log --max-parents=0 --oneline
0ca71b3 basic options parsing and whatnot.
16d6b8a Initial import of a python script...
cb07fc2 git-gui: Initial revision.
161332a first working version
1db95b0 Add initial version of gitk to the CVS repository
2744b23 Start of early patch applicator tools for git.
e83c516 Initial revision of "git", the information manager from hell
```

Searching changes in revisions

Sometimes, searching through commit messages and other revision metadata is not enough. Perhaps
descriptions of the changes were not detailed enough. Or, what if you are looking for a revision when
a function was introduced, or where variable started to be used?

Git allows you to look through the changes that each revision brought (the difference between a
commit and its parent). The faster option is called a **pickaxe** search.

With the -S<string> option, Git will look for differences that introduce or remove an instance
of a given string. Note that this is different from the string simply appearing in the diff output. You
can do a match using a regular expression with the --pickaxe-regex option. Git checks each
revision to see whether there are files whose *current* side and *parent* side have a different number of
the specified string, and show the revisions that match.

As a side effect, git log with the -S option would also show the changes that each revision made
(as if the --patch option were used), but only those differences that match the query. To show
differences for all the files and also differences where the change in number occurred, you need to
use the --pickaxe-all option:

```
$ git log -S'sub href'
commit 06a9d86b49b826562e2b12b5c7e831e20b8f7dce
Author: Martin Waitz <tali@admingilde.org>
Date:   Wed Aug 16 00:23:50 2006 +0200

    gitweb: provide function to format the URL for an action link.

    Provide a new function which can be used to generate an URL for
the CGI.
    This makes it possible to consolidate the URL generation in order
to make
    it easier to change the encoding of actions into URLs.
```

```
Signed-off-by: Martin Waitz <tali@admingilde.org>
Signed-off-by: Junio C Hamano <junkio@cox.net>
```

With -G<regex>, Git would literally look for differences whose added or removed line matches the given regular expression. Note that the unified diff format (that Git uses) considers the changed line to be a removal of the old version and adding of a new one; refer to *Chapter 2, Developing with Git* (the *Examining the changes to be committed* section) for an explanation of how Git describes changes.

To illustrate the difference between -S<regex> --pickaxe-regex and -G<regex>, consider a commit with the following diff:

```
     if (lstat(path, &st))
-        return error("cannot stat '%s': %s", path,
+        ret = error("cannot stat '%s': %s", path,
                           strerror(errno));
```

While git log -G"error\(" will show this commit (because the query matches both changed lines), git log -S"error\(" --pickaxe-regex will not (because the number of occurrences of that string did not change).

> **Tip**
>
> If you are interested in a single file, it is easier to use git blame (perhaps in a graphical blame browser, like with git gui blame) to check when the given change was introduced. However, git blame can't be used to find a commit that deleted a line — you need a pickaxe search for that.

Selecting types of changes

Sometimes, you might want to see only those changes that added or renamed files. With Git, you can do this with git log --diff-filter=AR. You can select any combination of types of changes; see the git-log(1) manpage for details. For example, to find all renames while listing all changed files, you can use --diff-filter=R*, such as in the following example:

```
$ git log --diff-filter=R* --oneline –stat
8b4dbde Rename random.js to gen_random.js
 index.html                      | 2 +-
 scripts/{random.js => gen_random.js} | 0
 2 files changed, 1 insertion(+), 1 deletion(-)
042a8af Directory structure
```

```
index.html                      | 2 +-
random.js => scripts/random.js | 0
2 files changed, 1 insertion(+), 1 deletion(-)
```

The mnemonics for types of changes are the same as those used by `git status --short` or `git log --name-status`:

```
$ git log -1 --diff-filter=R --oneline --name-status
8b4dbde Rename random.js to gen_random.js
R100    scripts/random.js       scripts/gen_random.js
```

Next, we will examine how to search the history based on which files were changed, and later, also how to format the `git log` output.

History of a file

As described in the *Whole-tree commits* section at the beginning of the previous chapter, Git revisions are about the state of the whole project as one single entity.

In many cases, especially with larger projects, we are interested only in the history of a single file, or the history limited to the changes in the given directory (in the given subsystem).

Path limiting

To examine the history of a single file, you can simply use `git log <pathname>`. Git will then only show all those revisions that affected the given pathname (a file or a directory), which means those revisions where there was a change to the given file or a change to a file inside the given subdirectory.

> **Disambiguation between branch names and path names**
>
> Git usually guesses what you mean by writing `git log foo`; did you mean to ask for the history of the `foo` branch (the line of development), or for the history of the `foo` file? However, sometimes, Git can get confused. To prevent confusion between pathnames and branch names, you can use `--` (two dashes) to separate filename arguments from other options. Everything after `--` would be taken to be a pathname, and everything before that would be taken to be the branch name or other option.
>
> For example, writing `git log -- foo` explicitly asks for the history of the `foo` path.
>
> One of the common situations where it is needed, besides when having the same name for a branch and a file, is when examining the *history of a deleted file* that is no longer present in a project.

You can specify more than one path; you can even look for changes that affect a given type of file with the help of wildcards (pattern match). For example, to find only changes to Perl scripts (files with the `*.pl` extension), you can use `git log -- '*.pl'`. Note that you need to protect the `*.pl` wildcard from being expanded by the shell before Git sees it — for example, via single quotes as shown here.

Pathspec magic

Most commands that accept `<path>` or `<pathspec>` as a parameter, such as `git log`, also support **pathspec magic**. A pathspec that starts with a colon ':' has a special meaning: it is expected that this colon is then followed by either one or more *magic signature* letters, or a comma-separated list of zero or more *magic words*. An optional colon ':' can be used to separate the magics from the pattern to match them. Here are a few examples: `:(top):data/` or `:/data/` would make the pattern match the data/directory at the top directory of the repository, regardless of where we are inside the repository (the current directory). The `git log :(exclude):*.html` or `git log :^*.html` command will list all revisions where there was at least one change to the file that is not an HTML file. You can find more magics in the `gitglossary(7)` manpage, in the `pathspec` entry.

However, because Git uses pathname parameters as *limiters* in showing the history of a project, querying for the history of a single file doesn't automatically *follow renames*. You need to use `git log --follow <file>` to continue listing the history of a file beyond renames. Unfortunately, it doesn't work in all cases. Sometimes, you need to use either the `git blame` command (see the *Blame – the line-wise history of a file* section), or examine the boundary commits with rename detection turned on (`git show -M -C --raw --abbrev <rev>`) and follow renames and file moving manually.

In modern Git, you can also trace *the evolution of the line range* within the file using `git log -L`, which is currently limited to the walk starting from a single revision (zero or one positive revision argument) and a single file. The range is given either denoting the start and end of the range with `-L <start>,<end>:<file>` (where either `<start>` or `<end>` can be a line number or a `/regexp/`), or a function to track with `-L :<funcname regexp>:<file>`. This technique cannot, however, be used together with the ordinary pathspec-based path limiting. For example, to see the history of the `index.html` file, limited to the changes in the `<head>` element, you can use the following command:

```
$ git log -L '/^<head>/','/^<\/head>/':index.html
```

History simplification

By default, when requested for the history of a path, Git would *simplify the history*, showing only those commits that are required (that are enough) to explain how the files that match the specified paths came to be. Git would exclude those revisions that do not change the given file. Additionally, for non-excluded merge commits, Git would exclude those parents that do not change the file (thus excluding lines of development).

You can control this kind of history simplification with the `git log` options such as `--full-history` or `--simplify-merges`. Check the Git documentation for more details, such as the *History simplification* section in the `git-log(1)` manpage.

Blame — the line-wise history of a file

The **blame** command is a version control feature designed to help you determine who made changes to a file. This command shows, for each line in the file, when the line was created, who authored the given line, and so on. It does this by finding the latest commit in which the current shape of each line was introduced. A revision introducing a given shape is the one where the given line has its current form, but where the line is different in its revision parents. The default output of `git blame` annotates each line with appropriate line-authorship information.

Git can start annotating from the given revision (useful when browsing the history of a file or examining how an older version of a file came to be), or even limit the search to a given revision range. You can also limit the range of lines annotated to make blame faster — for example, to check only the history of an `esc_html` function, you can use the following:

```
$ git blame -L '/^sub esc_html {/,/}/' gitweb/gitweb.perl
```

What makes the blame operation so useful is that it *follows the history* of a file across whole-file renames. It can optionally follow lines that were moved from one file to another (with the -M option), and even follow lines that were copied and pasted from another file (with the -C option); this includes internal code movement.

When following code movement, it is useful to ignore changes in whitespace to find when a given fragment of code was truly introduced and avoid finding when it was just re-indented (for example, due to refactoring repeated code into a function — code movement). This can be done by passing the diff-formatting option, -w / --ignore-all-space.

Rename detection

A good version control system should be able to deal with renaming files and other ways of changing the directory structure of a project. There are two ways to deal with this problem. The first is **rename tracking**, which means that the information about the fact that a file was renamed is saved at the commit time; the version control systems mark renames. This usually requires using the rename and move commands to rename files. For example, you cannot use a file manager that is not version-control aware to move files. However, you can detect the rename when creating the revision." It can involve some kind of **file identity** surviving across renames.

The second method, and the one used by Git, is **rename detection**. In this case, the mv command is only a shortcut for deleting a file with the old name and adding a file with the same contents and a new name. Rename detection means that the fact that the file was renamed is detected at the time it is needed: when doing a merge, when viewing the line-wise history of a file (if requested), or when showing a diff (if requested or configured). This has the advantage that the rename detection algorithm can be improved and is not frozen at the time of commit. It is a more generic solution, allowing it to handle not only the whole-file renames but also the code movement and copying within a single file and across different files, as can be seen in the description of git blame.

The disadvantage of rename detection, which in Git is based on the heuristic of the similarity of the file contents and the pathname, is that it takes resources to run and that in rare cases, it can fail: not detecting renames, or detecting a rename where there isn't one.

Note that, in modern Git, basic rename detection is turned on for diffs by default.

Many graphical interfaces for Git include a graphical version of the blame operation. The git gui blame is an example of such a graphical interface for the blame operation (it is a part of git gui, a Tcl/Tk-based graphical interface). Such graphical interfaces can show the full description of changes and simultaneously show the history with and without considering renames. From such a GUI, it is usually possible to go to a specified commit, browsing the history of the lines of a file interactively. In addition, the GUI blame tool makes it very easy to follow files across renames:

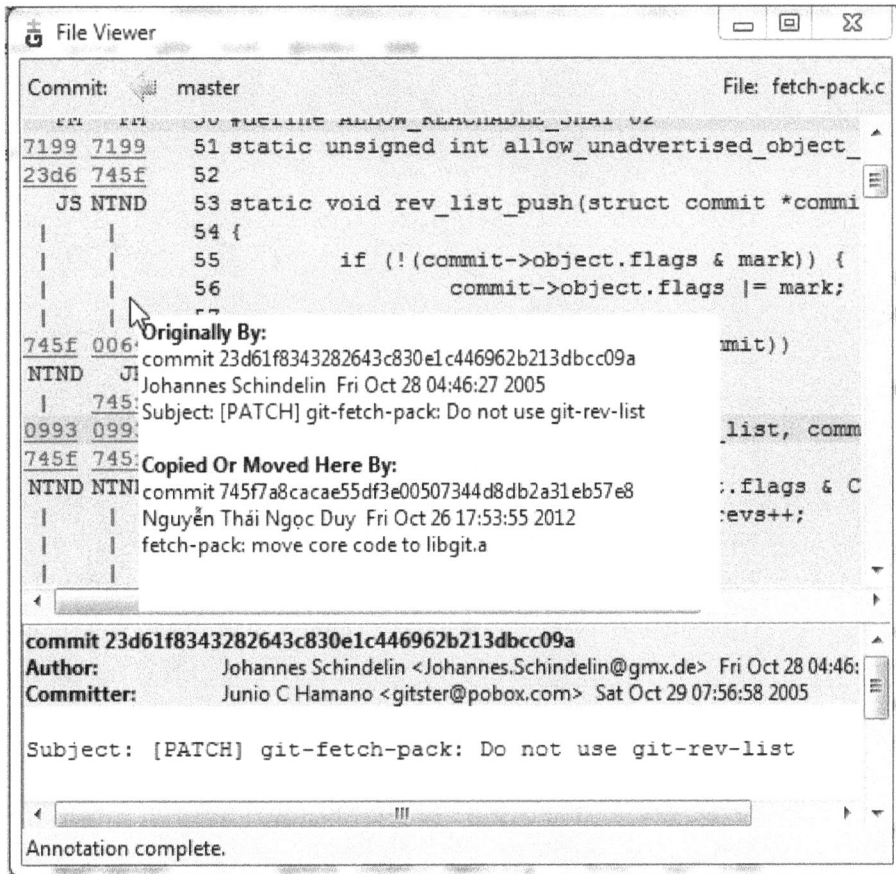

Figure 5.1 – 'git gui blame' in action, showing the detection of copying or moving fragments of code

Finding bugs with git bisect

Git provides a couple of tools to help you debug issues in your projects. These tools can be extremely useful, especially in the case of a software regression — that is, a software bug that makes a feature stop functioning as intended after a certain revision. If you don't know where the bug can be, and there have been dozens or hundreds of commits since the last state where you know the code worked, you'll likely turn to `git bisect` for help.

The **bisect** command searches semi-automatically, step by step, through project history, trying to find the revision that introduced the bug. In each step, it bisects the history into roughly equal parts and asks whether there is a bug in the dividing commit. It then uses the answer to eliminate one of the two sections and reduces the size of the revision range where there can be a commit that introduced the bug:

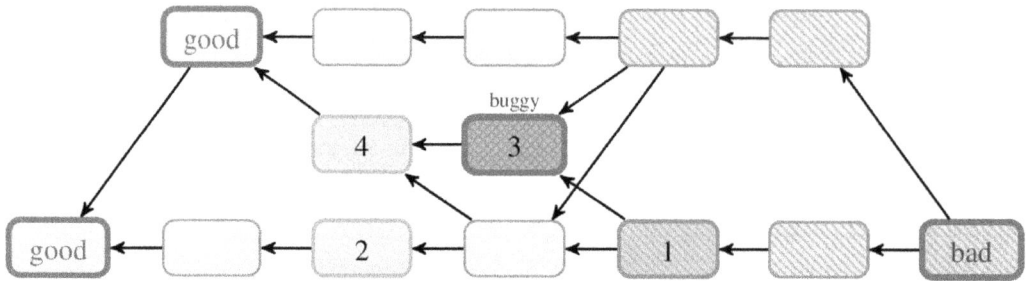

Figure 5.2 – An example of git bisect in action, finding the buggy commit after 4 steps

Starting the git bisect process

Suppose version 1.14 of your project worked, but the release candidate for the new version, 1.15-rc0, crashes. You go back to the 1.15-rc0 version, and it turns out you can *reproduce the issue* (this is very important!), but you can't figure out what is going wrong.

You can bisect the code history to find out. You need to start the bisection process with `git bisect start`, and then tell Git which version is broken with `git bisect bad`. Then, you must tell the bisect process the last known good state (or set of states) with `git bisect good`:

```
$ git bisect start
$ git bisect bad   v1.15-rc0
$ git bisect good v1.14
Bisecting: 159 revisions left to test after this (roughly 7 steps)
[7ea60c15cc98ab586aea77c256934acd438c7f95] Merge branch 'mergetool'
```

Finding the buggy commit

Git figured out that about 300 commits came between the commit you marked as the last good commit (`v1.14`) and the bad version (`v1.15-rc0`), and checked out the middle one (`7ea60c15`) for you. If you run `git branch` or `git status` at this point, you'll see that git has temporarily moved you to (`no branch`):

```
$ git branch
* (no branch, bisect started on master)
  master
$ git status
HEAD detached at 7ea60c15cc
You are currently bisecting, started from branch 'master'.
  (use "git bisect reset" to get back to the original branch)
```

At this point, you need to run your test to check whether the issue is present in the commit currently checked out by the bisect operation. If the program crashes, mark the current commit as bad with `git bisect bad`. If the issue is not present, mark it as correct with `git bisect good`. After about seven steps, Git would show the suspect commit:

```
$ git bisect good
b047b02ea83310a70fd603dc8cd7a6cd13d15c04 is first bad commit
commit b047b02ea83310a70fd603dc8cd7a6cd13d15c04
Author: PJ Hyett <pjhyett@example.com>
Date:   Tue Jan 27 14:48:32 2009 -0800

    secure this thing

:040000 040000 40ee3e7... f24d3c6... M  config
```

The last line in the preceding example output is in the so-called *raw* diff output, showing which files changed in a commit. You can then examine the suspected commit with `git show`. From there, you can find the author of the said commit, and ask them for clarification or ask them to fix it (by sending them a bug report). If the good practice of creating small, incremental changes was followed during the development of the project, the amount of code to examine after finding the bad commit should be small.

If, at any point, you land on a commit that broke something unrelated and is not a good one to test, you can skip such a commit with `git bisect skip`. You can even skip a range of commits by giving the revision range to the `skip` subcommand.

When you're finished, you should run `git bisect reset` to return you to the branch you started from:

```
$ git bisect reset
Previous HEAD position was b047b02... secure this thing
Switched to branch 'master'
```

To finish bisection while staying on the bad commit you found, you can use `git bisect reset HEAD`.

Automating testing during the git bisect process

You can even fully automate finding bad revisions with `git bisect run`. For this, you need to have a script that will test for the presence of a bug and exit with a value of 0 if the project works all right, or a non-0 value if there is a bug. The special exit code, 125, should be used when the currently checked-out code cannot be tested. In this case, you also start the `bisect` operation by providing the known bad and good commits. You can do this by simply listing them with the `bisect start` command if you want, listing the known bad commit first and the known good commit(s) second. You can even cut down the number of trials, if you know what part of the tree is involved in the problem you are tracking down, by specifying path parameters (the double-dash before the path is not strictly necessary, but is helpful). Then, you start automated bisection:

```
$ git bisect start v1.5-rc0 v1.4 -- arch/i386
$ git bisect run ./test-error.sh
```

Doing so automatically runs `test-error.sh` on each checked-out commit until Git finds the first broken commit.

If the bug is that the project stopped compiling (a broken build), you can use `make` as a test script (with `git bisect run make`).

Selecting and formatting the git log output

Now that you know how to select revisions to examine and limit which revisions are shown (selecting those that are interesting), it is time to see how to select which part of the information is associated with the queried revisions to show, and how to format this output. There is a huge number and variety of options for the `git log` command available for this.

Predefined and user-defined output formats

A very useful `git log` option is `--pretty`. This option changes the format of the log output. There are a few prebuilt formats available for you to use. The `oneline` format prints each commit on a single line, which is useful if you're looking at a lot of commits; there exists the `--oneline` shorthand for `--pretty=oneline --abbrev-commit` used together. In addition, the `short`, `medium` (the default), `full`, and `fuller` formats show the output in roughly the same format, but with less or more information, respectively. The `raw` format shows commits in internal Git representation, and `email` or `mboxrd` in a `git format-patch`-like format, as an email. The `reference` format is intended to refer to another commit in a commit message, per the following example:

```
$ git show --no-patch --pretty=reference master^
20cfc7c (Added COPYRIGHT, 2021-05-30)
```

It is possible to change the format of dates shown in those verbose, pretty formats with an appropriate `--date` option: make Git show relative dates such as, for example, **2 hours ago**, with `--date=relative`, dates in your local time zone with `--date=local`, and so on.

You can also specify your own log output format with `--pretty=format:<string>` (and its `tformat` variant, with terminator rather than separator semantics — output for each commit has the newline appended). This is especially useful when you're generating output for machine parsing for use in scripts because when you specify the format explicitly, you know it won't change with updates to Git. The format string works a little bit like in `printf`:

```
$ git log --pretty="%h - %an, %ar : %s"
50f84e3 - Junio C Hamano, 7 days ago : Update draft release notes
```

```
0953113 - Junio C Hamano, 10 days ago : Second batch for 2.1
afa53fe - Nick Alcock, 2 weeks ago : t5538: move http push tests out
```

There is a very large number of placeholders. Selected ones of those are listed in the following table:

Placeholder	Description of output
%H	Commit hash (full SHA-1 identifier of revision)
%h	Abbreviated commit hash
%an	Author name
%ae	Author email
%ar	Author date, relative
%cn	Committer name
%ce	Committer email
%cr	Committer date, relative
%s	Subject (first line of a commit message, describing revision)
%%	A raw %

Table 5.1 – Placeholders and their description

Author versus committer

The **author** is the person who originally wrote the patch (authored the changes), whereas the **committer** is the person who last applied the patch (created a commit object with those changes, representing the revision in the DAG). So, if you send in a patch to a project and one of the core members applies the patch, both of you get credit — you as the author and the core member as the committer. Also, after rebase, for rebased revisions the original author of the commit is kept, while the person performing the rebase is made the committer.

The --oneline format option is especially useful together with another git log option called --graph, though the latter can be used with any format. The latter option adds a nice little ASCII graph showing your branch and merge history. To see where tags and branches are, you can use an option named --decorate (which in modern Git is now the default):

```
$ git log --graph --decorate --oneline origin/maint
*   bce14aa (origin/maint) Sync with 1.9.4
|\
| * 34d5217 (tag: v1.9.4) Git 1.9.4
| *   12188a8 Merge branch 'rh/prompt' into maint
| |\
| * \   64d8c31 Merge branch 'mw/symlinks' into maint
```

```
|  |\ \
*  |  |  |  d717282 t5537: re-drop http tests
*  |  |  |  e156455 (tag: v2.0.0) Git 2.0
```

You might want to use a graphical tool to visualize your commit history. One such tool is a Tcl/Tk program called `gitk` that is distributed with Git. You can find more information about various types of graphical tools in *Chapter 13, Customizing and Extending Git*.

Including, formatting, and summing up changes

You can examine a single revision with the `git show` command, which, in addition to the commit metadata, shows changes in the unified diff format, described in *Chapter 2, Developing with Git*, in the *Unified Git diff format* subsection. Sometimes, however, you might want to display changes alongside the selected part of the history in the `git log` output. You can do this with the help of the `-p` option. This is very helpful for code review, or to quickly browse what happened during a series of commits that a collaborator has added.

Ordinarily, Git would not show the changes for a merge commit. To show changes from all parents, you need to use the `-c` option (or `-cc` for compressed output), while to show changes from each parent individually, use `-m`.

Sometimes, it's easier to review changes on the word level rather than on the line level. The `git log` accepts various options to change the format of the diff output. One of those options is `--word-diff` (with various variants, including `color`). This way of viewing differences is useful for examining changes in documents (for example, documentation):

```
commit 06ab60c06606613f238f3154cb27cb22d9723967
Author: Jason St. John <jstjohn@purdue.edu>
Date:    Wed May 21 14:52:26 2014 -0400

    Documentation: use "command-line" when used as a compound
adjective, and fix

    Signed-off-by: Jason St. John <jstjohn@purdue.edu>
    Signed-off-by: Junio C Hamano <gitster@pobox.com>


diff --git a/Documentation/config.txt b/Documentation/config.txt
index 1932e9b..553b300 100644
--- a/Documentation/config.txt
+++ b/Documentation/config.txt
@@ -381,7 +381,7
        Set the path to the root of the working tree.
        This can be overridden by the GIT_WORK_TREE environment
        variable and the '--work-tree' [-command line-]{+command-
line+} option.
```

```
        The value can be an absolute path or relative to the path to
        the .git directory, which is either specified by --git-dir
        or GIT_DIR, or automatically discovered.
```

Another useful set of options is about ignoring changes in whitespace, including –w / --ignore-all-space to ignore all whitespace changes, and -b / --ignore-space-change to ignore changes in the amount of whitespace.

With color support, you can ask Git to show moved code with --color-moved, possibly ignoring whitespace changes (with --color-moved-ws).

Sometimes, you are interested only in the summary of changes and not the details. There is a series of diff summarizing options that you can use. If you want to know only which files changed, use --names-only (or --raw --abbrev). If you also want to know how much those files changed, you can use the --stat option (or perhaps its machine-parse-friendly version, --numstat) to see some abbreviated stats. If you are interested only in a short summary of changes, use --shortstat or --summary.

Summarizing contributions

Ever wondered how many commits you've contributed to a project? Or, perhaps, who the most active developer was during the last month (with respect to the number of commits)? Well, wonder no more, because this is what git shortlog is good for:

```
$ git shortlog -s -n
  13885  Junio C Hamano
   1399  Shawn O. Pearce
   1384  Jeff King
   1108  Linus Torvalds
    743  Jonathan Nieder
```

The -s option squashes all of the commit messages into the number of commits; without it, git shortlog would list a summary of all the commits, grouped by the developer. The -n option sorts the list of developers by the number of commits; otherwise, it is sorted alphabetically. You can add an –e option to also show an email address; note, however, that with this option, Git will separate contributions made by the same author under different emails. The git shortlog output can be configured to some extent with a pretty-like --format option.

The git shortlog command accepts a revision range and other revision-limiting options such as --since=1.month.ago — anything that git log accepts and makes sense for shortlog. For example, to see who contributed what to the last release candidate, you can use the following command:

```
$ git shortlog -e v2.0.0-rc2..v2.0.0-rc3
Jonathan Nieder <jrnieder@gmail.com> (1):
      shell doc: remove stray "+" in example
```

```
Junio C Hamano <gitster@pobox.com> (14):
      Merge branch 'cl/p4-use-diff-tree'
      Update draft release notes for 2.0
      Merge branch 'km/avoid-cp-a' into maint
   ...
```

> **Tip**
>
> One needs to remember that the number of revisions authored is only one way of measuring contribution. For example, somebody who creates buggy commits only to fix them later would have a larger number of commits than a developer who doesn't make mistakes (or cleans the history before publishing changes).
>
> There are other measures of programmer productivity — for example, the number of changed lines in authored commits, or the number of surviving lines. Those can be calculated with the help of Git, but there is no built-in command to calculate them.

Mapping authors

One problem with running `git shortlog -s -n -e` or `git blame` in Git repositories of long-running projects is that an author may change their name or email, or both, during the course of the project, due to many reasons: changing work (and work email), misconfiguration, spelling mistakes, and others. For example, you might have **Bob Hacker <bob@example.com>** in one place, but **Bob <bob@example.com>** in the other. When that happens, you can't get proper attribution. Git allows you to coalesce author/email pairs with the help of the `.mailmap` file in the top directory of your project. This file allows you to specify *canonical* names for contributors, which in its simplest form looks like this:

```
Bob Hacker <bob@example.com>
```

(Actually, it allows you to specify a canonical name, canonical email, or both name and email, matching by email or by name and email.)

By default, those corrections are applied to all commands: `git blame`, `git shortlog`, and `git log`. With custom `log` output, you can use placeholders that output the original name or corrected name, and the original email or corrected email.

Viewing a revision and a file at revision

Sometimes, you might want to examine a single revision (for example, a commit suspected to be buggy, found with `git bisect`) in more detail, together with changes and their descriptions. Or, perhaps, you want to examine the tag message of an annotated tag together with the commit it points to. Git provides a generic `git show` command for this; it can be used for any type of object.

For example, to examine the grandparent of the current version, you can use the following command:

```
$ git show HEAD^^
commit ca3cdd6bb3fcd0c162a690d5383bdb8e8144b0d2
Author: Bob Hacker <bob@virtech.com>
Date:    Sun Jun 1 02:36:32 2014 +0200

    Added COPYRIGHT

diff --git a/COPYRIGHT b/COPYRIGHT
new file mode 100644
index 0000000..862aafd
--- /dev/null
+++ b/COPYRIGHT
@@ -0,0 +1,2 @@
+Copyright (c) 2014 VirTech Inc.
+All Rights Reserved
```

The git show command can also be used to display directories (trees) and file contents (blobs). To view a file (or a directory), you need to specify where it is from (from which revision) and the path to the file, using : to connect them. For example, to view the contents of the src/rand.c file as it was in the version tagged v0.1, use the following:

```
$ git show v0.1:src/rand.c
```

This might be more convenient than checking out the required version of the file into the working directory with git checkout v0.1 -- src/rand.c. Before the colon may be anything that names a commit (v0.1 here), and after that, it may be any path to a file tracked by Git (src/rand.c here). The pathname here is the full path from the top of the project directory, but you can use ./ after the colon for relative paths — for example, v0.1:./rand.c if you are in the src/ subdirectory.

You can use the same trick to compare arbitrary files at arbitrary revisions; on the other hand, the git show :src/rand.c command (as if with an empty revision) will show the state of the file at the time git add was run — the state of the chosen file in the index (in the staging area).

If you want to find out what files are present at a given revision (to select one to examine), you can use git ls-tree <revision>. To find out what files are present in the worktree and in the index, use git ls-files with the appropriate option to select what you want to see.

Summary

This chapter showed us the various ways of exploring project history: selecting and filtering revisions to display, searching through various parts of commit-related data, and formatting the output.

You have learned how to find all the revisions that were made by a given developer, how to search through the commit message and the changes made by the commit, and how to narrow the search to a specific range of time.

We can even try to find bugs in the code by exploring the history: finding when a function was deleted from the code with a *pickaxe search*, examining a file for how its code came to be and who wrote it with `git blame`, and utilizing semi-automatic or automatic search through the project history to find which version introduced a regression with `git bisect`.

When examining a revision, we can select the format in which the information is shown, even to the point of user-defined formats. There are various ways of summarizing the information, from the statistics of the changed files, to the statistics of the number of commits per author.

In the next chapter, we will examine how Git can help developers work together as a team on a single project.

Questions

Answer the following questions to test your knowledge of this chapter:

1. How would you list all commits made since yesterday on any remote-tracking branch?

2. How would you find out who the original author of a given function or class was, to ask for clarification or a code review?

3. How would you use Git to help find the source of regression — that is, a bug that is present in the new revision of the project, but was not there in older versions?

4. You have noticed that your colleague made a few commits with a misconfigured email, using `bob@laptop.company.com` instead of `bob@company.com`. How would you fix the attribution, assuming that it is not possible to rewrite those commits?

Answers

Here are the answers to the questions given above:

1. Combine time-limiting options with the `--remotes` option:

2. `git log --since=yesterday --remotes`.

3. Use the `git blame` command or an interactive GUI to do this, such as `git gui blame` (or an integration with your editor or **integrated development environment** (IDE); you can also search through the history of the relevant fragment of a file with `git log -L`.

4. Use `git bisect` to find the commit that introduced the bug, perhaps even by automating the search with `git bisect run`.

5. Add the correct name and email to the `.mailmap` file.

Further reading

To learn more about the topics that were covered in this chapter, take a look at the following resources:

- Scott Chacon, Ben Straub, *Pro Git*, 2nd Edition (2014), Apress, *Chapter 7.5 Git Tools – Searching*: `https://git-scm.com/book/en/v2/Git-Tools-Searching`

- Christian Couder, *Fighting regressions with git bisect* (slides from the Linux-Kongress 2009 conference): `http://www.linux-kongress.org/2009/slides/fighting_regressions_with_git_bisect_christian_couder.pdf`

- Junio C Hamano, *Fun with first parent history* (2013): `https://git-blame.blogspot.com/2013/09/fun-with-first-parent-history.html`

- Junio C Hamano, *Measuring Project Activities (2)* (2013): `https://git-blame.blogspot.com/2013/03/measuring-project-activities-2.html`

- Jan Goyvaerts, *Regular Expressions Tutorial - Learn How to Use and Get The Most out of Regular Expressions*: `https://www.regular-expressions.info/tutorial.html`

Part 2 -
Working with
Other Developers

In this part, you will learn how to choose the correct workflow for you (which includes a repository setup and branching model) and how to use Git to collaborate with other developers. You will also discover how to rewrite history and what to do if it is not possible.

This part has the following chapters:

- *Chapter 6, Collaborative Development with Git*
- *Chapter 7, Publishing Your Changes*
- *Chapter 8, Advanced Branching Techniques*
- *Chapter 9, Merging Changes Together*
- *Chapter 10, Keeping History Clean*

6

Collaborative Development with Git

Chapter 2, Developing with Git, and *Chapter 3, Managing Your Worktrees*, taught you how to make new contributions to a project, but limited this information to affecting only your own clone of the project's repository on your computer. *Chapter 2* described how to commit new revisions, while *Chapter 3* showed you how Git can help you prepare it.

This chapter and *Chapter 7, Publishing Your Changes*, present a bird's-eye view of the various ways to collaborate with others, showing centralized and distributed workflows. These two chapters will focus on the repository-level interactions in collaborative development, while the patterns of branches used will be covered in *Chapter 8, Advanced Branching Techniques*.

This chapter describes different collaborative workflows, explaining the advantages and disadvantages of each one. You will also learn about the chain of trust concept and how to use **signed tags**, **signed merges**, and **signed commits**.

The following topics will be covered in this chapter:

- Centralized and distributed workflows, and bare repositories
- Managing remotes and one-off single-shot collaboration
- How versions are addressed—the chain of trust
- Tagging; lightweight tags versus signed tags
- Signed tags, signed merges, and signed commits

Collaborative workflows

There are various levels of engagement while using a version control system. One might, for example, only be interested in using Git to examine how the project came to be. *Chapter 4*, *Exploring Project History*, and *Chapter 5*, *Searching Through the Repository*, covered this use of Git. Of course, examining a project's history is an important part of development, too.

One might use version control for one's private development, using it for a single developer project, on a single machine. *Chapter 2*, *Developing with Git*, and *Chapter 3*, *Managing Your Worktrees*, show how to do this with Git. Of course, people usually don't work in isolation, but in a team: one's own development is usually part of a collaboration.

But one of the main goals of version control systems is to help multiple developers work together on a project, collaboratively. Version control makes it possible for them to work simultaneously on a given piece of software in an effective way, ensuring that their changes do not conflict with each other, and thus helping with merging those changes.

One might work on a project together with a few other developers, or with many. One might be a contributor, or a project maintainer. Maybe the project is so large that it needs subsystem maintainers. One might work in tight software teams, or one might want to make it easy for external contributors to provide proposed changes (for example, to fix bugs, or fix an error in the documentation). Various workflows might be best suited for those different situations:

- Centralized workflow
- Peer-to-peer workflow
- Maintainer workflow
- Hierarchical workflow

Bare repositories

There are two types of repositories:

- One with a working directory, and a staging area (**non-bare**)
- A **bare repository**, without the working directory

The former type is meant for private solo development, and creating new history, while the latter is intended for collaboration and synchronizing development results.

By convention, **bare repositories** use the `.git` extension—for example, `project.git`—while **non-bare repositories** don't have it—for example, `project` (with the administrative area and the local repository in `project/.git`).

You can usually omit this extension when cloning the repository, pushing to it, or fetching from it; using either `https://github.com/git/git.git` as the repository URL or `https://github.com/git/git` will work the same.

To create the bare repository, you need to add the `--bare` option to the `git init` or `git clone` commands, as in the following example:

```
$ git init --bare project.git
Initialized empty Git repository in /home/user/project.git/
```

Interacting with other repositories

After creating a set of revisions and extending the project's history, you usually need to share it with other developers. You need to synchronize with other repository instances: publish your changes, and get changes from others.

From the perspective of the local repository instance – your own clone of the repository – you need to **push** your changes to the repository meant for publishing changes and **fetch** changes from other repositories. Often the only repository you need to interact with is simply the repository you cloned from. *Chapter 7, Publishing Your Changes,* will describe this process (and its alternatives) in more detail.

After fetching changes, you sometimes need to incorporate them into your work by **merging** two lines of development (or **rebasing**)—which you can do in one operation with **pull**. Merging and rebasing operations will be described in more detail in *Chapter 9, Merging Changes Together.*

Git assumes that you don't want your local repository to be visible to the public, because these repositories are intended for private work (which helps to keep work not yet ready for the public from being visible).

This means that there is an additional step required to make your finished work available: you need to **publish** your changes, for example with the `git push` command.

The diagram in *Figure 6.1*, which is an extension of the one in *Figure 2.2* in *Chapter 2, Developing with Git*, demonstrates the steps one can take when creating and publishing commits. The arrows in this diagram show the Git commands used to copy contents from one place to another, including to and from the remote repository.

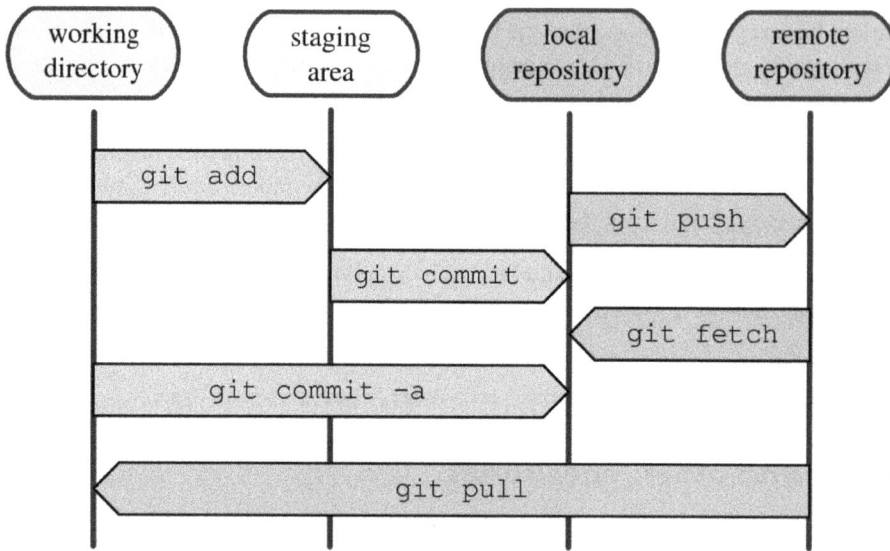

Figure 6.1 – Creating commits, publishing commits, and fetching changes
published by other developers into your local repository

Now, let us understand the centralized workflow.

The centralized workflow

With distributed version control systems, you can use different collaboration models, some more distributed, some less distributed. In a **centralized workflow**, there is one central hub: a shared repository, usually bare, that everyone uses to synchronize their work.

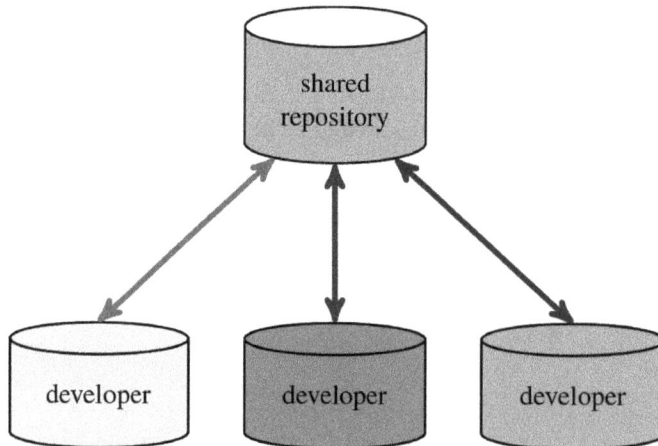

Figure 6.2 – Centralized workflow – the shared repository is bare

In this workflow, each developer has their own **non-bare clone** of the central shared repository, which they use to develop new revisions of software. When changes are ready, they push those changes to the central repository, and fetch (or pull) changes from other developers from it. One might have to merge changes before being able to push. In this workflow integration of changes is distributed. This workflow is shown in *Figure 6.2*.

Let us now look into the advantages and disadvantages of a *centralized workflow*.

Advantages of a centralized workflow

Some of the key advantages of centralized workflows include the following:

- This workflow has a simple setup; it is a familiar paradigm for people coming from centralized version control systems and used to working with centralized management. It provides centralized access control and easy backups.

- It makes it easy to set up **continuous integration (CI)**.

- The process of merging changes is shared among developers, with no person solely responsible for integrating changes.

- It might be a good setup for a private project with a small team, or where all developers are trusted and capable.

Disadvantages of a centralized workflow

Some of the disadvantages of centralized workflows are as follows:

- The shared repository is a single point of failure: if there are problems with the central repository, then there is no way to synchronize changes.

- Each developer pushing changes (making them available for other developers) might require updating one's own repository first, and merging changes from others. Shared integration means that each developer needs to know how to do it.

- You also need to trust developers with access to the shared repository in this setup, or to provide access controls.

The peer-to-peer or forking workflow

The opposite of a centralized workflow is a **peer-to-peer** or **forking workflow**. Instead of using a single central shared public repository, each developer has a public repository (which is bare), in addition to a private working repository (with a working directory), like in the *Figure 6.3*.

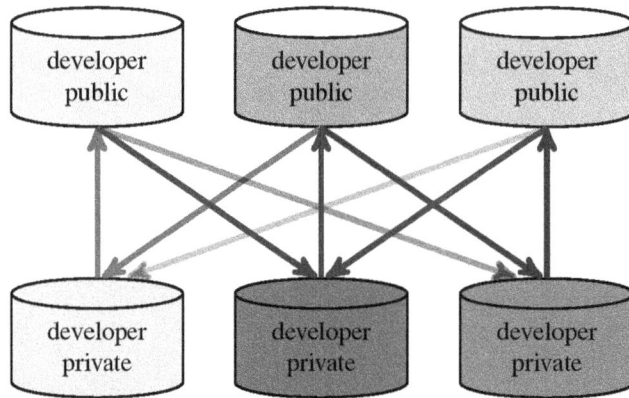

Figure 6.3 – Peer-to-peer workflow – here, lines pointing up represent push operation, while lines pointing down represent fetch/pull operation

When the changes are ready, developers push to their own public repositories. To incorporate changes from other developers, one needs to fetch them from the public repositories of each of the other developers.

The advantages and disadvantages of this rarely used peer-to-peer workflow, also called forking workflow, are as follows:

Advantages of the peer-to-peer workflow

- One advantage of the forking workflow is that contributions can be integrated without the need for a central repository; it is a fully distributed workflow

- Another advantage is that you are not forced to integrate if you want to publish your changes; you can merge at your leisure

- It is a good workflow for organic teams without requiring much setup

Disadvantages of the peer-to-peer workflow

- The disadvantages are a lack of the canonical version, no centralized management, and the fact that in the basic form of this workflow you need to interact with many repositories. Though the `git remote update` or `git fetch --multiple` commands can help here by doing multiple fetches with a single command.

- Setting up this workflow requires developers' public repositories to be reachable from other developers' workstations, which might not be as easy as using one's own machine as a server for one's public repositories

- Also, as can be seen in *Figure 6.3*, collaboration gets more complicated with the growing number of developers; this workflow does not scale well

The maintainer or integration manager workflow

One of the problems with peer-to-peer workflows is that there is no canonical version of a project, something that non-developers can use. Another is that each developer has to do their own integration (which was also the case for the centralized workflow). If we promote one of the public repositories in *Figure 6.3* to be the canonical (official) repository and make one of the developers responsible for integration, we arrive at the **integration manager workflow** (or **maintainer workflow**). The following diagram shows this workflow, with bare repositories at the top and non-bare at the bottom.

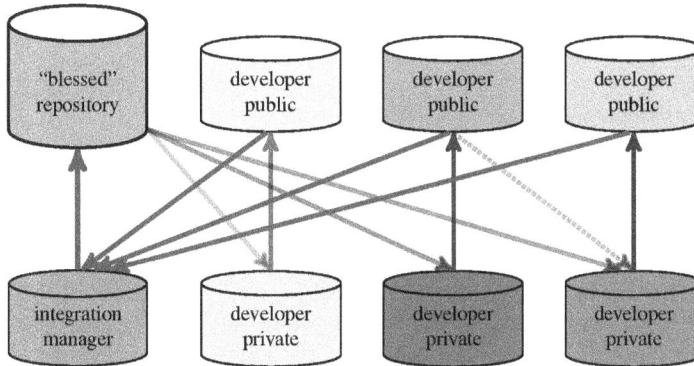

Figure 6.4 – Integration-manager (maintainer) workflow – lines pointing up are
push operations, while lines pointing down are fetch operations

In this workflow, when changes are ready, the developer pushes those changes to their own public repository and tells the maintainer (for example, via a **pull request**) that they are ready. The **maintainer** pulls changes from the developer's repository into their own working repository and integrates the changes. Then the maintainer pushes the merged changes to the **"blessed" repository**, for all to see, making them available to be fetched.

The advantages and disadvantages are as follows:

Advantages of the integration manager workflow

- The advantages are having an official version of a project, and that developers can continue to work without doing or waiting for integration, as maintainers can pull their changes at any time.

- It is a good workflow for a large organic team, as in open source projects.

- The fact that the blessed repository is decided by social consensus makes it easy to switch to other maintainers, either temporarily (for example, when one maintainer takes some time off) or permanently (such as when forking a project), without the need to hand out access rights.

- This setup makes it easy for a smaller group of developers to collaborate by simply denoting one of the repositories in the group as the one to fetch from. The dotted line in *Figure 6.4* shows this possibility of fetching from a non-official repository.

Disadvantages of the integration manager workflow

- The primary disadvantage is that the ability of the maintainer to integrate changes can be a bottleneck (as opposed to the centralized workflow, with distributed integration).

  This can happen especially for large teams and large projects. Thus, for very large organic teams, such as in Linux kernel development, it is better to use the hierarchical workflow, described in the next section.

- There needs to be dedicated person that does the merging and is responsible for the state of the "*blessed*" repository.

- Another disadvantage is that it is more difficult to set up continuous integration than in the centralized repository workflow.

The hierarchical or dictator-and-lieutenants workflow

The **hierarchical workflow** is a variant of the blessed repository workflow, generally used in huge projects with hundreds of collaborators. In this workflow, the project maintainer (sometimes called the **benevolent dictator**) is accompanied by additional integration managers, usually in charge of certain parts of the repository (subsystems). They are called **lieutenants**. The benevolent dictator's public repository serves as the *blessed* reference repository from which all the collaborators need to pull. Lieutenants pull from developers and the maintainer pulls from the lieutenants, as shown in *Figure 6.5*. (Note that in the following diagram, repositories shown with dashed patterns are actually pairs of private and public repositories of a developer or a lieutenant).

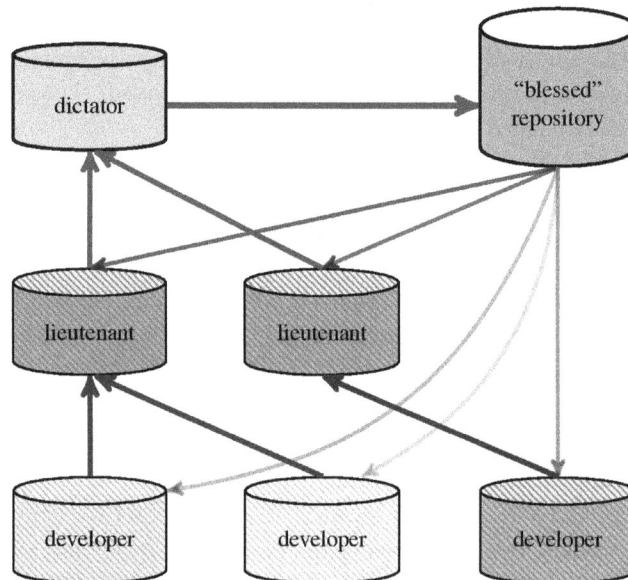

Figure 6.5 – Dictator and lieutenants workflow (hierarchical workflow)

In a **dictator and lieutenants** workflow, there is a hierarchy (a network) of repositories. Before starting work, either development or merging, one would usually pull updates from the canonical **(blessed) repository** for a project.

- **Developers** prepare changes in their own private repository, then send changes to an appropriate subsystem maintainer (lieutenant).

  Changes can be sent as patches in email, or by pushing them to the **developer's** public repository and sending a pull request to an appropriate integration manager (appropriate subsystem maintainer).

- **Lieutenants** are responsible for merging changes in their respective areas of responsibility.

- The master maintainer (**dictator**) pulls from the lieutenants (and occasionally directly from developers). The **dictator** is also responsible for pushing merged changes to the reference (canonical) repository, and usually also for release management (for example, creating tags for releases).

An overview of the advantages and disadvantages of this workflow follows.

Advantages of the hierarchical workflow

- The advantage of this workflow is that it allows the project leader (the dictator) to delegate much of the integration work.

- This can be useful in very big projects (concerning the number of developers and/or changes), or in highly hierarchical environments. Such a workflow is used, for example, to develop the Linux kernel.

Disadvantages of the hierarchical workflow

- Its complicated setup is a disadvantage of this workflow. It is usually overkill for an ordinary project.

- Almost all other disadvantages of the integration manager workflow are present in this workflow, which is its more complex variant.

Which workflow to choose, and how to set up repositories, depend on how the project is developed. You need to decide which drawbacks are acceptable and which advantages matter most.

Managing remote repositories

When collaborating on any project managed with Git, you will interact often with a constant set of other repositories. For example, using the integration-manager workflow will involve (at least) the canonical blessed repository of a project. In many cases, you will interact with more than one remote repository.

Git allows us to save the information about a remote repository (or just **remote** for short) in the config file by giving it a nickname (a shorthand name). This configuration can be managed with the `git remote` command.

> **Legacy mechanisms for storing remote repository information**
>
> There are also two legacy mechanisms to store the information about remote repositories.
>
> This first is a named file in `.git/remotes`—the name of this file will be the nickname of the remote. This file can contain information about the URL or URLs, and fetch and push refspecs.
>
> The second is a named file in `.git/branches`—the name of this file will also be the nickname of the remote. The contents of this file are just a URL for the repository, optionally followed by # and the branch name.
>
> Neither of those mechanisms is likely to be found in modern repositories. See the *Remotes* section in the `git-fetch(1)` manpage for more details.

The "origin" remote

When cloning a repository, Git will create one remote for you—the **origin remote**, which stores information about where you cloned from—that is the origin of your copy of the repository (hence the name). You can use this remote to fetch updates.

This is the default remote; for example, `git fetch` without the remote name will use the origin remote. You can change this using the `remote.default` configuration variable on a per-repository basis, or you can set up a default remote differently for a given branch with `branch.<branchname>.remote`.

Listing and examining remotes

To see which remote repositories you have configured, you can run the `git remote` command. It lists the short names of each remote you've got. In a cloned repository you will have at least one remote named `origin`:

```
$ git remote
origin
```

To see the URL together with remotes, you can use the `-v` or `--verbose` options:

```
$ git remote --verbose
origin  https://github.com/git/git.git (fetch)
origin  https://github.com/git/git.git (push)
```

From the output of this command, you can easily guess that the fetch and push URLs can be different (in a so-called **triangular workflow**).

If you want to inspect remotes to see more information about a particular remote, you can use the `git remote show <remote>` subcommand:

```
$ git remote show origin
remote origin
  Fetch URL: https://github.com/git/git.git
```

```
Push  URL: https://github.com/git/git.git
HEAD branch: master
Remote branches:
  maint   tracked
  master  tracked
  next    tracked
  pu      tracked
  todo    tracked
Local branch configured for 'git pull':
  master merges with remote master
Local ref configured for 'git push':
  master pushes to master (up to date)
```

Git will consult the remote configuration, the branch configuration, and the remote repository itself (for an up-to-date status). If you want to skip contacting the remote repository and use cached information instead, add the -n option to git remote show. If there is no internet connection,and you did not use '-n' option, Git will tell you that it was unable to contact the repository.

As the information about remotes is stored in the repository configuration file, you can simply examine .git/config:

```
[remote "origin"]
    fetch = +refs/heads/*:refs/remotes/origin/*
    url = git://git.kernel.org/pub/scm/git/git.git
```

The difference between local and remote branches (and **remote-tracking branches**: local representations of remote branches) will be described in *Chapter 8, Advanced Branching Techniques*, together with an explanation of **refspecs**. The refspec is a thing that is used to describe mapping between branches in remote repository and local remote-tracking branches, which looks like this: +refs/heads/*:refs/remotes/origin/* . You can see it in the second line in the preceding example.

Adding a new remote

To add a new remote Git repository and to store its information under a short name, run git remote add <shortname> <URL>:

```
$ git remote add alice \
  https://git.company.com/alice/random.git
```

Adding remote doesn't fetch from it automatically—you need to use the -f option for that (or run git fetch <shortname> afterwards).

This command has a few options that affect how Git creates a new remote. You can select which branches in the remote repository you are interested in with the -t <branch> option. You can change which branch is the default one in the remote repository (and which you can refer to by the remote name) using the `-m <branch>` option; otherwise, it would be the current branch in the remote repository. You can fetch all tags or no tags with `--tags` or `--no-tags`, respectively; otherwise, only tags on fetched branches would be imported. Or you can configure the remote repository for mirroring rather than for collaboration with `--mirror=push` or `--mirror=fetch`.

For example, running the following command:

```
$ git remote add -t master -t next -t maint github \
  https://github.com/jnareb/git.git
```

will result in the following configuration of the remote:

```
[remote "github"]
    url = https://github.com/jnareb/git.git
    fetch = +refs/heads/master:refs/remotes/github/master
    fetch = +refs/heads/next:refs/remotes/github/next
    fetch = +refs/heads/maint:refs/remotes/github/maint
```

Updating information about remotes

The information about the remote repository is stored in three places:

- In the remote configuration: `remote.<remote name>`,
- In remote-tracking branches and in the remote-HEAD (`refs/remotes/<remote name>/HEAD`)
- And optionally, in the per-branch configuration: `branch.<branch name>`

The remote-HEAD is a symbolic reference (`symref`) that defines the **default remote-tracking branch**. The remote-HEAD is a symbolic reference (symref) that defines the default remote-tracking branch. This means it determines which remote-tracking branch `<remote name>` refers to when used as a branch name, such as in the command `'git log <remote name>'`.

You could manipulate this information directly—either by editing the appropriate files or using manipulation commands such as `git config` and `git symbolic-ref`—but Git provides various `git remote` subcommands for this.

Renaming remotes

Renaming the remote—that is, changing its nickname—is quite a complicated operation. Running `git remote rename <old> <new>` will not only change the section name in `remote.<old>`, but also the remote-tracking branches and accompanying `refspec`, their reflogs (if there are any—see the `core.logAllRefUpdates` configuration variable), and the respective branch configurations.

Changing the remote URLs

You can add or replace the URL for a remote with `git remote set-url`, but it is also quite easy to simply directly edit the configuration.

You can also use the `insteadOf` (and `pushInsteadOf`) configuration variables. This can be useful if you want to temporarily use another server, for example, if the canonical repository is temporarily down. Say that you want to fetch Git from the repository on GitHub, because `https://www.kernel.org` that you cloned Git from is down. You can do this by adding the following text to the config file:

```
[url "https://github.com/git/git.git"]
    insteadOf = git://git.kernel.org/pub/scm/git/git.git
```

Another use case for this feature is handling repository migration. You can use `insteadOf` rewriting in the per-user configuration file, that is, in `~/.gitconfig` (or `~/.config/git/config`), without having to change the URL in each and every repository's `.git/config` file. In the case of more than one match, the longest match is used.

> **Tip – multiple URLs for a remote**
>
> You can set multiple URLs for a remote. Git will try all these URLs sequentially when fetching and use the first one that works. When pushing, Git will publish to all URLs (all servers) simultaneously.

Changing the list of branches tracked by remote

A similar situation to changing the URL occurs when changing the list of branches tracked by a remote (that is, the contents of `fetch` lines). You can use `git remote set-branches` (with a sufficiently modern Git client) or edit the config file directly.

> **Note – stale remote-tracking branches**
>
> Freeing a branch in a remote repository from being tracked does not remove the remote-tracking branch—the latter is simply no longer updated on fetch. This is explained in more detail in the *Deleting remote-tracking branches and remotes* section later in this chapter, which describes how to prune remote-tracking branches that correspond to branches deleted in remote the repository.

Setting the default branch of the remote

Having a **default branch on the remote** is not required, but it lets us specify the remote name (for example, `origin`) instead of a specific remote-tracking branch (for example, `origin/master`). This information is stored in the symbolic ref `<remote name>/HEAD` (for example, `origin/HEAD`).

You can set this with `git remote set-head` command. The `--auto` option does that based on what the current branch in the remote repository is:

```
$ git remote set-head origin master
$ git branch -r
  origin/HEAD -> origin/master
  origin/master
```

You can delete the default branch on the remote with the `--delete` option.

Deleting remote-tracking branches and remotes

When a public branch is deleted in the remote repository, Git nevertheless keeps the corresponding remote-tracking branch. It does that because you might want to do, or might have already done, your own work on top of it. You can, however, delete the remote-tracking branch with `git branch -r -d`, or you can ask Git to prune all stale remote-tracking branches under the remote with `git remote prune`. You can configure Git to do this automatically on every fetch, as if `git fetch` were run with the `--prune` option, by setting the `fetch.prune` and/or `remote.<name>.prune` configuration variables (the latter on a per-remote basis).

You can check which remote-tracking branches are stale with the `--dry-run` option to `git remote prune`, or with the `git remote show` command.

Deleting remote is as simple as running `git remote delete` (or its alias, `git remote rm`). It also removes remote-tracking branches for the deleted remote.

Support for triangular workflows

In many collaborative workflows, such as the maintainer (or integration manager) workflow, you fetch from one URL (from the blessed repository), but push to another URL (to your own public repository).

As shown in *Figure 6.4*, the developer interacts with three repositories: they fetch from the **blessed** repository (top left) into their own private repository (darker, at the bottom), then push their work into their own public repository (lighter, at the top).

In such a **triangular workflow** (three repositories), the remote you fetch or pull from is usually the default `origin` remote (or `remote.default`). One option for configuring which repository you push to is to add this repository as a separate remote, and perhaps also set it up as the default with `remote.pushDefault`:

```
[remote "origin"]
    url = https://git.company.com/project
    fetch = +refs/heads/*:refs/remotes/origin/*
[remote "myown"]
    url = git@work.company.com:user/project
```

```
fetch = +refs/heads/*:refs/remotes/myown/*
[remote]
    pushDefault = myown
```

You could also set it as `pushRemote` in the per-branch configuration:

```
[branch "master"]
    remote = origin
    pushRemote = myown
    merge = refs/heads/master
```

Another option is to use a single remote (perhaps even `origin`) but set it up with a separate `pushurl`. This solution, however, has the slight disadvantage that you don't have separate remote-tracking branches for the push repository (and thus there is no support `@{push}` notation in addition to having `@{upstream}` as a shortcut for specifying the appropriate remote-tracking branches):

```
[remote "origin"]
    url = https://git.company.com/project
    pushurl = git@work.company.com:user/project
    fetch = +refs/heads/*:refs/remotes/origin/*
```

Having separate remote-tracking branches for the push repository allows you to track which branches were pushed to the push remote, and which have local unpublished changes.

Chain of trust

An important part of collaborative efforts during the development of a project is ensuring the quality of its code. This includes protection against the accidental corruption of the repository, and also from **malicious intent**—a task that the version control system can help with. Git needs to ensure trust in the repository contents: both your own and other developers' (especially trust in the canonical repository of the project).

Content-addressed storage

In *Chapter 4*, *Exploring Project History*, in the *SHA-1 and the shortened SHA-1 identifier* section, we learned that Git currently uses SHA-1 hashes as a native identifier of commit objects (which represent revisions of the project and form the project's history). This mechanism makes it possible to generate commit identifiers in a distributed way, taking a cryptographic hash of the commit object. This hash is then used to link to the previous commit (to the parent commit or commits).

Moreover, all other data stored in the repository (including the file contents in the revision represented by the blob objects, and the file hierarchy represented by the tree objects) also use the same mechanism. All types of object are addressed by their contents, or to be more accurate, the hash function of the object. You can say that the base of a Git repository is the *content-addressed object database*.

Thus, Git provides a built-in **trust chain** through secure SHA-1 hashes, via a kind of a hash tree, also known as a Merkle tree. In one dimension, the SHA-1 hash of a commit depends on its contents, which includes the SHA-1 hash of the parent commit or commits, which depends on the contents of the parent commit, and so forth down to the initial root commit. In the other dimension, the content of a commit object includes the SHA-1 hash of the tree representing the top directory of a project, which in turn depends on its contents, and these contents include the SHA-1 hash of the subdirectory trees and blobs of file contents, and so forth down to the individual files.

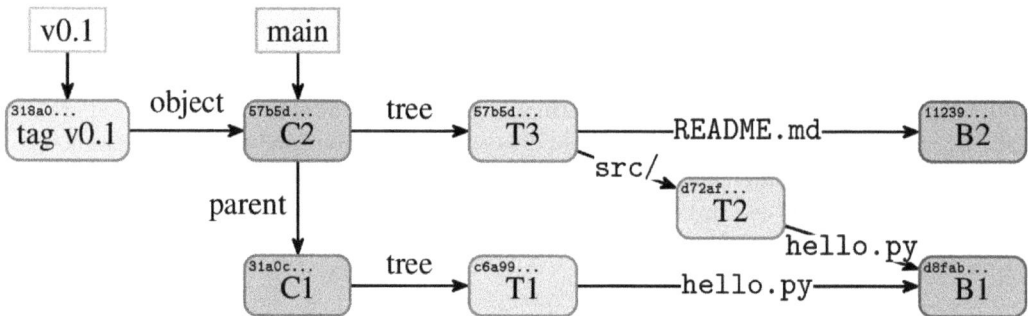

Figure 6.6 – Hash tree of a short history of a project, with a tag, two commits, and their contents. The SHA-1 hashes, shown in shortened form, depending on their contents

All of this allows SHA-1 hashes to be used to verify whether objects obtained from a (potentially untrusted) source have been corrupted or modified since they were created.

Lightweight, annotated, and signed tags

The trust chain allows us to verify the contents but does not verify the identity of the person who created the content (the author and committer name are fully configurable and under user control). This is the task for GPG/PGP signatures: signed tags, signed commits, and signed merges.

Since Git version 2.34, you can also use SSH keys for signing by setting the gpg.format configuration variable to the value ssh, for example with git config gpg.format ssh (you may also need to use your public key as the configuration value for the user.signingKey configuration variable).

Lightweight tags

Git uses two types of tags: lightweight tags and annotated tags (there are also signed tags, which are a special case of annotated tags).

A **lightweight tag** is very much like a branch that doesn't change – it's just a pointer (reference) to a specific commit in the graph of revisions, though in the refs/tags/ namespace rather than in refs/heads/.

Annotated tags

Annotated tags, however, involve **tag objects**. Here the tag reference (in `refs/tags/` namespace) points to a tag object, which in turn points to a commit. Tag objects contain a creation date, the tagger identity (name and e-mail), and a tagging message. You create an annotated tag with `git tag -a` (or `--annotate`). If you don't specify a message for an annotated tag on the command line (for example, with `-m "<message>"`), Git will launch your editor so you can enter it.

You can view the tag data along with the tagged commit with the `git show` command as follows (commit skipped):

```
$ git show v0.2
tag v0.2
Tagger: Joe R Hacker <joe@company.com>
Date:   Sun Jun 1 03:10:07 2014 -0700

random v0.2

commit 5d2584867fe4e94ab7d211a206bc0bc3804d37a9
```

Signed tags

Signed tags are annotated tags with a clear text PGP signature (or, with modern Git, an SSH signature) of the tag data attached. You can create them with `git tag -s` (which uses your committer identity to select the signing key, or `user.signingKey` if set), or with `git tag -u <key-id>`; both versions assume that you have a **private GPG key** (created, for example, with `gpg --gen-key`).

> **Lightweight tags versus annotated and signed tags**
>
> Annotated or signed tags are meant for marking a release, while lightweight tags are meant for private or temporary revision labels. For this reason, some Git commands (such as `git describe`) will ignore lightweight tags by default.

Of course in collaborative workflows, it is important that the signed tag is *made public*, and that there is a way to *verify it*; both of those operations will be described in the following sections.

Publishing tags

Git does not push tags by default: you need to do it explicitly. One solution is to individually **push a tag** with `git push <remote> tag <tag-name>` (here, `tag <tag>` is equivalent to the longer **refspec** (describing how refs on the remote translate to refs in the local repository), namely `refs/tags/<tag>:refs/tags/<tag>`); however, if you don't have the naming conflict between a branch and a tag (i.e., you don't have branch and tag with the same name), then you can skip the word `tag` here in this operation.

Another solution is to push tags en masse: either all the tags—both lightweight and annotated—with the use of the `--tags` option, or just all annotated tags that point to pushed commits with `--follow-tags`. This explicitness allows you to re-tag (using `git tag -f`) with impunity if it turns out that you tagged the wrong commit, or there is a need for a last-minute fix—but only if the tag was not made public. Git does not (and should not) change tags behind the user's back; thus, if you pushed the wrong tag, you need to ask others to delete this old tag to change it.

When fetching changes, Git automatically **follows tags**, downloading annotated tags that point to fetched commits. This means that downstream developers will automatically get signed tags, and will be able to verify releases.

Tag verification

To verify a signed tag, you use `git tag --verify <tag-name>` (or -v for short). You need the signer's **public GPG key** in your keyring for this (imported using `gpg --import` or `gpg --keyserver <key-server> --recv-key <key-id>`), and of course the tagger's key needs to be vetted in your chain of trust. For **SSH keys** there is no *web of trust*; you need to specify the trusted public keys with the `gpg.ssh.allowedSignersFile` configuration variable.

```
$ git tag --verify v0.2
object 1085f3360e148e4b290ea1477143e25cae995fdd
type commit
tag signed
tagger Joe Random   1411122206 +0200

project v0.2
gpg: Signature made Fri Jul 19 12:23:33 2014 CEST using RSA key ID
A0218851
gpg: Good signature from "Joe Random <jrandom@example.com>"
```

Signed commits

Signed tags are a good solution for users and developers to verify that the tagged release was created by the maintainer. But how do we make sure that a commit purporting to be by somebody named Jane Doe, with the `jane@company.com` e-mail, is *actually* a commit from her? How can we make it so anybody can check this?

One possible solution is to sign individual commits. You can do this with `git commit --gpg-sign[=<keyid>]` (or -S for short). The key identifier is optional—without this, Git would use your identity as the author. Note that -S (capital S) is different from -s (small s); the latter adds a *Signed-off-by* line at the end of the commit message for the *Digital Certificate of Ownership*:

```
$ git commit -a --gpg-sign

You need a passphrase to unlock the secret key for
```

```
user: "Jane Doe <jane@company.com>"
2048-bit RSA key, ID A0218851, created 2014-03-19

[master 1085f33] README: eol at eof
 1 file changed, 1 insertion(+), 1 deletion(-)
```

To make commits available for verification, just push them. Anyone can then verify them with the `--show-signature` option to `git log` (or `git show`), or with one of the `%Gx` placeholders in `git log --format=<format>`:

```
$ git log -1 --show-signature
commit 1085f3360e148e4b290ea1477143e25cae995fdd
gpg: Signature made Wed Mar 19 11:53:49 2014 CEST using RSA key ID
A0218851
gpg: Good signature from "Jane Doe <jane@company.com>
Author: Jane Doe <jane@company.com>
Date:   Wed Mar 19 11:53:48 2014 +0200

    README: eol at eof
```

You can also use the `git verify-commit` command for this.

Merging signed tags (merge tags)

The *signed commit* mechanism, described in the previous section, may be useful in some workflows, but it is inconvenient in an environment where you push commits out early, and only after a while do you decide whether they are worth including in the upstream. In such cases, you would want to sign only those parts that are ready to be published.

This situation can happen if you follow the recommendations in *Chapter 10, Keeping History Clean*; you know only after the fact (long after the commit was created) that the given iteration of the commit series passes code review. Commits need to be signed at commit creation time, but you can create a signed tag after the fact, after the series of commits gets accepted.

You can deal with this issue by rewriting the whole commit series after its shape is finalized (after passing the review), signing each rewritten commit, or just by amending and signing only the top commit. Both of those solutions would require a forced push to replace the old history where commits were not signed. You can always sign every commit, or you can create an empty commit (with `--allow-empty`), sign it, and push it on top of the series. But there is a better solution: requesting the pull of a signed tag.

In this workflow, you work on your changes and, when they are ready, you create and push a signed tag (tagging the last commit in the series). You don't have to push your working branch—pushing the tag is enough. If the workflow involves sending a pull request to the integrator, you create it using a signed tag instead of the end commit:

```
$ git tag -s 1253-for-maintainer
$ git request-pull origin/master public-repo \
   1253-for-maintainer >msg.txt
```

The signed tag message is shown between the dashed lines in the pull request, which means that you may want to explain your work in the tag message when creating the signed tag. The maintainer, after receiving such a pull request, can copy the repository line from it, fetching and integrating the named tag. When recording the merge result of pulling the named tag, Git will open an editor and ask for a commit message. The integrator will see a template starting with the following:

```
Merge tag '1252-for-maintainer'

Work on task tsk-1252

# gpg: Signature made Wed Mar 19 12:23:33 2014 CEST using RSA key ID
A0218851
# gpg: Good signature from "Jane Doe <jane@company.com>"
```

This commit template includes the commented-out output of the verification of the signed tag object being merged (so it won't be in the final merge commit message). The tag message helps describe the merge better.

The signed tag being pulled is *not* stored in the integrator's repository, not as a tag object. Its content is stored, hidden, in a merge commit. This is done to avoid polluting the tag namespace with a large number of such working tags. The developer can safely delete the tag (`git push public-repo --delete 1252-for-maintainer`) after it gets integrated.

Recording the signature inside the merge commit allows for after-the-fact verification with the `--show-signature` option:

```
$ git log -1 --show-signature
commit 0507c804e0e297cd163481d4cb20f3f48ceb87cb
merged tag '1252-for-maintainer'
gpg: Signature made Wed Mar 19 12:23:33 2014 CEST using RSA key ID
A0218851
gpg: Good signature from "Jane Doe <jane@company.com>"
Merge: 5d25848 1085f33
```

```
Author: Jane Doe <jane@company.com>
Date:   Wed Mar 19 12:25:08 2014 +0200

    Merge tag 'for-maintainer'

    Work on task tsk-1252
```

Summary

Through this chapter, we learned how to use Git for collaborative development and how to work together in a team on a project. We got to know different collaborative workflows, that is, different ways of setting up repositories for collaboration. Which one to use depends on circumstances: how large the team is, how diverse, and so on. This chapter focuses on repository-to-repository interaction; the interplay between branches and remote-tracking branches in those repositories is left for *Chapter 8, Advanced Branching Techniques*.

We learned how Git can help manage information about remote repositories involved in the chosen workflow. We were shown how to store, view, and update this information. This chapter explains how one can manage triangular workflows, in which you fetch from one repository (canonical), and push to the other (public).

We learned about the chain of trust: how to verify that a release comes from the maintainer, how to sign your work so that the maintainer can verify that it comes from you, and how the Git architecture helps with this.

The next chapter, namely *Chapter 7, Publishing Your Changes*, will talk about how to get your contribution to other remote repositories. The two further following chapters will expand on the topic of collaboration: *Chapter 8, Advanced Branching Techniques*, will explore relations between local branches and branches in a remote repository and how to set up branches for collaboration, while *Chapter 9, Merging Changes Together*, will talk about the opposite issue—how to join the results of parallel work.

Questions

Answer the following questions to test your knowledge of this chapter:

1. What operation do you need to publish your changes to your public remote repository, and what operation do you need to get changes from a remote?

2. What is the difference between `git fetch` and `git pull`?

3. How can you remove stale remote-tracking branches (that is, remote-tracking branches where the corresponding branch on the remote was deleted)?

Answers

Here are the answers to the questions given above:

1. Use `git push` to publish your changes, and use `git fetch` or `git pull` (or `git remote update`) to get changes from the remote repository.

2. The `fetch` operation only downloads changes and updates the remote-tracking branches, while the `pull` operation also tries to update the current branch with `merge` or `rebase` (if it is configured as tracking some branch in the remote repository).

3. You can use `git branch -d -r` to delete individual remote-tracking branches, or `git remote prune` to delete all stale remote-tracking branches.

Further reading

To learn more about the topics that were covered in this chapter, take a look at the following resources:

- Scott Chacon and Ben Straub: *Pro Git, 2nd Edition* (2014) `https://git-scm.com/book/en/v2`

  - *Chapter 5.1 Distributed Git - Distributed Workflows*

  - *Chapter 2.5 Git Basics - Working with Remotes*

  - *Chapter 7.4 Git Tools - Signing Your Work*

- Ryan Brown: *gpg-sign releases* (2014) `https://gitready.com/advanced/2014/11/02/gpg-sign-releases.html`

- Danilo Bargen: *Signing Git Commits with SSH Keys* (2021) `https://blog.dbrgn.ch/2021/11/16/git-ssh-signatures/`

- Carl Tashian: *SSH Tips & Trick – Add a second factor to your SSH login* (2020) `https://smallstep.com/blog/ssh-tricks-and-tips/#add-a-second-factor-to-your-ssh`

- Junio C Hamano: *Git Blame: Fun (?) with GnuPG* (2014) `https://git-blame.blogspot.com/2014/09/fun-with-gnupg.html`

7

Publishing Your Changes

Chapter 6, Collaborative Development with Git (the previous chapter), taught you how to use Git to work together as a team, focusing on repository-to-repository interaction. It described different ways of setting up repositories for collaboration, presenting different collaborative workflows, such as centralized and integration manager workflows. It also showed how Git manages information about remote repositories.

In this chapter, you will find out how you can exchange information between your local repository and remote repositories, and how Git can manage credentials that might be needed to access remote repositories.

This chapter will also teach you how to provide your changes upstream, so that they appear in the official history of the project, in its canonical repository. This can be done by pushing your changes to a central repository, pushing them to your own publishing repository and sending some kind of a pull request to the integration manager, or even exchanging patches.

This chapter will cover the following topics:

- Transport protocols used by Git and their advantages and disadvantages
- Managing credentials (passwords, keys) for remote repositories
- Publishing changes: push and pull requests, and exchanging patches
- Using bundles for offline transfer and speeding up the initial clone
- Remote transport helpers and their use

Transport protocols and remote helpers

In general, URLs in the configuration of remote repository contain information about the transport protocol, the address of the remote server (where appropriate), and the path to the repository. Sometimes, the server that provides access to the remote repository supports various transport protocols; you need to select which one to use. This section is intended to help with this choice.

Local transport

If the remote repository is on the same local filesystem, you can use either the path to the repository or the `file://` schema to specify the repository URL:

```
/path/to/repo.git/
file:///path/to/repo.git/
```

The former implies the `--local` option to the Git clone, which bypasses the smart Git-aware mechanism and simply makes a copy (or a hard link for immutable files under `.git/objects`, though you can avoid this with the `--no-hardlinks` option); the latter is slower but can be used to get a clean copy of a repository (for example, after history rewriting done to remove an accidentally committed password or another secret; which is described in *Chapter 10*, *Keeping History Clean*, in the *Rewriting history* section).

This transport is a nice option for quickly grabbing work from someone else's working repository, or for sharing work using a shared filesystem with the appropriate permissions.

As a special case, a single dot (`.`) denotes the current repository. This means that

```
$ git pull . next
```

is, assuming that `pull.rebase` is set to false, roughly equivalent to

```
$ git merge next
```

Smart transports

When the repository you want to fetch from is on another machine, you need to access the Git server. Nowadays, Git-aware smart servers are most commonly encountered. The smart downloader negotiates which revisions are necessary, and creates a customized `packfile` to send to a client. Similarly, during the push, the Git server talks to Git on the user's machine (to the client) to find which revisions to upload.

Git-aware smart servers use the `git upload-pack` downloader for fetching and `git receive-pack` for pushing. You can tell Git where to find them if they are not in `PATH` (but, for example, are installed in one's home directory) with the `--upload-pack` and `--receive-pack` options for fetching and pushing, or the `uploadpack` and `receivepack` configuration variables in the `remote.<name>` section.

With very few exceptions (such as the repository using submodules accessed by an ancient Git instance that does not understand them), Git transport is backward- and forward-compatible—the client and server negotiate what capabilities they can both use.

The native Git protocol

The native transport, using `git://` URLs, provides read-only anonymous access (though you could, in theory, configure Git to allow pushing by enabling the `receive-pack` service, either from the command line via the `--enable=receive-pack` option, or via the `daemon.receivePack` boolean-valued configuration variable—using this mechanism is not recommended at all, even in a closed local network).

The Git protocol does no authentication, including no server authentication, and should be used with caution on unsecured networks. The `git daemon` TCP server for this protocol normally listens on port `9418`; you need to be able to access this port (through the firewall) to be able to use the native Git protocol.

> Trivia
>
> There is no secure version of the `git://` protocol. There is no `git://` over TLS like there is for the FTP and HTTP protocols—namely, FTPS and HTTPS. On the other hand, one can consider SSH transport, as used by Git, to be the `git://` protocol over SSH.

The SSH protocol

The **Secure Shell (SSH)** transport protocol provides authenticated read-write access. Git simply runs `git upload-pack` or `git receive-pack` on the server, using SSH to execute the remote command. There is no possibility for anonymous, unauthenticated access, though you could, as a workaround, set up a guest account for it (passwordless or with an empty password).

Using public-private key authentication allows access without requiring you to provide a password on every connection. You might, however, need to provide it once to unlock a password-protected private key. You can read more about authentication in the *Credentials/password management* section in this chapter. Many Git hosting sites and software forges require key authentication for accessing repositories via SSH.

For the SSH protocol, you can use the URL syntax with `ssh://` as the protocol part:

```
ssh://[user@]host.example.com[:port]/path/to/repo.git/
```

Alternatively, you can use the `scp`-like syntax:

```
[user@]host.example.com:path/to/repo.git/
```

The SSH protocol additionally supports the `~username` expansion, just like the native Git transport (`~` is the home directory of the user you log in as, and `~user` is the home directory of `user`), in both syntax forms:

```
ssh://[user@]host.example.com/~[user]/path/to/repo.git/
[user@]host.example.com:~[user]/path/to/repo.git/
```

SSH uses the first contact authentication for servers (**TOFU**—short for **Trust On First Use**)—it remembers the key that the server side previously used, and warns the user if it has changed, asking for confirmation (the server key could have been changed legitimately, for example, due to an SSH server reinstall). You can check the server key fingerprint on the first connection.

The smart HTTP(S) protocol

Git also supports the smart HTTP(S) protocol, which requires a Git-aware CGI or server module—for example, `git-http-backend` (itself a CGI module). This protocol uses the following URL syntax:

```
http[s]://[user@]host.example.com[:port]/path/to/repo.git/
```

By default, without any other configuration, Git allows anonymous downloads (`git fetch`, `git pull`, `git clone`, and `git ls-remote`), but requires that the client is authenticated for upload (`git push`).

Standard HTTP authentication is used if authentication is required to access a repository, which is done by the HTTP server software. Using SSL/TLS with HTTPS ensures that if the password is required (for example, if the server uses basic HTTP authentication), then it is sent encrypted and the server identity is verified (using server CA certificate).

Legacy (dumb) transports

Some transports do not require any Git-aware smart server—they don't need Git installed on the server (for smart transports, at least `git-upload-pack` and/or `git-receive-pack` is needed). Those are the FTP(S) and dumb HTTP(S) protocol transports (nowadays, implemented using the `remote-curl` helper).

These transports need only the appropriate stock server (an FTP server, or a web server), and up-to-date data from `git update-server-info`. When fetching from such a server, Git uses the so-called **commit walker** downloader: going down from fetched branches and tags, Git walks down the commit chain, and downloads objects or packs containing missing revisions and other data (for example, file content at revision).

This transport is inefficient (in terms of bandwidth, but especially in terms of latency), but on the other hand, it can be resumed if interrupted. Nevertheless, there are better solutions than using dumb protocols—namely, involving bundles (see the *Offline transport with bundles* section in this chapter), when the network connection to the server is unreliable enough that you can't get the clone operation to finish.

Pushing to a dumb server is possible only via the HTTP and HTTPS protocols. It requires the web server to support WebDAV, and Git has to be built with the *expat* library linked. The FTP and FTPS protocols are read-only (supporting only `clone`, `fetch`, and `pull`).

As a design feature, Git can automatically upgrade dumb protocol URLs to smart URLs. Conversely, a Git-aware HTTP server can downgrade to the backward-compatible dumb protocol (at least for fetching: smart HTTP servers don't support WebDAV-based dumb HTTP push operation). This feature allows the use of the same HTTP(S) URL for both dumb and smart access:

```
http[s]://[user@]host.example.com[:port]/path/to/repo.git/
```

Offline transport with bundles

Sometimes, there is no direct connection between your machine and the server holding the Git repository that you want to fetch from. Or, perhaps there is no server running, and you want to copy changes to another machine anyway. Maybe your network is down. Perhaps you're working somewhere on-site and don't have access to the local network for security reasons. Maybe your wireless/Ethernet card just broke.

Enter the `git bundle` command. This command will package up everything that would normally be transferred over the wire, putting objects and references into a special binary archive file called `bundle` (like `packfile`, only with branches and so on). You need to specify which commits are to be packed—something that network protocols do automatically for you for online transport.

> **Trivia**
>
> When you are using one of the smart transports, a **want/have negotiation** phase takes place, where the client tells the server what it does have in its repository and which advertised references on the server it wants, to find common revisions. This is then used by the server to create a packfile, and to send the client only what's necessary, minimizing the bandwidth use.

Next, you need to move this bundle (this archive) by some means to your machine. It can be done, for example, by so-called **sneakernet**, which means saving the bundle to removable storage and physically moving the media. You can then incorporate the bundle contents by using `git clone` or `git fetch` with the filename of the bundle in place of the repository URL.

Proxies for Git transports

When direct access to the server is not possible, for example, from within a firewalled LAN, sometimes you can connect via a proxy.

For the native Git protocol (`git://`), you can use the `core.gitProxy` configuration variable, or the `GIT_PROXY_COMMAND` environment variable to specify a proxy command—for example, `ssh`. This can be set on a per-remote basis with special syntax for the `core.gitProxy` value—namely, `<command> for <remote>`; for example, `"ssh" for kernel.org`.

You can use the `http.proxy` configuration variable or appropriate *curl* environment variables such as `http_proxy` to specify the HTTP proxy server to use for the HTTP(S) protocol (`http(s)://`). This can be set on a per-remote basis with the `remote.<remote name>.proxy` configuration variable.

You can configure SSH (using its configuration files—for example, `~/.ssh/config`) to use tunneling (port forwarding) or a proxy command (for example, the `netcat/nc`; or `netcat` mode of SSH—that is, `ssh -W -`—if your SSH implementation supports this feature). It is a recommended solution for the SSH proxy; if neither tunneling nor using a proxy is possible, you can use the `ext::` transport-helper, as shown later in this chapter, in the *Transport relay with remote helpers* section.

Cloning and updating with bundles

Let's assume that you want to transfer the history of a project (say, limited to the `master` branch for simplicity) from `machineA` (for example, your work computer) to `machineB` (for example, an onsite computer). There is, however, no direct connection between those two machines.

First, we create a bundle that contains the whole history of the `master` branch (see *Chapter 4, Exploring Project History*), and tag this point of history to know what we bundled, which will be needed later:

```
user@machineA ~$ cd repo
user@machineA repo$ git bundle create ../repo.bundle master
user@machineA repo$ git tag -f lastbundle master
```

Here, the bundle file was created outside the working directory. This is a matter of choice; storing it outside of the repository means that you don't have to worry about accidentally adding it to your project history, or having to add a new `ignore` rule. The `*.bundle` file extension is OR simply a matter of the naming convention.

> **Important note**
>
> For security reasons, to avoid information disclosure about the parts of history that were deleted but not purged (for example, an accidentally committed file with a password), Git only allows fetching from `git show-ref`-compatible references: branches, remote-tracking branches, and tags.
>
> The same restrictions apply when creating a bundle. This means, for example, that (for implementation reasons) you cannot run `git bundle creates master^1`. Though, of course, because you control the server end, as a workaround you can create a new branch for `master^`, (temporarily) rewind `master`, or check out the detached HEAD at `master^`.

Then you transfer the just-created `repo.bundle` file to `machineB` (via email, on a USB pen drive, and so on). Because this bundle consists of a self-contained, whole subset of the history, down to the first (parent-less) root commit, you can create a new repository by cloning from it, by simply putting the bundle filename in place of the repository URL:

```
user@machineB ~$ git clone repo.bundle repo
Cloning into 'repo'...
```

```
warning: remote HEAD refers to non-existent ref, unable to checkout.

user@machineB ~$ cd repo
user@machineB repo$ git branch -a
   remotes/origin/master
```

Oops. We didn't bundle HEAD, so the Git clone didn't know which branch is current and therefore should be checked out:

```
user@machineB repo$ git bundle list-heads ../repo.bundle
5d2584867fe4e94ab7d211a206bc0bc3804d37a9 refs/heads/master
```

> **Tip**
>
> Because a bundle can be treated as a remote repository, we could have simply used the `git ls-remote ../repo.bundle` command here instead of `git bundle list-heads ../repo.bundle`.

Therefore, with this bundle being as it is, we need to specify which branch to check out to avoid the problem (this would not be necessary if we had bundled HEAD too):

```
user@machineB ~$ git clone repo.bundle --branch master repo
```

Instead of cloning again, we can fix the problem with the failed checkout by selecting the current branch:

```
user@machineB repo$ git switch master
Already on 'master'
Branch 'master' set up to track remote branch 'master' from 'origin'.
```

As you can see, here, Git guessed that when trying to switch to a non-existent local branch, `master`, what we actually wanted was to create a local branch to create new commits to submit to the remote `master` branch. In other words, create a local branch following (tracking) the remote branch with the same name existing in the `origin` remote. What Git did is the same as if we ran the following command:

```
$ git switch --create master --track origin/master
```

To update the repository on `machineB` cloned from the bundle, you can fetch or pull after replacing the original bundle stored at `/home/user/repo.bundle` with the one with incremental updates.

To create a bundle containing changes since the last transfer in our example, go to `machineA` and run the following command:

```
user@machineA repo$ git bundle create ../repo.bundle \
   lastbundle..master
user@machineA repo$ git tag -f lastbundle master
```

This will bundle all changes since the `lastbundle` tag; this tag denotes what was copied with the previous bundle (see *Chapter 4, Exploring Project History*, the *Double-dot notation* section, for an explanation of double-dot syntax). After creating a bundle, this will update the tag (using `-f` or `--force` to replace it), like was done the first time when creating a bundle, so that the next bundle can also be created incrementally from the now current point.

Then, you need to copy the bundle to `machineB`, *replacing* the old one. At this point, one can simply perform the pull operation to update the repository, as shown in the following example:

```
user@machineB repo$ git pullFrom /home/user/repo.bundle


   ba5807e..5d25848  master      -> origin/master
Updating ba5807e..5d25848
Fast-forward
```

Using a bundle to update an existing repository

Sometimes, you might have a repository cloned already, only for the network to fail. Or, perhaps you moved outside the **local area network** (**LAN**), and now you have no access to the server. The end result is that you have an existing repository, but no direct connection to the upstream (to the repository we cloned from).

Now, if you don't want to bundle up the whole repository, which is wasteful, like in the *Cloning and updating with bundles* section, you need to find some way to specify the cut-off point (the base) in such a way that it is surely present in the target repository (which you want to update). You can specify the range of revisions to pack into the bundle using almost any technique from *Chapter 4, Exploring Project History*. The only limitation is that the history, as was said earlier, must start at a branch or tag (anything that `git show-ref` accepts). You can, of course, check the range with the `git log` command.

Commonly used solutions for specifying the range of revisions to pack into a bundle are as follows:

- Use the tag that is present in both repositories:

  ```
  git bundle create ../repo.bundle v0.1..master
  ```

- Create a cut-off based on the time of commit creation:

  ```
  git bundle create ../repo.bundle --since=1.week master
  ```

- Bundle just the last few revisions, limiting the revision range by the number of commits:

  ```
  git bundle create ../repo.bundle -5 master
  ```

> **Tip**
>
> It's better to pack in too much than too little. You can check whether the repository has the requisite commits to fetch from the bundle with `git bundle verify`. If you pack in too little, you'll get the following error:
>
> ```
> user@machineB repo$ git pull ../repo.bundle master
> error: Repository lacks these prerequisite commits:
> error:  ca3cdd6bb3fcd0c162a690d5383bdb8e8144b0d2
> ```

Then, after transporting it to `machineB`, you can use the bundle file just like a regular repository to do a one-off pull (putting the bundle filename in place of the URL or the remote name):

```
user@machineB repo$ git pull ../repo.bundle master
From ../repo.bundle
 * branch              master       -> FETCH_HEAD
Updating ba5807e..5d25848
```

If you don't want to deal with the merge, you can fetch into the remote-tracking branch (the `<remote branch>:<remote-tracking branch>` notation used here, which is known as *refspec*, will be explained in *Chapter 8*, *Advanced Branching Techniques*):

```
user@machineB repo$ git fetch ../repo.bundle \
    refs/heads/master:refs/remotes/origin/master
From ../repo.bundle
    ba5807e..5d25848  master       -> origin/master
Updating ba5807e..5d25848
```

Alternatively, you can use `git remote add` to create a new shortcut, using the path to the bundle file in place of the repository URL. Then, you can simply deal with bundles as described in the previous section.

Utilizing a bundle to help with the initial clone

Smart transports provide much more effective transport than dumb ones. On the other hand, the concept of a resumable clone using smart transport remains elusive to this day (it is not available in Git version 2.34, though perhaps somebody will implement it in the future). For large projects with a long history and with a large number of files, the initial clone might be quite large (for example, `linux-next` is more than 2.7 GB) and take a pretty long time. This might be a problem if the network is unreliable.

> **Tip – workaround**
>
> You can work around the issue of an unreliable network by using a shallow clone or a sparse clone (see *Chapter 12, Managing Large Repositories*) and widening it step by step until you arrive at the full repository. There are some third-party tools that do this automatically.

You can create a bundle from the source repository, for example, with the following command (which needs to run on the server):

```
$ git –git-dir=/dir/repo.git bundle create -- all HEAD
```

Some servers may offer such bundles to help with the initial clone. There is a practice where a bundle intended for cloning is available at the same URL as the repository, but with a `.bundle` suffix instead of `.git`. For example, `https://git.example.com/user/repo.git` has its bundle available at `https://git.example.com/user/repo.bundle`.

You can then download such a bundle, which is an ordinary static file, using any resumable transport: HTTP(S), FTP(S), rsync, or even BitTorrent (with the appropriate client that supports resuming the download).

With modern Git, the user can specify the bundle URI with the `--bundle-uri` command-line option, or a bundle list can be advertised by a Git server. A list of bundle URIs can also be saved in a config file. Fetching from bundle servers (such as `https://github.com/git-ecosystem/git-bundle-server`) is then automatic.

Remote transport helpers

When Git doesn't know how to handle a certain transport protocol (when one tries to use a protocol that doesn't have built-in support), it attempts to use the appropriate **remote helper** for a protocol, if one exists. That's why if there is an error within the protocol part of the repository URL—Git responds with an error message that looks like the following:

```
$ git clone shh://git@example.com:repo
Cloning into 'repo'…
fatal: Unable to find remote helper for 'shh'
git: 'remote-shh' is not a git command. See 'git --help'.
```

This error message means that Git tried to find `git-remote-shh` to handle the `shh` protocol (actually a typo for `ssh`), but didn't find an executable with such a name.

You can explicitly request a specific remote helper with the `<transport>::<address>` syntax, where `<transport>` defines the helper (`git remote-<transport>`), and `<address>` is a string that the helper uses to find the repository.

Modern Git implements support for the dumb HTTP, HTTPS, FTP, and FTPS protocols with the `curl` family of remote helpers: `git-remote-http`, `git-remote-https`, `git-remote-ftp`, and `git-remote-ftps`, respectively.

Transport relay with remote helpers

Git includes two generic remote helpers that can be used to proxy smart transports: the `git-remote-fd` helper to connect to the remote server via either a bidirectional socket or a pair of pipes, and the `git-remote-ext` helper to use an external command to connect to the remote server.

In the case of the latter, which uses the `"ext::<command> <arguments">"` syntax for the repository URL, Git runs the specified command to connect to the server, passing data for the server to the standard input of the command, and receiving a response on its standard output. This data is assumed to be passed to a `git://` server, `git-upload-pack`, `git-receive-pack`, or `git-upload-archive` (depending on the situation).

For example, let's assume that you have your repository on a LAN host where you can log in using SSH. However, for security reasons, this host is not visible on the internet, and you need to go through the gateway host: `login.example.com`:

```
user@home ~$ ssh user@login.example.com
user@login ~$ ssh work
user@work ~$ find . -name .git -type d -print
./repo/.git
```

The trouble is that—also for security reasons—this gateway host either doesn't have Git installed (reducing the attack surface) or doesn't have your repository present (it uses a different filesystem). This means that you cannot use the ordinary SSH protocol—not unless you can set up an **SSH tunnel** from your home via a gateway to your work computer (with `ssh -L`). The SSH transport is just `git-receive-pack`/`git-upload-pack` accessed remotely via SSH, with the path to the repository as a parameter. This means that you can use the `ext::` remote helper:

```
user@home ~$ git clone \
    "ext::ssh -t  ssh work %S 'repo'" repo
Cloning into 'repo'...
Checking connectivity... done.
```

Here, `%S` will be expanded by Git into the full name of the appropriate service—`git-upload-pack` for fetching and `git-receive-pack` for pushing. The `-t` option is needed if logging to the internal host uses interactive authentication (for example, a password). Note that you need to give the name (`repo`, here) to the result of cloning; otherwise, Git will use the command (`ssh`) as the repository name.

> **Tip**
>
> You can also use `"ext::ssh [<parameters>...] %S '<repository>'"` to use specific options for SSH transport—for example, selecting the keypair to use without needing to edit `.ssh/config`.

This is not the only possible solution—though there is no built-in support for sending the SSH transport through a proxy like there is for the native `git://` protocol (among others, `core.gitProxy`) and for HTTP (among others, `http.proxy`), you can do it by configuring the SSH using the `ProxyCommand` config option, or by creating an SSH tunnel.

On the other hand, you can also use the `ext::` remote helper to proxy the `git://` protocol—for example, with the help of `socat`—including using a single proxy for multiple servers. See the `git-remote-ext(1)` manpage for details and examples.

Using foreign SCM repositories as remotes

The remote helper mechanism is very powerful. It can also be used to interact with other version control systems, transparently using their repositories as if they were native Git repositories. Though there is no such built-in helper (unless you count the `contrib/` area in the Git sources), you can find the `git-remote-hg`, `gitifyhg`, or `git-cinnabar` helper to access Mercurial repositories, and `git-remote-bzr` to access Bazaar repositories.

Once installed, those remote helper bridges will allow you to clone, fetch, and push to and from the Mercurial or Bazaar repositories as if they were Git ones, using the `<helper>::<URL>` syntax. For example, to clone the Mercurial repository, you can simply run the following command:

```
$ git clone "hg::https://hg.example.com/repo"
```

There is also the `remote.<remote name>.vcs` configuration variable, if you don't like using the `<helper>::` prefix in the repository URL. With this method, you can use the same URL for Git as for the original **version control system (VCS)**.

> **Foreign version control system clients**
>
> The alternative approach to using remote helper bridges is to use a specialized client, such as `git-svn` for Subversion, or `git-p4` for Perforce. Those clients interact with the foreign VCS (usually a centralized VCS), manage and update the Git repository based on this interaction, and update the foreign repository based on changes present in the Git repository.

Of course, one needs to remember impedance mismatches between different version control systems, and the limitations of the remote helper mechanism. Some features do not translate at all or do not translate well—for example, octopus merges (with more than two parent commits) in Git, or multiple anonymous branches (heads) in Mercurial.

With remote helpers, there is also no place to fix mistakes, replace references to other revisions with target native syntax, and otherwise clean up artifacts created by repository conversions—as can and should be done with one-time conversion when changing version control systems. (Such a cleanup can be done with, for example, the help of the `reposurgeon` third-party tool.)

With remote helpers, you can even use things that are not version control repositories in the strict sense; for example, with the *Git-Mediawiki* project, you can use Git to view and edit a MediaWiki-based wiki (for example, Wikipedia), treating the history of pages as a Git repository:

```
$ git clone "mediawiki::https://wiki.example.com"
```

Besides that, there are remote helpers that allow additional transport protocols, or storage options—such as `git-remote-s3bundle` to store the repository as a bundle file on Amazon S3, or `git-remote-codecommit` for AWS CodeCommit (if you don't want to or cannot use HTTPS authentication with static credentials). There is also `git-ssb` to encode repositories in a peer-to-peer log store via the Secure ScuttleButt protocol.

Credentials/password management

In most cases, with the exception of the local transport (where filesystem permissions control access), publishing changes to the remote repository requires authentication (the user identifies itself) and authorization (the given user has permission to perform the push operation). Sometimes, fetching the repository also requires authentication and authorization, like with private repositories.

Commonly used **credentials** for authentication are *username* and *password*. You can put the username in the HTTP and SSH repository URLs if you are not concerned about information leakage (in respect of leaking the information about valid usernames), or you can use the **credential helper** mechanism. You should *never* put passwords in URLs, even though it is technically possible for HTTP ones—the password can be visible to other people, for example when they are listing processes.

Besides the mechanism inherent in the underlying transport engine, be it SSH_ASKPASS for SSH or the ~/.netrc file for curl-based transport, Git provides its own integrated solutions.

Asking for passwords

Some of the Git commands that interactively ask for a password (and a username if it is not known)—such as `git svn`, the HTTP interface, or IMAP authentication—can be told to use an external program. The program is invoked with a suitable prompt (a so-called **authentication domain**, describing what the password is for), and Git reads the password from the standard output of this program.

Git will try the following places to ask the user for usernames and passwords; see the `gitcreden-tials(7)` manpage:

- The program specified by the `GIT_ASKPASS` environment variable, if set (Git-specific environment variables always have higher precedence than configuration variables)

- Otherwise, the `core.askpass` configuration variable is used, if set

- Otherwise, the `SSH_ASKPASS` environment variable is used, if set (it is not Git-specific, that is why it is consulted later in the sequence)

- Otherwise, the user is prompted on the terminal

This `askpass` external program is usually selected according to the desktop environment of the user (after installing it, if necessary):

- `(x11-)ssh-askpass` provides a plain X-window dialog asking for the username and password

- There is `ssh-askpass-gnome` for GNOME and `ksshaskpass` for KDE

- `mac-ssh-askpass` can be used for macOS

- `win-ssh-askpass` can be used for MS Windows

Git comes with a cross-platform password dialog in Tcl/Tk—`git-gui--askpass`—to accompany the `git gui` graphical interface and the `gitk` history viewer.

The Git configuration precedence (that we have seen an example of here) will be described in more detail in *Chapter 13, Customizing and Extending Git*.

Public key authentication for SSH

For the SSH transport protocol, there are additional authentication mechanisms besides passwords. One of them is **public key authentication**. This method is very useful to avoid being asked for a password over and over. Also, the repository hosting service may require using it when providing SSH access—possibly because identifying a user based on their public key doesn't require an individual account (that's what, for example, `gitolite` uses—`https://gitolite.com`).

The idea of public key authentication is that the user creates a **public/private key pair** by running, for example, `ssh-keygen`. The public key is then sent to the server, for example, by using `ssh-copy-id` (which also adds the public key, `*.pub`, at the end of the `~/.ssh/authorized_keys` file on the remote server), or by pasting it into a web form on a hosting service. You can then log in with your private key that is on your local machine, for example, as `~/.ssh/id_rsa`. You might need to configure SSH (in `~/.ssh/config` on Linux, and a similar configuration file on MS Windows) to use a specific identity file for a given connection (hostname) if it is not the default identity key.

Another convenient way to use public key authentication is with an authentication agent such as `ssh-agent` (or Pageant from PuTTY on MS Windows). Utilizing an agent also makes it more convenient to work with passphrase-protected private keys—you need to provide the password only once, to the agent, at the time of adding the key (which might require user action, for example, running `ssh-add` for `ssh-agent`).

Credential helpers

It can be cumbersome to input the same credentials over and over. For SSH, you can use public key authentication, but there is no true equivalent for other transports. Git credential configuration provides two methods to at least reduce the number of questions.

The first is the static configuration of default usernames (if one is not provided in the URL) for a given **authentication context**—for example, hostname:

```
[credential "https://git.example.com"]
    username = user
```

It helps if you don't have secure storage for credentials.

The second is to use external programs from which Git can request both usernames and passwords— **credential helpers**. These programs usually interface with secure storage (a keychain, keyring, wallet, credentials manager, and so on) provided by the desktop environment or the operating system.

Git, by default, includes at least the `cache` and `store` helpers. The `cache` helper (`git-credential-cache`) stores credentials in memory for a short period of time; by default, it caches usernames and passwords for 15 minutes. The `store` helper (`git-credential-store`) stores *unencrypted* credentials for an indefinitely long time on disk, in files readable only by the user (similar to `~/.netrc`); there is also a third-party `netrc` helper (`git-credential-netrc`) for GPG-encrypted `netrc/authinfo` files.

Selecting a credential helper to use (and its options) can be configured either globally or per authentication context, as in the previous example. Global credentials configuration looks like the following:

```
[credential]
    helper = cache --timeout=300
```

This will make Git use the `cache` credential helper, which will then cache credentials for 300 seconds (5 minutes). If the credential helper name is not an absolute path (for example, `/usr/local/bin/git-kde-credentials-helper`), Git will prepend the `git credential-` prefix to the helper's name. You can check what types of credential helpers are available with `git help -a | grep credential-`. Git for Windows also includes, optionally, `git credential-helper-selection`.

There exist credential helpers that use secure storage of the desktop environment. When you are using them, you need to provide the password only once, to unlock the storage (some helpers can be found in the `contrib/` area in Git sources). There is `git-credential-libsecret` for GNOME and KDE, `git-credential-osxkeychain` for the macOS Keychain, and `git-credential-manager` for Microsoft's cross-platform **Git Credential Manager (GCM)**.

You can also use `git-credential-oauth` to avoid having to set up personal access tokens or SSH keys. With this solution, the first time you authenticate, the helper opens a browser window to the host. Subsequent access uses cached credentials. Here, one can use the fact that Git supports multiple credential helpers. GitHub, GitLab, and Bitbucket are among the Git hosting services that support OAuth authentication.

Git will use credential configuration for the most specific authentication context, though if you want to distinguish the HTTP URL by pathname (for example, providing different usernames to different repositories on the same host), you need to set the `useHttpPath` configuration variable to `true`. If there are multiple helpers configured for context, each will be tried in turn, until Git acquires both a username and a password.

> **Historical note**
>
> Before the introduction of credential helpers, one could use *askpass* programs that interface with the desktop environment keychain—for example, `kwalletaskpass` (for KDE Wallet) or `git-password` (for the macOS Keychain).

Publishing your changes upstream

The *Collaborative workflows* section in *Chapter 6, Collaborative Development with Git* explained various repository setups. Here, we'll learn about a few common patterns for contributing to a project. We'll see what our main options for publishing changes are.

Before starting work on new changes, you should usually sync with the main development, incorporating the official version into your repository. This, and the work of the maintainer, is left to be described in *Chapter 9, Merging Changes Together*.

Pushing to a public repository

In a **centralized workflow**, publishing your changes consists simply of **pushing** them to the central server, as shown in *Figure 6.2*. Because you share this central repository with other developers, it can happen that somebody has already pushed to the branch you are trying to update (the non-fast-forward case). In this scenario, you need to pull (fetch and merge, or fetch and rebase) others' changes, before being able to push yours. This case is shown at the start of the *Updating your repository (with merge)* section in *Chapter 1, Git Basics in Practice*.

Another possible system with a similar workflow is when your team submits each set of changes to the code review system—for example, Gerrit. One available option is to push to a special ref, `refs/for/<branchname>` (which is named after a target branch), in a special repository. Then, the change review server makes each set of changes land automatically on a separate per-set ref (for example, `refs/changes/<change-id>` for commits belonging to a series with the given change ID).

> **Important note**
>
> In both peer-to-peer (see *Figure 6.3*) and maintainer workflows, or the hierarchical workflow variant (*Figure 6.4* and *Figure 6.5*, respectively), the first step in getting your changes included in the project is to perform the push operation, but pushing to *your own* "public" repository (visible to the appropriate group) of your fork of the project. Then, you need to ask your co-developers, or the project maintainer, to merge your changes. You can do this, for example, by generating a **pull request**, as described below.

Generating a pull request

In all workflows other than the centralized workflow, one needs to send a notification that changes are available in the public repository to co-developers, to the maintainer, or to integration managers. The `git request-pull` command can help with this step. Given the starting point (the bottom of the revision range of interest), the URL or the name of the remote public repository, and optionally, the commit to end at (if it is not the `HEAD`), this command will generate a summary of changes:

```
$ git request-pull origin/master publish
The following changes since
commit  ba5807e44d75285244e1d2eacb1c10cbc5cf3935:

  Merge: strtol() + checks (2014-05-31 20:43:42 +0200)

are available in the Git repository at:

  https://git.example.com/random master

for you to fetch changes up to
82006acd359717624fb33a7ae554cba6be717911:

  Merge branch 'master' of https://git.company.com/random (2021-05-30
00:58:23 +0200)
----------------------------------------------------------
Alice Developer (1):
      Support optional <count> parameter

  src/rand.c |   26 +++++++++++++++++++++-----
  1 files changed, 21 insertions(+), 5 deletions(-)
```

The pull request contains the SHA-1 of the base of the changes (which is the revision just before the first commit, in the series proposed for pull), the title of the base commit, the URL, the branch of the public repository (suitable as `git pull` parameters), the title of the final commit, the **shortlog** (the name of the `git shortlog` output), and **diffstat** (the name of the `git diff --stat` output) of the changes. This output can be sent to the maintainer—for example, by email.

Figure 7.1 – "New pull request" action shown in a list of branches on GitHub

A lot of Git hosting software and services include a built-in equivalent of `git request-pull` (for example, the **Create pull request** action in GitHub).

Exchanging patches

Many larger projects (and many open source projects) have established procedures for accepting changes in the form of **patches**, for example, to lower the barrier to entry for contributing. If you want to send a one-off code proposal to a project but do not plan to be a regular contributor, sending patches might be easier than a full collaboration setup (acquiring permission to commit in the centralized workflow, setting up a personal public repository for forking and similar workflows—on GitHub, that would consist of **forking** the project). Besides, one can generate patches with any compatible tool, and the project can accept patches no matter what is the version control setup.

> **Tip**
>
> Nowadays, with the proliferation of various free Git hosting services, it might be more difficult to set up an email client for sending properly formatted patch emails—though services such as *GitGitGadget* (for submitting patches to the Git project mailing list), or the older *submitGit* service, could help. Git itself also includes commands for sending mail, namely `git send-email` and `git imap-send`, both of which need configuration.

Additionally, patches, being a text representation of changes, can be easily understood by computers and humans alike. This makes them universally appealing, and very useful for *code review* purposes. Many open source projects historically used the public mailing list for that purpose: you can email a patch to this list and the public can review and comment on your changes (with services such as *public-inbox* and *lore+lei*, it is possible even without subscribing to the mailing list).

To generate email versions of each commit series, turning them into mbox-formatted patches, you can use the `git format-patch` command, as follows:

```
$ git format-patch -M -1
0001-Support-optional-count-parameter.patch
```

You can use any revision range specifier with this command. The most commonly used is limiting by the number of commits, as in the preceding example, or by using the double-dot revision range syntax—for example, `@{u}..` (see *Chapter 4*, *Exploring Project History*, the *Double-dot notation* section). When generating a larger number of patches, it is often useful to select a directory to save generated patches. This can be done with the `-o <directory>` option. The `-M` option for `git format-patch` (passed to `git diff`) turns on rename detection; this can make patches smaller and easier to review.

The patch files end up looking like this:

```
From db23d0eb16f553dd17ed476bec731d65cf37cbdc Mon Sep 17 00:00:00 2001
From: Alice Developer <alice@company.com>
Date: Sat, 31 May 2014 20:25:40 +0200
Subject: [PATCH] Initialize random number generator

Signed-off-by: Alice Developer
---
 random.c |    2 ++
 1 files changed, 2 insertions(+), 0 deletions(-)

diff --git a/random.c b/random.c
index cc09a47..5e095ce 100644
--- a/random.c
+++ b/random.c
@@ -1,5 +1,6 @@
 #include <stdio.h>
 #include <stdlib.h>
+#include <time.h>

 int random_int(int max)
@@ -15,6 +16,7 @@ int main(int argc, char *argv[])
 int max = atoi(argv[1]);


+    srand(time(NULL));

     int result = random_int(max);

     printf("%d\n", result);


--
2.42.0
```

It is actually a complete email in the mbox format. The subject (after stripping the [PATCH] prefix) and everything up to the three-dash line (- - -) forms the commit message—the description of the change. To email this to a mailing list or a developer, you can use either git send-email or git imap-send, or any email client capable of sending plain text email. The maintainer can then use git am to apply the patch series, creating commits automatically; there's more about this in *Chapter 9, Merging Changes Together*, in the *Applying a series of commits from patches* section.

> **Email subject convention for patches**
>
> The [**PATCH**] prefix is there to make it easier to distinguish patches from other emails. This prefix can—and often does—include additional information, such as the number in the series (set) of patches, the revision of the series, the information about it being a **work in progress (WIP)**, or the **Request For Comments (RFC)** status—for example, [RFC/PATCHv4 3/8].

You can also edit these patch files to add more information for prospective reviewers—for example, information about alternative approaches, the differences between previous revisions of the patch (previous attempts), or a summary and/or references to the discussion on implementing the patch (for example, on a mailing list). You add such text between the - - - line and the beginning of the patch, before the summary of changes (diffstat); it will be ignored by git am.

> **Tip – range diff**
>
> If the series of patches is undergoing revision and needs to be redone in a different way, it is recommended practice to provide in the cover letter the git range-diff output, showing the differences between one iteration of the series and the other.

Summary

In this chapter, we have learned how to choose a transport protocol (if the remote server offers such a choice), and a few tricks such as using foreign repositories as if they were native Git repositories and offline transport with bundles.

Contact with remote repositories can require providing credentials—usually, the username and password, to be able to, for example, push to the repository. This chapter described how Git can help make this part easier thanks to credential helpers.

Publishing your changes and sending them upstream may involve different mechanisms, depending on the workflow. This chapter described push, pull request, and patch-based techniques.

The two following chapters expand on the topic of collaboration: *Chapter 8, Advanced Branching Techniques*, explores relations between local branches and branches in a remote repository, and how to set up branches for collaboration, while *Chapter 9, Merging Changes Together*, talks about the opposite issue—how to join the results of parallel work.

Questions

Answer the following questions to test your knowledge of this chapter:

1. How can one clone a large repository when the connection to the host is quite unreliable, but you can log in to the host with the remote repository?

2. What do you need to get your changes into the canonical repository in the centralized workflow, and what do you need to do in the integration manager workflow?

3. How can you set up Git so you would need to provide a password only once, and not for each contact with the remote?

4. Can you use Git to interact with foreign version-control system repositories, to submit commits and download updates?

Answers

Here are the answers to the questions given above:

1. One possible solution is to use `git bundle` on the remote host, and send the generated file via resumable transport such as HTTPS, rsync, or BitTorrent, or simply transport it via removable media such as a USB stick.

2. In the centralized workflow, you need to push to said central canonical repository, which might require merging changes from others first; in the integration manager workflow, you need to either push to your public repository and send some kind of pull request (for example, with `git request-pull` and email) against the canonical repository, or send patches by email to the maintainer.

3. You can set up a credential helper appropriate for the operating system and desktop environment used; for SSH transport, you can also use `ssh-agent` or the equivalent.

4. With appropriate tools, you can either use Git to work as a client for a foreign version control system (for example, `git svn`) or use a remote transport helper to treat a foreign repository as a Git remote (for example, `git-cinnabar`).

Further reading

To learn more about the topics that were covered in this chapter, take a look at the following resources:

- Scott Chacon and Ben Straub: *Pro Git, 2nd Edition* (2014) `https://git-scm.com/book/en/v2`

  - *Chapter 7.12 Git Tools – Bundling*

  - *Chapter 7.14 Git Tools - Credential Storage*

- Bundle URI: `https://git-scm.com/docs/bundle-uri`

- Anthony Heddings: *Should You Use HTTPS or SSH For Git?* (2021) `https://www.howtogeek.com/devops/should-you-use-https-or-ssh-for-git/#why-use-https`

- *A visual guide to SSH tunnels* `https://robotmoon.com/ssh-tunnels/`

- Carl Tashian: *SSH Tips & Trick – Add a second factor to your SSH login* (2020) `https://smallstep.com/blog/ssh-tricks-and-tips/#add-a-second-factor-to-your-ssh`

- Greg Kroah-Hartman: *"Patches carved into stone tablets", or why the Linux kernel developers rely on plain text email*, a Kernel Recipes 2016 talk `https://kernel-recipes.org/en/2016/talks/patches-carved-into-stone-tablets/`

8

Advanced Branching Techniques

Chapter 6, Collaborative Development with Git, described how to arrange teamwork while focusing on repository-level interactions. In that chapter, you learned about various centralized and distributed workflows, as well as their advantages and disadvantages.

This chapter will dive deeper into the details of collaboration in distributed development. We'll explore the relationships between local branches and branches in remote repositories. Then, we'll introduce the concept of remote-tracking branches, branch tracking, and upstream. This chapter will also teach us how to specify the synchronization of branches between repositories by using *refspecs* and *push modes*.

You will also learn branching techniques: how branches can be used to create new features, prepare new releases, and fix bugs. You will learn about the advantages and disadvantages of different branching patterns. Among other things, this chapter will show you how to use branches so that it would be easy for you to select which features will go into the next version of the project.

In this chapter, we will cover the following topics:

- Different kinds of branches, both long-lived and short-lived, and their purpose
- Various branching patterns, and how they can be composed into workflows
- Release engineering for different branching models
- Using branches to fix a security issue in more than one released version
- Remote-tracking branches and refspecs
- Rules for fetching and pushing branches and tags
- Selecting a push mode to fit the chosen collaboration workflow

The purpose of branching

A **branch** in a version control system is an active parallel line of development (also called a **codeline**). They are used to isolate, separate, and gather different types of work. For example, branches can be used to prevent your current unfinished work on a feature in progress from interfering with the management of bug fixes (isolation), or to gather fixes for an older version of the developed software (gathering and integration).

A single Git repository can have an arbitrarily large number of branches. Moreover, with a distributed version control system, such as Git, there could be many repositories for a single project (known as **forks** or **clones**), some public and some private; each such repository will have their own local branches. This can be considered **source branching**. Each developer would have at least one private clone of the project's public repository to work in.

> **A bit of history – a note on the evolution of branch management**
>
> Early distributed version control systems used one branch per repository model. Both *Bazaar* (then Bazaar-NG) and *Mercurial documentation*, at the time when they began their existence, recommended cloning the repository to create a new branch.
>
> Git, on the other hand, had good support for multiple branches in a single repository almost from the start. However, in the beginning, it was assumed that there would be one central multibranch repository interacting with many single-branch repositories (see, for example, the legacy `.git/branches` directory, which was used to specify URLs and fetch branches, as described in the `gitrepository-layout(7)` man page), though with Git it was more about defaults than capabilities.
>
> Because branching is cheap in Git (and merging is easy), and collaboration is quite flexible, people started using branches more and more, even for solitary work. This led to the wide use of the extremely useful topic branch workflow (also known as feature branching).

Isolation versus integration

Version control systems such as Git allow different people to work on the same code base without interfering with each other. They also makes it easy to switch between different types of work. But this separated work would then need to be merged back together into some integration target to be useful and to be later included in a release.

We need *isolation*, but we also need to *integrate* changes, combining work into a coherent whole. To avoid conflicts as best as possible, our changes need to be visible to others, or even better be integrated. For example, if we change a calling convention of some API, but our work remains isolated, others cannot easily adjust to those changes. They would use the old version of the API in their work – leading to merge conflicts and a more difficult integration in the future. So, from this point of view, earlier and more *frequent integration* is something to be desired.

However, some features are more involved, and their development consists of many steps. The goal of frequent integration conflicts with the need to *isolate unfinished work*, and to prevent such work from being visible. If we want frequent integration, we need to be able to handle such issues.

The path to production release

The main goal of software development is to deploy code into production, to create a usable release of the project, and to have something to be used. A proper branching technique helps us achieve a stable base for creating such a release.

What branching pattern to use depends on the particularities of the project. For example, the team may need to isolate a work in progress from a stable base. There can also be more or less friction in the release process. Additionally, you might need to manage multiple versions of releases, or multiple versions of the project in production.

There are specific branching patterns to help you handle such issues.

Long-running and short-lived branches

Branches whose main purpose is to gather and integrate changes need to be long-lived, or even permanent. They are intended to last indefinitely, or at least for a very long time; they are rarely deleted.

From a collaboration point of view, a **long-lived branch** can be expected to be there when you are next updating data or publishing changes. This means that you can safely start work by forking off any of the long-lived branches in the remote repository, and be assured that there should be no problems with integrating that work. This means that at least one such branch must exist. Branches that people usually base their work on, that define the *current version* of the project, are sometimes called **mainlines**.

While long-lived branches stay forever, **short-lived** or **temporary branches** are created to deal with single issues, and are usually removed (deleted) after dealing with said issue (after the branch is merged or the feature is dropped). They are intended to last only as long as the issue is present. Their purpose is time-limited.

Having a separate branch for a separate issue helps us isolate and gather subsequent steps in the process of resolving a problem, whether it's adding a new feature or creating an urgent bug fix. Those branches are usually named after their topic.

Visibility of branches

What you can find in *public repositories* are usually only *long-lived branches*. In most cases, these branches should never rewind (the new version is always a descendant of the old versions). This makes it possible for you to safely build your work on top of the public branch.

There are some special cases here, though; there can be branches that are rebuilt after each new release (requiring forced fetch at that time), and there can be branches that do not fast forward. Each such case should be explicitly mentioned in the developer documentation to help you avoid unpleasant surprises.

Because of their provisional nature, *short-lived branches* are usually only present in the *local private repository* of a developer or integration manager (maintainer), and are not pushed to public distribution repositories. If they appear in public repositories, they are often only present in a public repository of an individual contributor as a target for a pull request (see the blessed repository workflow in *Chapter 6, Collaborative Development with Git*).

Alternatives to branching

With frequent integration, potential conflicts are discovered early. However, some features take longer to develop, and they are simply not ready when the time comes to push them to the mainline. But teams don't want to expose half-developed features. With a branching workflow geared toward integration rather than isolation, there is often a need for some kind of mechanism to hide unfinished work.

One technique is to build backend code first, and only create the user interface for it when it is ready, like a keystone. On the other hand, changing the existing code can be done by creating a temporary abstraction layer, which would then allow you to switch the underlying implementation to the new one when it is ready.

Another useful method is to hide different unfinished implementations behind **feature switches** or **feature toggles**. This technique is useful outside providing separation for integrated but unfinished features. For example, with runtime feature toggles, you can compare two different algorithms on live production data, or you can perform A/B tests.

Visibility without integration

An alternative to frequent integration into the mainline could be to use outside channels. It can be done by creating a `proposed-updates` type of branch, which would be used to merge all feature branches. This improves the visibility of changes and provides a place to test branch integrations.

Tools and services such as *GitLive* (available as a VS Code extension and as a JetBrains IDE plugin) exist that can show who is working on which branch, on which issue, and even show working copy local changes of teammates.

Branching patterns

In many cases, the choice of branching pattern (of the branching technique) depends on how stable the branch is, or in other words how healthy it is. A **stable branch**, or a **healthy branch** is a branch in such a state that the current tip commit on that branch always builds and deploys successfully, and the software runs with zero or at most a few bugs.

Ensuring that a branch is healthy pretty much requires doing daily builds and having a comprehensive suite of automated tests that are run frequently – if not at each commit, then at least at each integration (merge). However, explaining how to do this is outside the scope of this book.

Integration patterns

Deciding what branching strategy to use to integrate individual changes into a coherent and healthy mainline depends on various factors. Techniques that tend toward frequent integration, such as continuous integration, require the branch being merged into to be healthy. This requires a disciplined team, where each developer can make sure that each change is well-tested and does not break the developed application.

On the other hand, if it isn't certain that the feature being developed is of a good enough quality, and we want it assessed as a unit only after it is finished, then integrating less frequently might make more sense. Requiring pre-integration code reviews also drives you toward specific branching patterns.

Mainline integration

The simplest possible branching strategy is to work directly out of the mainline (the **trunk**) and merge your changes (commits) directly into it. In this workflow, the developer starts from the mainline and creates their work on top of it.

This strategy is called **mainline integration** or **trunk-based development** (the name depends on how the main branch is called).

After the developer reaches a point where they want to integrate, they start by fetching the current state of the mainline. If other developers published their changes when they were working on the project, they would need to combine those changes, using either a merge or a rebase operation – see *Chapter 9, Merging Changes Together* for more detail. Then, they verify that the code is healthy and push integrated changes into the mainline.

Topic branches-based development

In the **topic branching** pattern (also called **feature branching**), the idea is to make a new separate branch for each topic. This might be creating a new feature or a bug fix. This type of branch intends to gather subsequent development steps of a feature (where each step – a commit – should be a self-contained piece, easy to review), and to isolate the work on one feature from work on other topics. Using a feature branch allows topical changes to be kept together, and not mixed with other commits. It also makes it possible for a whole topic to be dropped (or reverted) as a unit, be reviewed as a unit, and be accepted (integrated) as a unit.

The end goal for the commits on a topic branch is to have them included in a released version of a product. This means that, ultimately, the short-lived topic branch is to be merged into the long-lived branch, which is gathering stable work, and must be deleted.

To make it easier to integrate topic branches, the recommended practice is to create such branches by forking off the oldest, most stable integration branch that you will eventually merge into. Usually, this means creating a branch, starting from the stable-work graduation branch. However, if a given feature does depend on a topic not yet in the stable line, you need to fork off the appropriate topic branch containing the dependency you need.

Note that if it turns out that you forked off the wrong branch, you can always fix it by rebasing (see *Chapter 9*, *Merging Changes Together*, and *Chapter 10*, *Keeping History Clean*), because topic branches are not public.

Continuous integration

When using the mainline integration pattern, integrations are as frequent as possible: after each commit. Feature branching implies a lower bound to the period of integrations – you integrate fully developed cohesive features.

With the **continuous integration pattern** (which is also called **scaled trunk-based development**), you try to integrate as frequently as possible – that is, whenever you have made a worthwhile number of changes and the branch is still healthy. The work is best done with short-lived feature branches, just integrated more frequently. The recommended practice is to integrate at least daily, with the feature branch living a few days at most.

When using this pattern, you need to be able to deal with a partially built feature. If mainline code runs in production (continuous delivery), you need to consider how to avoid exposing such unfinished features in the running code. This was described in the *Alternatives to branching* section.

Release engineering

If the mainline is kept healthy enough and is in an always-releasable state (following the tenets of continuous delivery), you can mark revision for release simply by creating a Git tag from the current tip. This simple branching pattern is called the **release-ready mainline**.

But if this is not the case, or if you need to manage more than one version of the product, more complex branching patterns are needed. In that case, specialized branches are required on the path from the integration branch to the production release.

Progressive-stability branches

One possible solution to the problem of ongoing development not being stable enough to be always ready for the release (temporarily including some unstable code), is to put less mature and more mature code in separate **maturity branches**. Thus, the latest possibly unstable version is kept isolated from the one that is always ready for release. The intent of each of these branches is to integrate the development work of the respective degree of stability, from maintenance work (which accepts only relevant bugfixes), through stable work (production-ready), to unstable or development work (planning for the future) – for example, from `maint`, to `master`, to `next`:

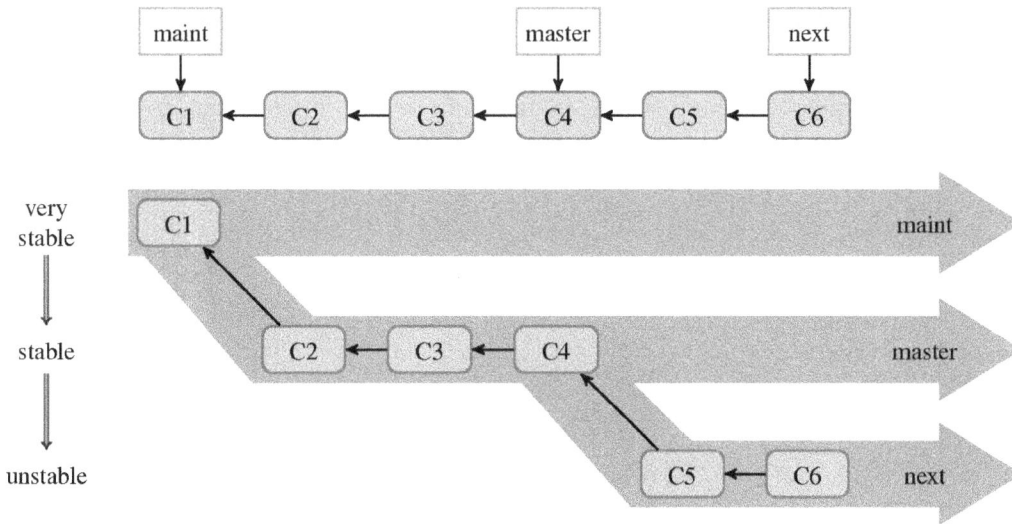

Figure 8.1 – A linear view and a "silo" view of the maturity branches
(also called progressive-stability branches)

These branches form a hierarchy with a decreasing level of **graduation** or **stability** of work, as shown in *Figure 8.1*. In the linear view (top of the figure), the stable revisions are further down the line in your commit history, and the cutting-edge unstable work is further up the history. Alternatively, we can think of branches as work silos (bottom of the figure), where work goes depending on the stability (graduation) of changes. Note that, in real development, progressive-stability branches would not keep being this simple. There would be new revisions on the branches after the forking points. Nevertheless, the overall shape will be the same, even in the presence of merging.

Here and in the following figures, the chosen commit names (C1, C2, C3, and so on) are only to distinguish commits, and in some cases also to make it easy to see which commit corresponds to another.

With maturation branches, the rule is to always merge more stable branches into less stable ones – that is, **merge upwards**. This would preserve the overall shape of branch silos (see *Figure 8.3* in the *Graduation or progressive-stability branches workflow* section of this chapter). This is because merging means including all the changes from the merged branch.

Therefore, merging a less stable branch into a more stable one would bring unstable work to the stable branch, violating the purpose and the contract of a stable branch.

Often, we see the graduation branches of the following levels of stability:

- The `maint`, `maintenance`, or `fixes` branch only contains bug fixes to the last major release; minor releases are done with the help of this branch.

- The `main`, `master`, `trunk`, or `stable` branch, with the development intended for the next major release; the tip of this branch should always be in the production-ready state.

- The next `devel`, `development`, `unstable` branch, where the new development goes to test whether it is ready for the next release; the tip can be used for nightly builds.

- The `pu` or `proposed` branch for the proposed updates. This is the integration testing branch and is meant for checking compatibility between different new features.

Having multiple long-running branches is not necessary, but it's often helpful, especially in very large or complex projects. Often, in operations, each level of stability corresponds to its own platform or deployment environment.

You don't need to – and probably shouldn't – use every type of branch listed here. Pick only what is needed for your project.

Per-release branches and per-release maintenance

Preparing for the new release of a project can be a lengthy and involved process. **Per-release branches** can help with this. The release branch is meant to separate the ongoing development from preparing for the new release. It allows other developers to continue working on writing new features and on integration testing, while the quality assurance team, with the help of the release manager, takes time to test and stabilize the release candidate.

After creating a new release, keeping such per-release branches allows us to support and maintain older released versions of the software. At these times, such branches work as a place to gather bug fixes (for their software versions) and create minor releases.

Not all projects need to utilize per-release branches. You can prepare a new release on the stable-work graduation branch, or use a separate repository in place of using a separate branch. Also, not all projects must provide support for more than the latest version.

This type of branch is often named after the release it is intended for – for example, `release-v1.4`. It is better not to give the branch the same name that the tag has for the release.

Release train with feature-freeze

If your project is doing releases on a regular cadence (such as every 2 weeks or every 6 months), and the release process is complex and involved (for example, there is external testing or a verification process), then it might be beneficial to use a release train branching pattern. It can be considered a variant of the per-release branch pattern. It is depicted in *Figure 8.2*:

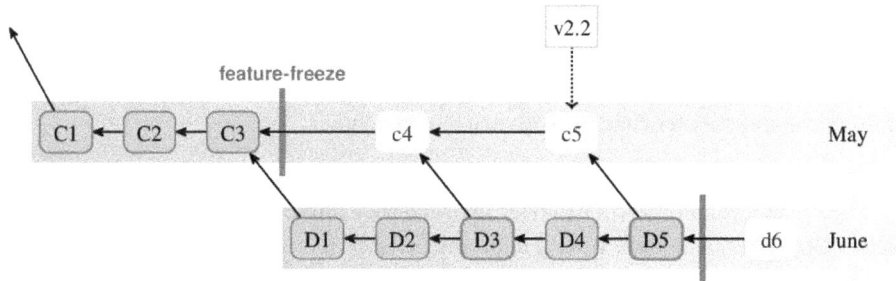

Figure 8.2 – Release train branching pattern for monthly releases, with the May "train" tagged and released into production and the June "train" in a state just after a feature freeze

In this approach, each per-release branch is coupled with a feature-freeze date (in advance of the planned release date). A new branch for the new release is created no later than the feature-freeze date for the previous release. After the feature freeze, an integration branch turns into a release branch, accepting only bug fixes and changes that prepare the project for release. This approach is often used with feature branching.

If there's more than one per-release branch active and accepting features, the developer can estimate how long would it take to finish the new feature and push it to the "train" corresponding to the later date (loading future trains). Earlier departing "trains" can be regularly merged into later departing ones.

This pattern can be transitioned into the continuous delivery pattern (production-ready mainline) by increasing the frequency of "trains" and reducing friction in the release process.

Hotfix branches for security fixes

Hotfix branches are like release branches but for unplanned releases. Their purpose is to act upon the undesired state of a live production or a widely deployed version, usually to resolve some critical bug in production (usually a severe security bug). This type of branch can be considered a longer-lived equivalent of the bugfix topic branches (see the *Bugfix branches* section of this chapter).

Other branching patterns involving long-lived branches

The main purpose of the different types branches is to isolate and/or integrate lines of development. However, branching patterns exist that do not fit around the themes of team integration or the path to production (to release).

Per-customer or per-deployment branches

Let's say that some of your project's customers require a few customization tweaks since they do things differently, or perhaps some deployment sites have special requirements. Suppose that these customizations cannot be done by simply changing the configuration. In this case, you would need to create separate lines of development for these customers or customizations.

But you don't want these lines of development to remain separate. You expect that there will be changes that apply to all of them. One solution is to use one branch for each customization set, per customer, or per deployment. Another would be to use separate repositories. Both solutions help maintain parallel lines of development and transfer changes from one line to another.

Such environment branches can be considered an *anti-pattern*. With this approach, it is very easy to introduce changes that lead to different behavior in production and on the developer's workstation, or end with having to maintain wildly divergent products for each customer.

Automation branches

Say that you are working on a web application, and you want to automate its deployment using a version control system. One solution would be to set up a daemon to watch a specific branch (for example, the one named `deploy`) for changes. Updating this branch would automatically update and reload the application.

This is, of course, not the only possible solution. Another possibility would be to use a separate `deploy` repository and set up hooks there, so pushing would cause the web application to refresh. Alternatively, you could configure a hook in a public repository so that pushing to a specific branch triggers redeployment (this mechanism will be described in *Chapter 14, Git Administration*).

These techniques can also be used for continuous integration; instead of deploying the application, pushing it into a specific branch would cause the test suite to run (the trigger could create a new commit on this branch or merge it).

Mob branches for anonymous push access

Having a branch in a remote repository (on a server) with special treatment on push is a technique that has many uses, including helping to collaborate. It can be used to enable *controlled anonymous push access* for a project.

Let's assume that you want to allow random contributors to push into the central repository. However, you would want to do this in a managed way: one solution is to create a special `mob` branch or a `mob/*` namespace (set of branches) with relaxed access control.

You'll learn how to set this up in *Chapter 14, Git Administration*.

The orphan branch trick

The different types of branches described up to this point differed in their purpose and management. However, from a technical point of view (that is, from the point of view of the graph of commits), they all look the same. This is not the case with the so-called orphan branches.

An **orphan branch** is a parallel disconnected (orphaned) line of development that shares no revisions with the main history of a project. It is a reference to a disjoint subgraph in the DAG of revisions, without any intersection with the main DAG graph. In most cases, their checkout is also composed of different files.

Such branches are sometimes used as a trick to store tangentially related contents in a single repository, instead of using separate repositories. (When using separate repositories to store related content, you might want to use some naming convention to denote this fact – for example, a common prefix.) They can be used to do the following:

- Store the project's web page files. For example, GitHub uses a branch named `gh-pages` for the project's pages.

- Store generated files when the process of creating them requires some nonstandard toolchain. For example, the project documentation can be stored in the `html`, `man`, and `pdf` orphan branches (the `html` branch can be also used to deploy the documentation). This way, the user can get specific format of the documentation without needing to install the toolchain required to generate it.

- Store the project TODO notes (for example, in the `todo` branch), perhaps together with storing there some specialized maintainer tools (scripts).

- Have deployment configuration for GitOps in the same repository as the source code, instead of having two separate repositories – one for code and one for deployment configuration.

You can create such a branch with `git checkout --orphan <new branch>` or by pushing into (or fetching into) a specific branch from a separate repository, as follows:

```
$ git fetch repo-htmldocs master:html
```

This command fetches the `master` branch from the unrelated `repo-htmldocs` repository into the unconnected `html` "orphan" branch.

Trivia

Creating an orphan branch with `git checkout --orphan` does not technically create a branch – that is, it does not make a new branch reference. What it does is point the symbolic HEAD reference to an unborn branch. The reference is created after the first commit on a new orphan branch. That's why there is no option to create an orphan branch with the `git branch` command.

Other types of short-lived branches

While long-lived branches stay forever, short-lived or temporary branches are created to deal with single issues, and are usually removed after dealing with said issue. They are intended to last only as long as the issue is present.

Because of their provisional nature, they are usually only present in the local private repository of a developer or integration manager (maintainer), and are not pushed to public distribution repositories. If they appear in public repositories, they are there only in a public repository of an individual contributor (see the blessed repository workflow in *Chapter 6, Collaborative Development with Git*), as a target for a pull request.

Bugfix branches

We can distinguish a special case of a topic branch whose purpose is fixing a bug. Such a branch should be created starting from the oldest integration branch it applies to (the most stable branch that contains the bug). This usually means forking off the maintenance branch or the divergence point of all the integration branches rather than the tip of the stable branch. A bugfix branch's goal is to be merged into relevant long-lived integration branches.

Bugfix branches can be thought of as a short-lived equivalent of a long-lived hotfix branch. Using them is a better alternative to simply committing fixes on the maintenance branch (or another appropriate integration branch).

Detached HEAD – the anonymous branch

You can think of the **detached HEAD** state (described in *Chapter 2, Developing with Git*) as the ultimate in temporary branches – so temporary that it even doesn't have a name. Git uses such anonymous branches automatically in a few situations, such as during bisection and rebasing.

Because, in Git, there is only one anonymous branch and it must always be the current branch, it is usually better to create a true temporary branch with a temporary name; you can always change the name of the branch later.

One possible use of the detached HEAD is for proof-of-concept work. However, you need to remember to set the name of the branch if the changes turn out to be worthwhile (or if you need to switch branches). It is easy to go from an anonymous branch to a named branch. You simply need to create a new branch from the current detached HEAD state.

Branching workflows and release engineering

Now that we know about the different branching patterns and their purposes, let's examine how they can be composed into different branching workflows. Different situations call for different uses of branches, as well as different policies. For example, smaller projects are better suited for simpler branching workflows, while larger projects might need more advanced ones.

In this section, we'll describe how to use a few common workflows. Each workflow is distinguished by the various types of branches it uses. In addition to getting to know what the ongoing development looks like for a given workflow, we'll also examine what it recommends doing at the time of the new release (major and minor, where relevant).

The release and trunk branches workflow

One of the simplest workflows is to use just a single integration branch. Such branches are sometimes called **trunk** branches; in Git, it would usually be the main or master branch (it is the default branch when creating a repository). In a pure version of this workflow, you would commit everything to the said branch, at least during the normal development stage. This way of working comes from the times of centralized version control, when branching and especially merging were more expensive, and people avoided branch-heavy workflows.

This workflow is well-suited for continuous integration. If you can maintain a healthy trunk, new releases can be cut directly from it via tagging.

On the other hand, if the release process is more involved, this workflow can be used together with per-release branches. In this case, when we decide to cut the new release, we create the new release branch out of the trunk. This is done to avoid the interference between stabilizing for release and the ongoing development work. The rule is that all the stabilization work goes on the release branch, while all the ongoing development goes to the trunk. **Release candidates** are cut (tagged) from the release branch, as is the final version of a **major release**. The release branch for a given version can be later used to gather bug fixes and to cut **minor releases** from it.

The disadvantage of such a simple workflow is that it requires maintaining a healthy branch. Otherwise, if we get in an unstable state during development, it can be hard to come up with a good starting point for a new release. An alternative solution is to create revert commits on the release branch, undoing the work that isn't ready. However, this can be a lot of work, and it would make the history of a project hard to follow.

Another difficulty with this workflow is that a feature that looks good at first glance might cause problems later. This is something this workflow has trouble dealing with. If it turns out during development that some feature created with multiple commits feature is not a good idea, reverting it can be difficult. This is true especially if its commits are spread across the timeline (across the history).

Despite these problems, this simple workflow can be a good fit for a small or well-disciplined team.

The graduation branches workflow

To be able to provide a stable line of the product, and to be able to test it in practice as a kind of floating beta version, you need to separate work that is stable from the work that is ongoing and might destabilize code. That's what **graduation branches** are for – to integrate revisions with different degrees of maturation and stability (this type of long-running branch is also called an **integration** branch or **progressive-stability** branch). See *Figure 8.1* in the *Maturity or progressive-stability branches* section, which shows a graph view and a silo view of a simple case with progressive-stability branches and linear history. Let's call the technique that utilizes mainly (or only) this type of branch the **graduation branches workflow**.

Besides keeping stable and unstable development separate, there is also a need for ongoing maintenance. If there is only one version of the product to support, and the process of creating a new release is simple enough, you can also use the graduation-type branch for this.

Here, simple enough means that you can just create the next major release out of the stable branch.

In such a situation, you would have at least three integration branches. There would be one branch for the ongoing maintenance work (containing only bug fixes to the last version), to create minor releases, another branch for stable work to create major releases (this branch can also be used for nightly stable builds), and another branch for ongoing development, possibly unstable:

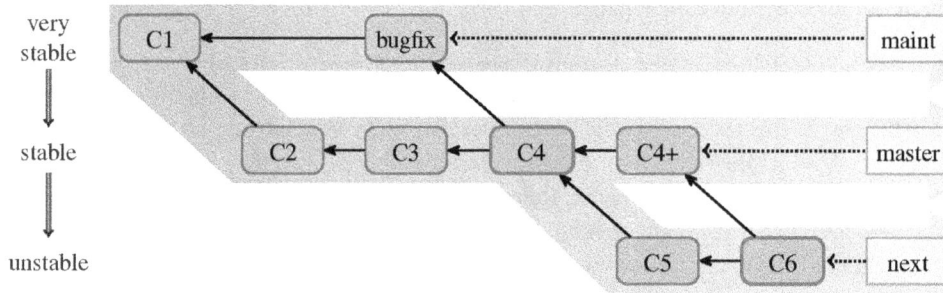

Figure 8.3 – The graduation or progressive-stability branches workflow. You should never merge a less stable branch into a more stable one as it would bring all the unstable history

You can use this workflow as-is, with only graduation branches, and no other types of branches:

- You commit bug fixes on the maintenance branch and merge it into the stable branch and development branch, if necessary

- You create revisions with the well-tested work on the stable branch, merging it into the development branch when needed (for example, if the new work depends on them)

- You put the work in progress, which might be unstable, on the development branch

During normal development, you should never merge less stable branches into more stable ones as this decreases their stability. This simple version of this workflow is shown in *Figure 8.3*.

This, of course, requires that you know upfront whether the feature that you are working on should be considered stable or unstable. There is also an underlying assumption that different features work well together from the start. In practice, however, you would expect that each piece of the development matures from the proof of concept, through being a work in progress during possibly several iterations, before it stabilizes. This problem can be solved by using topic branches, as described next.

In the pure graduation branches workflow, you would create minor releases (with bug fixes) out of the maintenance branch. Major releases (with new features) are created out of the stable-work branch. After a major release, the stable-work branch is merged into the maintenance branch to begin supporting the new release that was just created. At this point, an unstable (development) branch can be merged into a stable one. This is the only time when merging upstream – that is, merging fewer stable branches into more stable branches – should be done.

The topic branches workflow

The idea behind the topic branches workflow is to create a separate short-lived branch for each topic or feature so that all the commits belonging to a given topic (all the steps in its development) are kept together. The purpose of each **topic branch** is to develop a new feature or create a bug fix:

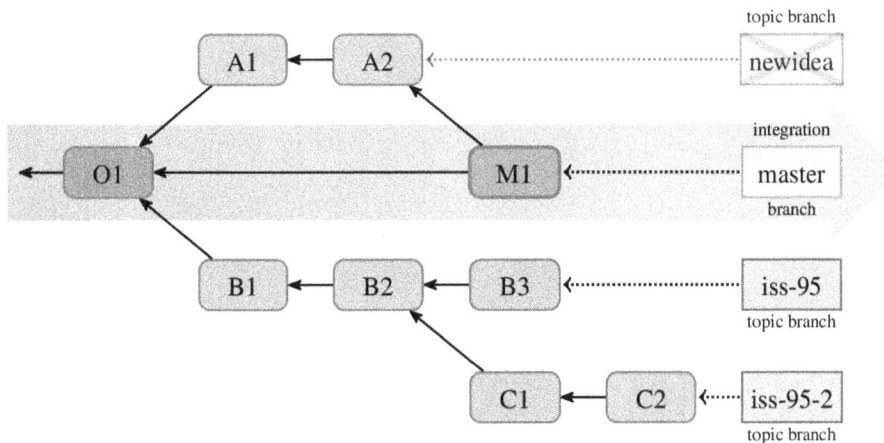

Figure 8.4 – The topic branches workflow with one integration branch (master) and three topic or feature branches. One of the topic branches is merged into the integration branch and deleted

In the **topic branches workflow** (also called the **feature branches workflow**), you have at least two different types of branches. First, there needs to be at least one permanent (or just long-lived) *integration branch*. This type of branch is used purely for merging. Integration branches are public.

Second, there are separate short-lived temporary *feature branches*, each intended for the development of a topic or the creation of a bug fix. They are used to carry all the steps, and only the steps required in the development of a feature or a fix – this is a unit of work for a developer. These branches can be deleted after the feature or the bug fix is merged. Topic branches are usually private and are often not present in public repositories.

When a feature is ready for review, its topic branch is often rebased to make integration easier, and optionally to make the history clearer. It is then sent for review as a whole. The topic branch can be used in a **pull request** or can be sent as a series of patches (for example, using git format-patch and git send-email). It is then often saved as a separate topic branch in a maintainer's working repository (for example, using git am --3way if it was sent as patches) to help in examining and managing it.

Then, the integration manager (the maintainer in the blessed repository workflow, or simply another developer in the central repository workflow) reviews each topic branch and decides whether it is ready for inclusion in the selected integration branch. If it is, then it will be merged (perhaps with the --no-ff option).

Graduation branches in a topic branch workflow

The simplest variant of the topic branches workflow uses only one integration branch. Usually, however, you would combine the graduation branches workflow with topic branches.

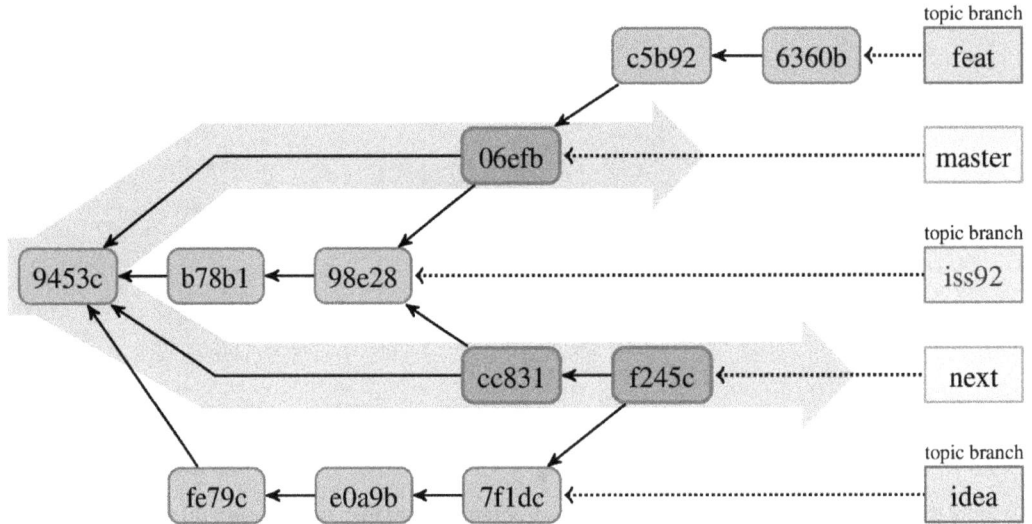

Figure 8.5 – The topic branches workflow with two graduation branches. Among topic branches, there is one that is stable enough to be merged into both graduation branches

In this often-used variant, the feature branch is started from the tip of a given stable branch (usually) or the last major release, unless the branch requires some other feature. In the latter case, the branch needs to be forked from (created from) the topic branch it depends on, such as the feat branch in *Figure 8.5*. Bugfix topic branches are created on top of the maintenance branch.

When the topic is considered done, it is merged into the development-work integration branch (for example, next) to be tested. For example, in *Figure 8.5*, topic branches idea and iss92 are both merged into next, while feat is not considered ready yet. Adventurous users can use builds from a given unstable branch to exercise the feature, though must take into account the possibility of crashes and data loss.

After this examination, when the feature is considered to be ready to be included in the next release, it is merged into the stable-work integration branch (for example, master). *Figure 8.5* includes one such branch: iss92. At this point, after merging it into the stable integration branch, the topic branch can be deleted.

Using a feature branch allows topical revision to be kept together and not mixed with other commits. The topic branch workflow allows you to easily undo topics as a whole and remove all bad commits together (removing a series of commits as a whole unit), instead of using a series of reverts.

If the feature turns out not to be ready, it is simply not merged into the stable branch, and it remains present only in the development-work branch. However, if we realize that it wasn't ready too late, after the topic was merged into the stable branch, we would need to revert the merge. This is a slightly more advanced operation than reverting a single commit, but it is less troublesome than reverting commits one by one, while ensuring that all the commits get correctly reverted. Problems with reverting merges will be covered in *Chapter 10, Keeping History Clean*.

The workflow for topic branches containing bug fixes is similar. The only difference is that you need to consider which of the integration branches the bugfix branch is to be merged into. This, of course, depends on the situation. Perhaps the bugfix applies only to the maintenance branch, because it was accidentally fixed by a new feature in the stable-work and development-work branches; then, it is merged only to this branch. Perhaps the bug only applies to the stable-work and development-work branches because it is about the feature that was not present in the previous version, thus the maintenance branch is excluded from being merged into.

Using a separate topic branch for bug fixing, instead of committing a bugfix directly, has an additional advantage: it allows us to easily correct the misstep if it turns out after the fact that the fix applies to more branches than we thought.

For example, if it turns out that the fix needs to also be applied to the maintained version and not only to the current work, with the topic branch you can simply merge the fix into additional branches. This is not the case if we were to commit the fix directly to the stable branch. In the latter situation, you cannot use merging as it would destabilize the maintenance branch. You would need to copy the revision with the fix by **cherry-picking** it from the branch it was committed to into the maintenance branch (see *Chapter 9, Merging Changes Together*, for a detailed description of this operation). But it means that duplicated commits are present in the history of the project, and cherry-picked commits can sometimes interact wrongly with the act of merging.

The topic branches workflow also allows us to check whether the features conflict with each other, and then fix them as necessary. You can simply create a throw-away integration branch and merge into it topic branches containing these features, to test the interaction between them. You can even publish such branches meant for integration testing (named `proposed-updates` or just `pu`, for example) to allow other developers to examine the works in progress. However, you should state explicitly in the developer documentation that said branch should not be used as a basis to work on, as it is recreated each time from scratch.

Branch management for a release in a topic branch workflow

Let's assume that we're using three graduation (integration) branches: `maint` for maintenance work on the last release, `master` for stable work, and `next` for development.

The first thing that the maintainer (the release manager) needs to do before creating a new release is to verify that `master` is a superset of `maint` – that is, all the bugs are also fixed in the version being considered for the next release. You can do this by checking whether the following command gives an empty output (see *Chapter 4, Exploring Project History*):

```
$ git log master..maint
```

If the preceding command shows some unmerged commits, the maintainer needs to decide what to do with them. If these bug fixes don't break anything, they can simply merge `maint` into `master` (as it's merging the more stable branch into the less stable one).

Now that the maintainer knows that `master` is a superset of `maint`, they can create the new release from the remote `master` branch by tagging it and push the just-created tag to the distribution point (to the public repository) with the following command:

```
$ git tag -s -m "Foo version 1.4" v1.4 master
$ git push origin v1.4 master
```

The preceding command assumed that the public repository of the Foo project is the one described by `origin` and that we use the double-digit version for major releases (following the semantic versioning specification).

> **Tip**
>
> If the maintainer wants to support more than one older version, they would need to copy an old maintenance branch, as the next step would be to prepare it for maintaining the just-released revision: `git branch maint-1.3.x maint`.

Then, the maintainer updates `maint` to the new release, advancing the branch (note that step one ensured that `maint` was a subset of `master`):

```
$ git checkout maint
$ git merge --ff-only master
```

If the second command fails, it means that some commits on the `maint` branch are not present in `master`, or to be more exact that `master` is not a strict descendant of `maint`.

Because we usually consider features for inclusion in `master` one by one, there might be some topic branches that are merged into `next`, but they were abandoned before they were merged into `master` (or they are not merged because they were not ready). This means that though the `next` branch contains a superset of topic branches that compose the `master` branch, `master` is not necessarily the ancestor of `next`.

That's why advancing the `next` branch after a release can be more complicated than advancing the `maint` branch. One solution is to rewind and rebuild the `next` branch:

```
$ git checkout next
$ git reset --hard master
$ git merge ai/topic_in_next_only_1...
```

You can find unmerged topics to be merged to rebuild `next` by running the following command:

```
$ git branch --no-merged next
```

After creating the release after rebuilding `next`, other developers would have to force fetch the `next` branch (see the next section) as it would not fast-forward if it were not already configured to force fetch:

```
$ git pull
From git://git.example.com/pub/scm/project
   62b553c..c2e8e4b  maint       -> origin/maint
   a9583af..c5b9256  master      -> origin/master
 + 990ffec...cc831f2 next        -> origin/next   (forced update)
```

Notice the forced update for the `next` branch here.

git-flow – a successful Git branching model

The more advanced version of the topic branching workflow builds on top of the graduation branch's one. In some cases, an even more involved branching model might be necessary, utilizing more types of branches: graduation branches, release branches, hotfix branches, and topic branches. One such model is `git-flow`.

This development model uses two main long-running **graduation branches** to separate the production-ready stable state from the work involved with integrating the latest delivered ongoing development. Let's call these branches `master` (stable work) and `develop` (gathers changes for the next release). The latter can be used for nightly builds. These two integration branches have an infinite lifetime.

These branches are accompanied by **supporting branches** – that is, *feature branches*, *release branches*, and *hotfix branches*.

Each new feature is developed on a **topic branch**. Such branches are forked off the tip of either the `devel` or `master` branch, depending on the requirements of the feature in question. When work on a feature is finished, its topic branch is merged with the `--no-ff` option (so that there is always a merge commit where a feature can be described) into `devel` for integration testing. When they are ready for the next release, they are merged into the `master` branch. A topic branch exists only as long as a feature is in development and is deleted when merged (or when abandoned).

The purpose of a **release branch** is twofold. When created, the goal is to prepare a new production release. This means doing last-minute cleanup, applying minor bug fixes, and preparing metadata for a release (for example, version numbers, release names, and so on). All but the last should be done using topic branches; metadata can be prepared directly on the release branch. This use of the release branch allows us to separate the quality assurance for the upcoming release from the work developing features for the next big release.

Such release branches are forked off when the stable state reflects, or is close to, the desired state planned for the new release. Each such branch is named after a release, usually something such as `release-1.4` or `release-v1.4.x`. You would usually create a few release candidates from this branch (tagging them `v1.4-rc1` and so on) before tagging the final state of the new release (for example, `v1.4`).

The release branch might exist only until the time the project release it was created for is rolled out, or it might be left to gather maintenance work: bug fixes for the given. In the latter situation, it replaces the `maint` branch of other workflows.

Hotfix branches are like release branches, but for an unplanned release usually connected with fixing serious security bugs. They are usually named `hotfix-1.4.1` or something similar. A hotfix branch is created out of an old release tag if the respective release (maintenance) branch does not exist. The purpose of this type of branch is to resolve critical bugs found in a production version. After putting a fix on such branches, the minor release is cut (for each such branch).

Ship/Show/Ask – a modern branching strategy

This approach tries to provide a balance between the advantages of doing pre-integration code reviews with pull requests and feature branches, and high-frequency integration and release-ready mainline that scaled trunk-based development provides.

In this workflow, you choose one of the three options – *Ship*, *Show*, or *Ask* – every time you make a change. With **Ship**, you add a change directly into the mainline (like in trunk-based development). This is useful if you want fast integration while being sure that the change is healthy – for example, if you add a feature using an established pattern, fix a simple bug, or update documentation.

With **Show**, you open a pull request but merge it straight away (if the automated checks pass). This allows for easy post-integration review while not making the feature wait.

Finally, with **Ask**, you follow the topic branch workflow and wait for the code review before integration.

Fixing a security issue

Let's examine another situation: how we can use branches to fix a bug, such as a security issue. This requires a slightly different technique than in ordinary development.

As explained in the *The topic branches workflow* section, while it is possible to create a bugfix commit directly on the most stable of the integration branches that is affected by the bug, it is usually better to create a separate bugfix branch.

You start by forking from the oldest (most stable) integration branch the fix needs to be applied to, perhaps even at the branching point of all the branches it would apply to. You put the fix (perhaps consisting of multiple commits) on the branch that you have just created. After testing it, you simply merge the bugfix branch into each of the integration branches that need the fix.

This model can be also used to resolve conflicts (dependencies) between branches at an early stage. Let's assume that you are working on some new feature (on a topic branch) that is not ready yet. While writing it, you noticed some bugs in the development version and you know how to fix them. You want to work on top of the fixed state, but you realize that other developers would also want the bugfix. Committing the fix on top of the feature branch takes the bugfix hostage. Fixing the bug directly on an integration branch has a risk of forgetting to merge the bugfix into the feature in progress.

The solution is to create a fix on a separate topic branch and merge it into both the topic branch for the feature being developed and into the test integration branch (and possibly the graduation branches).

> **Tip**
> You can use similar techniques to create and manage some features that are requested by a subset of customers. You need to simply create a separate topic branch for each such feature and merge it into the individual, per-customer branches.

The matter is a bit more complicated if there is security involved. In the case of a severe security bug, you would want to fix it not only in the current version but also in all the widely used versions.

To do this, you need to create a hotfix branch for various maintenance tracks (forking it from the specified version):

```
$ git checkout -b hotfix-1.9.x v1.9.4
```

Then, you need to merge the topic branch with the fix in question into the just created hotfix branch, to finally create the bugfix release:

```
$ git merge CVE-2014-1234
$ git tag -s -m "Project 1.9.5" v1.9.5
```

Interacting with branches in remote repositories

As we've seen, having many branches in a single repository is very useful. Easy branching and merging allow for powerful development models that utilize advanced branching techniques, such as topic branches. This means that remote repositories will also contain many branches. Therefore, we have to go beyond just the repository to the repository interaction, as described in *Chapter 6, Collaborative Development with Git*. We have to consider how to interact with multiple branches in the remote repositories.

We also need to think about how many local branches in our repository relate to the branches in the remote repositories (or, in general, other refs). The other important knowledge is how the tags in the local repository relate to the tags in other repositories.

Understanding the interaction between repositories, the branches in these repositories, and how to merge changes (as described in *Chapter 9, Merging Changes Together*) is required to truly master collaboration with Git.

Upstream and downstream

In software development, **upstream** refers to a direction toward the original authors or the maintainers of the project. We can say that the *repository* is upstream from us if it is closer (in the repository-to-repository steps) to the blessed repository – the canonical source of the software. If a change (a patch or a commit) is accepted upstream, it will be included either immediately or in a future release of an application, and all the people **downstream** will receive it.

Similarly, we can say that a given *branch* in a remote repository (the maintainer repository) is an **upstream branch** for a given local branch if changes in that local branch are to be ultimately merged and included in the remote branch.

Configuring what is considered upstream

A quick reminder: the upstream repository and the upstream branch in the said remote repository for a given branch are defined by the `branch.<branchname>.remote` and `branch.<branchname>.merge` configuration variables, respectively. The upstream branch can be referred to with the `@{upstream}` or `@{u}` shortcut.

The upstream branch is usually set while creating a branch out of the remote-tracking branch, and it can be modified using either `git branch --set-upstream-to` or `git push --set-upstream`.

The upstream branch does not need to be a branch in the remote repository. It can be a local branch, though we usually say that it is a **tracked branch** rather than saying that it is an upstream one. This feature can be useful when one local branch is based on another local branch, such as when a topic branch is forked from another topic branch (because it contains the feature that is a prerequisite for the latter work).

Remote-tracking branches and refspec

While collaborating on a project, you will be interacting with many repositories (see *Chapter 6, Collaborative Development with Git*). Each of these remote repositories you are interacting with will have a notion of the position of the branches. For example, the `master` branch in the remote repository, `origin`, doesn't need to be at the same place as your local `master` branch in your clone of the repository. In other words, they don't need to point to the same commit in the graph of revisions.

Remote-tracking branches

To be able to check the integration status to see what changes there are in the `origin` remote repository that are not yet in yours, or what changes you made in your working repository that you have not published yet, you need to know where the branches in the remote repositories are (well, where they were the last time you contacted these repositories). This is the task of **remote-tracking branches** – the references that track where the branch was in the remote repository:

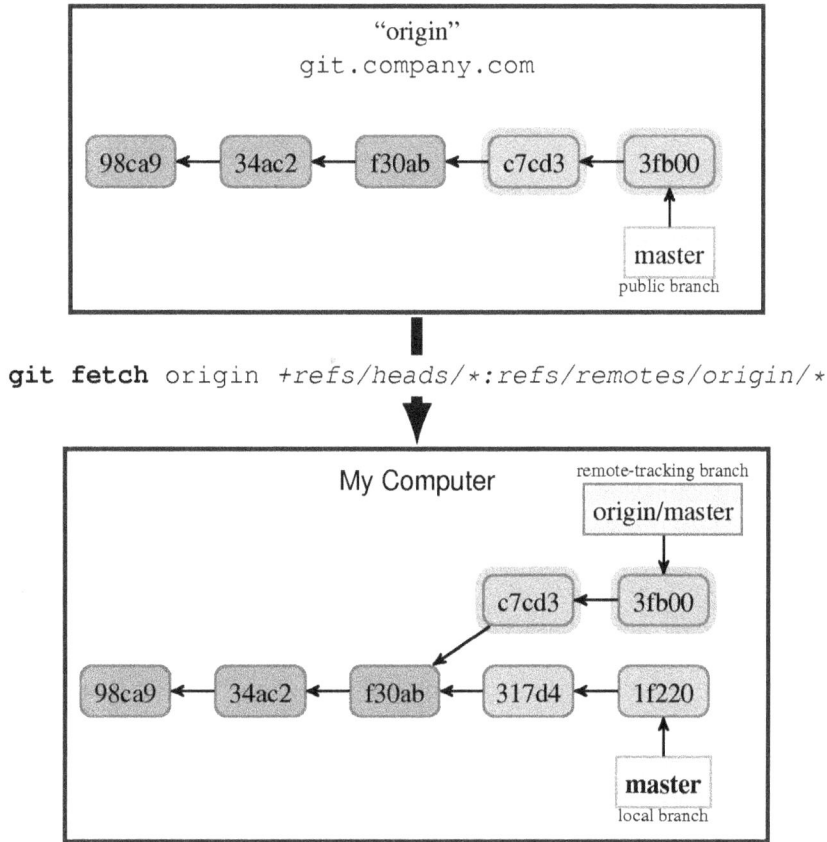

Figure 8.6 – Remote repository and local repository with local branches and remote-tracking branches. The grayed-out text in the fetch command denotes the default implicit parameters.

To track what happens in the remote repository, remote-tracking branches are updated automatically; this means that you cannot create new local commits on top of them (as you would lose these commits during updates). You need to create a local branch for it. This can be done, for example, by running `git checkout <branchname>`, assuming that the local branch with the given name does not already exist. This command creates a new local branch out of the remote branch's `<branchname>` and sets the upstream information for it.

Refspec – remote to local branch mapping specification

As described in *Chapter 4*, *Exploring Project History*, local branches are in the `refs/heads/` namespace, while remote-tracking branches for a given remote are in the `refs/remotes/<remote name>/` namespace. But that's just the default. The `fetch` (and `push`) lines in the `remote.<remote name>` configuration describe the mapping between branches (or refs in general) in the remote repository and the remote-tracking branches (or other refs) in the local repository.

This mapping is called **refspec**; it can be either explicit, mapping branches one by one, or globbing, describing a mapping pattern.

For example, the default mapping for the `origin` repository is as follows:

```
[remote "origin"]
    fetch = +refs/heads/*:refs/remotes/origin/*
```

This says that, for example, the content of the `master` branch (whose full name is `refs/heads/master`) in the remote repository, `origin`, is to be stored in the local clone of a repository in the remote-tracking branch, `origin/master` (whose full name is `refs/remotes/origin/master`). The plus (+) sign at the beginning of the pattern tells Git to accept the updates to the remote-tracking branch that are not fast-forwarded – that is, they are not descendants of the previous value.

The mapping can be given using the fetch lines in the configuration for the remote, as shown previously, or can be also passed as arguments to a command (it is often enough to specify just the short name of the reference instead of the full refspec). The configuration is only taken into account if there are no refspecs on the command line.

Fetching and pulling versus pushing

Sending changes (publishing) to the remote repository can be done with `git push`, while getting changes from it can be done with `git fetch`. These commands send changes in the opposite direction. However, note that your local repository has a very important difference – it has you sitting next to you're keyboard so that you're available to run other Git commands.

That's why there is no equivalent in the local-to-remote direction to `git pull`, which combines getting and integrating changes (see the next section). There is simply nobody there to resolve possible conflicts (problems occurring during doing automated integration).

In particular, there is a difference between how branches and tags are fetched and how they are pushed. This will be explained in detail later.

Pull – fetching and updating the current branch

Many times, you want to incorporate changes from a specific branch of a remote repository into the current branch. The **pull** command downloads changes (running `git fetch` with the parameters given); then, it automatically integrates the retrieved branch head into the current branch. Depending on the configuration, it either calls `git merge` or `git rebase` to do that. You can use the `--rebase=false` or `--rebase` option to override the default, something that can be configured globally with the `pull.rebase` configuration option or `branch.<branch name>.rebase` per-branch configuration option.

Note that if there is no configuration for the remote (you are doing the pull by URL), Git uses the `FETCH_HEAD` ref to store tips of the fetched branches.

There is also the `git request-pull` command to create information about published or pending changes for the pull-based workflows – for example, for a variant of the blessed repository workflow. It creates a plain text equivalent of the GitHub merge requests, one that is particularly suitable to send by email.

Pushing to the current branch in a non-bare remote repository

Usually, the repositories you push to are created for synchronization and are **bare** – that is, without a working area. A bare repository doesn't even have the concept of the current branch (`HEAD`) – there is no worktree, so there is no checked-out branch.

Sometimes, however, you might want to push to the non-bare repository. This may happen, for example, as a way of synchronizing two repositories, or as a mechanism for deployment (for example, of a web page or a web application). By default, Git on the server will deny the ref update to the currently checked-out branch. This is because it brings `HEAD` out of sync with the worktree and the staging area, which is very confusing if you don't expect it. You can, however, enable such a push by setting `receive.denyCurrentBranch` to `warn` or `ignore` (changing it from the default value of `refuse`). You can even make Git update the working directory (which must be clean – that is, without any uncommitted changes) by setting the said configuration variable to `updateInstead`.

An alternative and a more flexible solution to using `git push` for deployment is to configure appropriate hooks on the receiving side – see *Chapter 13, Customizing and Extending Git*, for information on hooks in general, and *Chapter 14, Git Administration*, for details on their use on the server.

The default fetch refspec and push modes

We usually fetch from public repositories with all the branches made public. Often, we want to get a full update of all the branches. That's why `git clone` sets up the default **fetch refspec** in the way shown earlier in this chapter. The common exception to the "fetch all" rule is following a pull request. But in this case, we have the repository and the branch (or the signed tag) stated explicitly in the request, and we will run the pull command with the provided parameters: `git pull <URL> <branch>`.

On the other side, in the private working repository, there are usually many branches that we don't want to publish or, at least, we don't want to publish them yet. In most cases, we would want to publish a single branch: the one we were working on and the one we know is ready. However, if you are the integration manager, you would want to publish a carefully selected subset of the branches instead of just one single branch.

This is yet another difference between fetching and pushing. That's why Git doesn't set up push refspec by default (you can configure it manually nonetheless), but instead relies on the so-called **push modes** to decide what should be pushed where. Note that the `push.default` configuration variable used to configure this applies only while running the `git push` command without branches to push stated explicitly.

> **Using git push to sync out of a host that one cannot pull from**
>
> When you work on two machines, machineA and machineB, each with its own worktree, a typical way to synchronize between them is to run git pull from each other. However, in certain situations, you may only be able to make the connection in one direction (for example, because of a firewall or intermittent connectivity). Let's assume that you can fetch and push from machineB, but you cannot fetch from machineA.
>
> You want to perform a push from machineB to machineA in such a way that the result of the operation is practically indistinguishable from doing a fetch while being on machineA. For this, you need to specify, via refspec, that you want to push the local branch into its remote-tracking branch:
>
> ```
> machineB$ git push machineA:repo.git \
>     refs/heads/master:refs/remotes/machineB/master
> ```
>
> The first parameter is the URL in the scp-like syntax, while the second parameter is refspec. You can set this in the config file, in case you need to do something like this more often.

Fetching and pushing branches and tags

The next section will describe which push modes are available, and when to use them (for which collaboration workflows). But first, we need to know how Git behaves concerning tags and branches while interacting with remote repositories.

Because pushing is not the exact opposite of fetching, and because branches and tags have different objectives (branches point to the lines of development, and tags point to name-specific revisions), their behavior is subtly different.

Fetching branches

Fetching branches is quite simple. With the default configuration, the git fetch command downloads changes and updates remote-tracking branches (if possible). The latter is done according to the fetch refspec for the remote.

There are, of course, exceptions to this rule. One such exception is **mirroring** the repository. In this case, all the refs from the remote repository are stored under the same name in the local repository. Here, git clone --mirror would generate the following configuration for origin:

```
[remote "origin"]
    url = https://git.example.com/project
    fetch = +refs/*:refs/*
    mirror = true
```

The names of refs that are fetched, together with the object names they point at, are written to the `.git/FETCH_HEAD` file. This information is used, for example, by `git pull`; this is necessary if we are fetching via URL and not via a remote name. This is done because, when we fetch by the URL, there are simply no remote-tracking branches to store the information about the fetched branch to be integrated.

You can delete remote-tracking branches on a case-by-case basis with `git branch -r -d`; you can also remove them on a case-by-case basis for which the corresponding branch in the remote repository no longer exists with `git remote prune` (or with `git fetch -- prune` in modern Git).

Fetching tags and automatic tag following

The situation with tags is a bit different. While we would want to make it possible for different developers to work independently on the same branch (for example, an integration branch such as `master`), in different repositories, we would need for all developers to have one specific tag to always refer to the same specific revision. That's why the position of branches in remote repositories is stored using a separate per-remote namespace, `refs/remotes/<remote name>/*`, in remote-tracking branches, but tags are mirrored – each tag is stored with the same name in the `refs/tags/*` namespace.

> **Tip**
> Note that the positions of tags in the remote repository can be configured with the appropriate fetch refspec; Git is that flexible. One example where it might be necessary is fetching a subproject, where we want to store its tags in a separate namespace (more information on this issue in *Chapter 11, Managing Subprojects*).

This is also why, by default, while downloading changes, Git will also fetch and store all the tags that point to the downloaded objects locally. You can disable this **automatic tag following** with the `--no-tags` option. This option can be set on the command line as a parameter, or it can be configured with the `remote.<remote name>.tagopt` setting.

You can also make Git download all the tags with the `--tags` option, or by adding the appropriate fetch refspec value for tags:

```
fetch = +refs/tags/*:refs/tags/*
```

Pushing branches and tags

Pushing is different. Pushing branches is (usually) governed by the selected push mode. You push a local branch (usually just a single current branch) to update a specific branch in the remote repository, from `refs/heads/` locally to `refs/heads/` in remote. It is usually a branch with the same name, but it might be a differently named branch configured as upstream – details will be provided later.

You don't need to specify the full refspec: using the ref name (for example, the name of a branch) means pushing to the ref with the same name in the remote repository, creating it if it doesn't exist. Pushing HEAD means pushing the current branch into the branch with the same name (not to HEAD in remote – it usually doesn't exist).

Usually, you push tags explicitly with `git push <remote repository> <tag>` (or `tag <tag>` if there is both a tag and branch with the same name – both mean the `+refs/tags/<tag>:refs/tags/<tag>` refspec). You can push all the tags with `--tags` (and with appropriate refspec) and turn on the automatic tag with `--follow-tags` (it is not turned on by default as it is for fetch).

As a special case of refspec, pushing an "empty" source into some ref in remote deletes it. The `--delete` option to `git push` is just a shortcut for using this type of refspec. For example, to delete a ref matching `experimental` in the remote repository, you can run the following command:

```
$ git push origin :experimental
```

Note that the remote server might forbid the deletion of refs with `receive.denyDeletes` configuration option or with hooks.

Push modes and their use

The behavior of `git push`, in the absence of the parameters specifying what to push, and in the absence of the configured push refspec, is specified by the **push mode**.

Different modes are available, each suitable for different collaborative workflows, which was shown in *Chapter 6, Collaborative Development with Git*.

The "simple" push mode – the default

The default push mode in Git 2.0 and later is the so-called `simple` mode. It was designed with the idea of *minimum surprise*: the idea that it is better to prevent publishing a branch than to make some private changes accidentally public.

With this mode, you always push the current local branch into the same named branch in the remote repository. If you push into the same repository you fetch from (the centralized workflow), it requires the upstream to be set for the current branch. The upstream is named the same as the branch.

This means that in the centralized workflow (push into the same repository you fetch from), it works like `upstream` with the additional safety that the upstream must have the same name as the current (pushed) branch. With a triangular workflow, while pushing to a remote that is different from the remote you normally pull from, it works like `current`.

This is the safest option; it is well-suited for beginners, which is why it is the default mode. You can turn it on explicitly with `git config push.default simple`.

The "matching" push mode for maintainers

Before version 2.0 of Git, the default push mode was `matching`. This mode is most useful for the maintainer (also known as the integration manager) in a blessed repository workflow. But most Git users are not maintainers; that's why the default push mode was changed to `simple`.

The maintainer would get contributions from other developers, be it via pull requests or patches sent in an email, and put them into topic branches. They could also create topic branches for their own contributions. Then, the topic branches considered to be suitable were merged into the appropriate integration branches (for example, `maint`, `master`, and `next`) – merging will be covered in *Chapter 9, Merging Changes Together*. All this is done in the maintainer's private repository.

The public blessed repository (one that everyone fetches from, as described in *Chapter 6, Collaborative Development with Git*) should only contain long-running branches (otherwise, other developers could start basing their work on a branch that suddenly vanishes). Git cannot know by itself which branches are long-lived and which are short-lived.

With the matching mode, Git will push all the local branches that have their equivalent with the same name in the remote repository. This means that only the branches that are already published will be pushed to the remote repository. To make a new branch public, you need to push it explicitly the first time, like so:

```
$ git push origin maint-1.4
```

> **Important note**
> Note that with this mode, unlike with other modes, using the `git push` command without providing a list of branches to push can publish multiple branches at once, and may not publish the current branch.

To turn on the matching mode globally, you can run the following command:

```
$ git config push.default matching
```

If you want to turn it on for a specific repository, you need to use a special refspec composed of a sole colon. Assuming that the said repository is named `origin` and that we want a not forced push, it can be done with the following command:

```
$ git config remote.origin.push :
```

You can, of course, push matching branches by using the following refspec on the command line:

```
$ git push origin :
```

The "upstream" push mode for the centralized workflow

In the centralized workflow, there is a single shared central repository that every developer with commit access pushes to. This shared repository will only have long-lived integration branches, usually only `maint` and `master`, and sometimes only `master`.

You should rather never work directly on `master` (perhaps except for simple single-commit topics), but rather fork a topic branch for each separate feature out of the remote-tracking branch:

```
$ git checkout -b feature-foo origin/master
```

In the centralized workflow, the integration is distributed: each developer is responsible for merging changes (in their topic branches), and publishing the result to the `master` branch in the central repository. You would need to update the local `master` branch, merge the topic branch to it, and push it:

```
$ git checkout master
$ git pull
$ git merge feature-foo
$ git push origin master
```

An alternate solution is to rebase the topic branch on the top of the remote-tracking branch rather than merging it. After rebasing, the topic branch should be an ancestor of `master` in the remote repository, so we can simply push it into `master`:

```
$ git checkout feature-foo
$ git pull --rebase
$ git push origin feature-foo:master
```

In both cases, you are pushing the local branch (`master` in the merge-based workflow and the feature branch in the rebase-based workflow) into the branch it tracks in the remote repository – in this case, origin's `master`.

That is what the `upstream` push mode was created for:

```
$ git config push.default upstream
```

This mode makes Git push the current branch to the specific branch in the remote repository – the branch whose changes are usually integrated into the current branch. This branch in the remote repository is the upstream branch (and can be referenced as `@{upstream}`). Turning this mode on makes it possible to simplify the last command in both examples to the following:

```
$ git push
```

The information about the upstream is created either automatically (while forking off the remote-tracking branch), or explicitly with the `--track` option. It is stored in the configuration file and it can be edited with ordinary configuration tools.

Alternatively, it can be changed later with the following:

```
$ git branch --set-upstream-to=<branchname>
```

The "current" push mode for the blessed repository workflow

In the blessed repository workflow, each developer has a private and public repository. In this model, you fetch from the blessed repository and push it to your public repository.

In this workflow, you start working on a feature by creating a new topic branch for it:

```
$ git checkout -b fix-tty-bug origin/master
```

When the features are ready, you push it into your public repository, perhaps rebasing it first to make it easier for the maintainer to merge it:

```
$ git push origin fix-tty-bug
```

Here, it is assumed that you used `pushurl` to configure the triangular workflow, and the push remote is `origin`. You would need to replace `origin` here with the appropriate name of the publishing remote if you are using a separate remote for your public repository (using a separate repository makes it possible to use it not only for publishing but also for synchronization between different machines).

You can configure Git in such a way that when you're on the `fix-tty-bug` branch, it is enough to just run `git push`. To do this, you need to set up Git to use the `current` push mode, like so:

```
$ git config push.default current
```

This mode will push the current branch to the branch with the same name at the receiving end.

Note that if you're using a separate remote for the publishing repository, you would need to set up the `remote.pushDefault` configuration option to be able to use just `git push` for publishing.

Summary

This chapter has shown how to effectively use branches for development and collaboration. You also got to know a few useful tricks.

First, we learned about the various uses of branches, from integration, through release management and the parallel development of features, to fixing bugs. You learned about different branching patterns and branching workflows. This knowledge should help you branch and customize workflows so that they fit the needs of the project and your team's preferences.

You also learned how to deal with multiple branches per repository while downloading or publishing changes. Git provides flexibility in how the information on branches and other refs in the remote repository is managed using the so-called refspecs to define a mapping to local refs: remote-tracking branches, local branches, and tags. Usually, fetching is governed by fetch refspec, but pushing is managed by the configured push mode. Various collaborative workflows require branch publishing to be handled differently; this chapter described which push mode to use with which workflow and explains why.

The next chapter, *Chapter 9*, *Merging Changes Together*, will explain how to integrate changes from other branches and other developers. You will learn about merging and rebasing, and how to deal with situations where Git can't do this automatically (how to handle various types of merge conflicts). You will also learn about cherry-picking and reverting commits.

Questions

Answer the following questions to test your knowledge of this chapter:

1. What are the advantages of frequent integration?

2. What are the advantages of topic branches?

3. How can you store a project web page or its GitOps configuration in the same repository as the code, while keeping their histories and files separate?

4. How can you synchronize the working directory of the Git repository hosted on other computers?

Answers

Here are the answers to this chapter's questions:

1. More frequent integration leads to easier integration because with smaller differences, there is less chance of conflict, and because conflicts are discovered earlier. It also makes it easier to maintain a production-ready mainline, decreasing the time it takes to put the feature into the production environment.

2. Using topic branches makes it easier to review and examine the steps it took to create a feature and remove it if needed. The use of topic branches also plays nicely with the requirement of pre-integration code review.

3. You can use the "orphan" branch trick – for example, with `git checkout -- orphan` – to have two or more unrelated histories in a single repository.

4. Log in to the other computer and use `git pull`; if this is not possible, you can `git push` into a non-bare repository (configuring what should happen to checked-out branches).

Further reading

To learn more about the topics that were covered in this chapter, take a look at the following resources:

- Martin Fowler, *Patterns for Managing Source Code Branches* (2020): `https://martinfowler.com/articles/branching-patterns.html`

- Rouan Wilsenach, *Ship / Show / Ask: A modern branching strategy* (2021): `https://martinfowler.com/articles/ship-show-ask.html`

- Vincent Driessen, *git-flow - A successful Git branching model* (2010): `https://nvie.com/posts/a-successful-git-branching-model/`

- *gitworkflows - An overview of recommended workflows with Git*: `https://git-scm.com/docs/gitworkflows`

- Paul Hammant and others, *Trunk Based Development*: `https://trunkbaseddevelopment.com/`

- Junio C Hamano: *Resolving conflicts/dependencies between topic branches early* (2009): `https://gitster.livejournal.com/27297.html`

- Junio C Hamano, *Fun with various workflows 1* and *2* (2013): `https://git-blame.blogspot.com/2013/06/fun-with-various-workflows-1.html` and `https://git-blame.blogspot.com/2013/06/fun-with-various-workflows-2.html`

Merging Changes Together

The previous chapter, *Advanced Branching Techniques*, described how to use branches effectively for collaboration and development.

This chapter will teach you how to integrate changes from different parallel lines of development together (that is, branches) by creating a merge commit, or by reapplying changes with the rebase operation. Here, the concepts of merge and rebase are explained, including the differences between them and how they can be used. This chapter will also explain the different types of merge conflicts and teach you how to avoid them, examine them, and resolve them.

In this chapter, we will cover the following topics:

- Merging, merge strategies, and merge drivers
- Cherry-picking and reverting a commit
- Applying a patch and a patch series
- Rebasing a branch and replaying its commits
- A merge algorithm at file and contents level
- Three stages in the index
- Merge conflicts – how to examine and resolve them
- Reusing recorded [conflict] resolutions with `git rerere`
- An external tool – `git-imerge`

Methods of combining changes

Now that you have changes from other people in the remote-tracking branches (or in the series of emails), you need to combine them, perhaps also with your changes.

Alternatively, your work on a new feature, created and performed on a separate topic branch, is now ready to be included in the long-lived development branch and made available to other people. Maybe you have created a bug fix and want to include it in all the long-lived graduation branches. In short, you want to join two divergent lines of development by integrating their changes.

Git provides a few different methods to combine changes and variations of these methods. One of these methods is a *merge* operation, joining two lines of development with a two-parent commit. Another way to copy introduced work from one branch to another is via cherry-picking, which is creating a new commit with the same changeset on another line of development (this is sometimes necessary to use). Alternatively, you can reapply changes, transplanting one branch on top of another with *rebase*. We will now examine all these methods and their variants, see how they work, and when they can be used.

In many cases, Git will be able to combine changes automatically; the next section will talk about what you can do if it fails and if there are merge conflicts.

Merging branches

The **merge operation** joins two (or more) separate branches together, including *all* the changes since the point of divergence into the current branch. You do this with the `git merge` command:

```
$ git switch master
$ git merge bugfix123
```

Here, we first switched to a branch we want to merge into (in this example, `master`) and then provided the branch to be merged (here, `bugfix123`).

No divergence – fast-forward and up-to-date cases

Let's say that you need to create a fix for a bug somebody found. Let's assume that you have followed the recommendations of the topic branch workflow from *Chapter 8, Advanced Branching Techniques*, and created a separate bugfix branch, named `i18n`.

In such cases, there is often no real divergence, which means that there were no commits on the maintenance branch (the branch we are merging into), since a bugfix branch was created. Because of this, Git would, by default, simply move the branch pointer of the current branch forward:

```
$ git switch maint
Switched to branch 'maint'
$ git merge i18n
Updating f41c546..3a0b90c
Fast-forward
  src/random.c | 2 ++
  1 file changed, 2 insertions(+)
```

You have probably seen this **Fast-forward** phrase among output messages during `git pull`, when there are no changes on the branch you pull into. The fast-forward merge situation is shown in *Figure 9.1*.

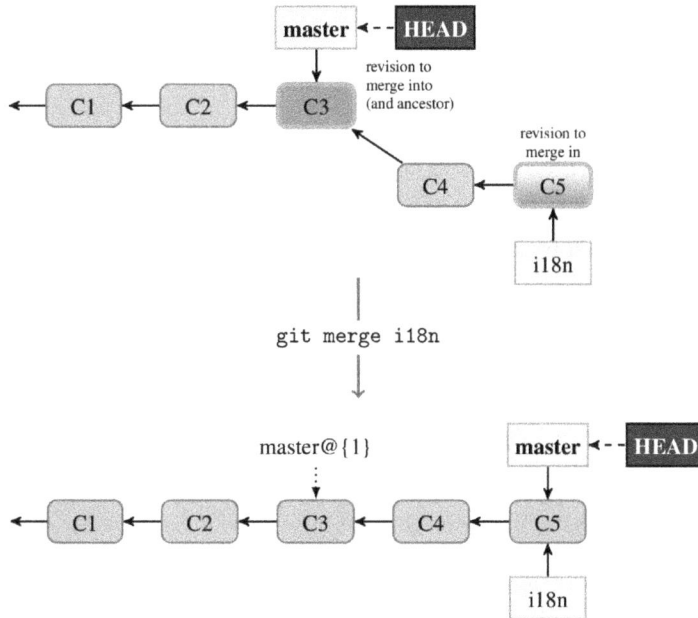

Figure 9.1 – The master branch is fast-forwarded to i18n during the merge

This case is important for the centralized and the peer-to-peer workflows (described in *Chapter 6, Collaborative Development with Git*), as it is the fast-forward merge that allows you to ultimately push your changes forward.

In some cases, that is not what you want. For example, note that after the fast-forward merge in *Figure 9.1*, we have lost the information that the **C4** and **C5** commits were done on the i18n topic branch. We can force the creation of a merge commit (which is described in the next section), even in a case where there are no changes to the current branch, using the `git merge --no-ff` command. The default is `--ff`; to fail instead of creating a merge commit, you can use `--ff-only` (ensuring fast-forward only).

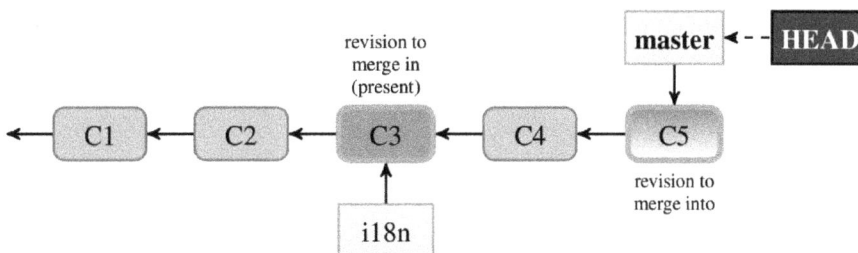

Figure 9.2 – The master branch is up to date with respect to the i18n branch (i.e., it includes it)

There is another situation where the head (tip) of one branch is the ancestor of the other – namely, the up-to-date scenario where the branch we try to merge is already included (merged) in the current branch (*Figure 9.2*). Git doesn't need to do anything in this case; it just informs the user about it:

```
$ git merge i18n
Already up to date.
```

Creating a merge commit

When you are merging fully fledged feature branches, rather than merging bugfix branches as in the previous section, the situation is usually different from the previously described fast-forward case. In the case of feature branch workflow, the development of the feature branch and integration branch would usually have diverged.

Suppose that you have decided that your work on a feature (for example, work on adding support for internationalization on the i18n topic branch) is complete and ready to be included in the master stable branch. In order to do so with a merge operation, you need to first check out the branch you want to merge into and then run the git merge command, with the branch being merged as a parameter:

```
$ git checkout master
Switched to branch 'master'
$ git merge i18n
Merge made by the 'ort' strategy.
 src/random.c |
 2 ++
 1 file changed, 2 insertions(+)
```

Because the top commit on the branch you are on (and merging into) is not a direct ancestor or a direct descendant of the branch you merge in, Git has to do more work than just moving the branch pointer. In this case, Git does a merge of changes since the divergence and stores it as a **merge commit** on the current branch. This commit has two parents, denoting that it was created based on more than one commit (more than one branch); the first parent is the previous tip of the current branch, and the second parent is the tip of the branch you merge in.

Note that Git does start committing the result of merge if it can be done automatically, and if there are no conflicts. However, the fact that the merge succeeded at the text level doesn't necessarily mean that the merge result is correct. You can either ask Git to not automatically commit a merge with git merge --no-commit to examine it first, or you can examine the merge commit and then use the git commit --amend command if it is incorrect (see *Figure 2.4*).

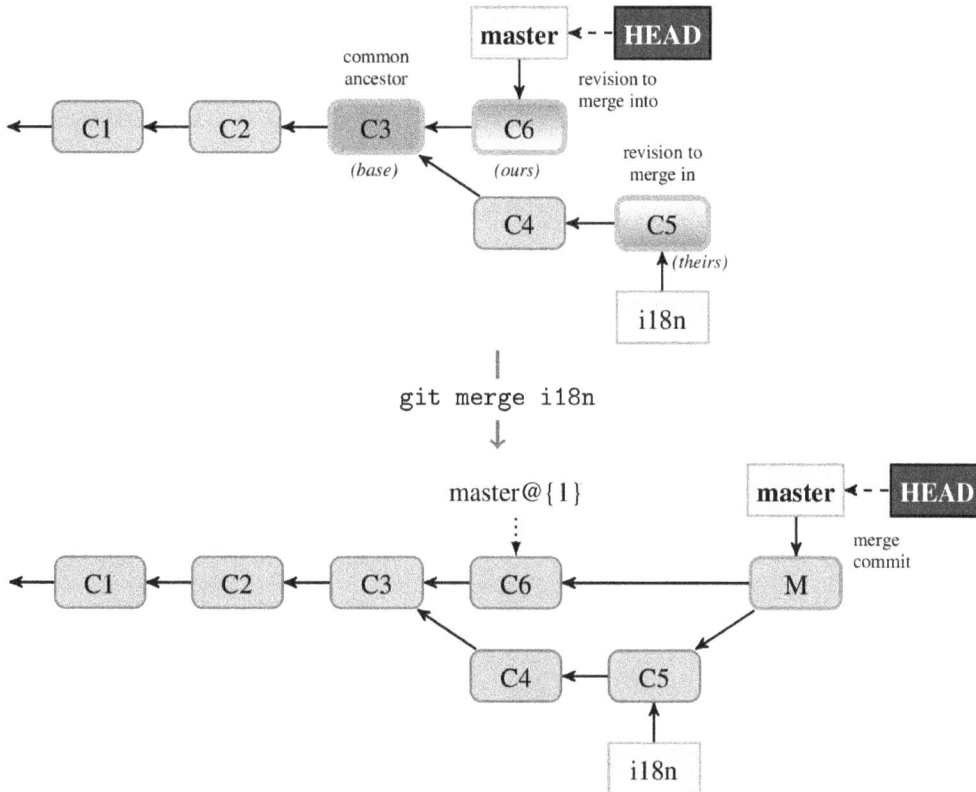

Figure 9.3 – Three revisions used in a typical merge and the resulting merge commit

Git creates contents of a merge commit (**M** in *Figure 9.3*), using by default (and in most cases) the three-way merge, which in turn uses the snapshots pointed to the tips of the branches being merged (master, **C6**, and i18n, **C5**) and the common ancestor of the two (**C3** here, which you can find with the git merge-base command).

A very important issue is that Git creates the merge commit contents based usually only on the three revisions – merged into (*ours*), merged in (*theirs*), and the common ancestor (*merge base*). It does not examine what happened on the divergent parts of the branches; this is what makes merging fast. However, because of this, Git also does not know about the cherry-picked or reverted changes on the branches being merged, which might lead to surprising results (see, for example, the section about reverting merges in *Chapter 10, Keeping History Clean*).

Merge strategies and their options

In the merge message, we have seen that it was made by the 'ort' strategy (known as **recursive** in older Git). The **merge strategy** is an algorithm that Git uses to compose the result of joining two or more lines of development.

There are a few merge strategies that you can select to use with the `--strategy`/`-s` option to the `git merge` command. By default, Git uses the *ort* merge strategy when joining two branches and a very simple *octopus* merge strategy when joining more than two branches. You can also choose the *resolve* merge strategy if the default one fails; it is fast and safe, although less capable of merging.

The two remaining merge strategies are special-purpose algorithms. The *ours* merge strategy can be used when we want to abandon changes in the merged-in branch but keep them in the history of the merged-into branch – for example, for documentation purposes. This strategy simply repeats the current snapshot (the *ours* version) as a merge commit. Note that this merge strategy, invoked with `--strategy=ours` or `-s ours`, should be not confused with the *ours* option to the default *ort* merge strategy, `--strategy=ort --strategy-option=ours`, or just *-Xours*, which means something different.

The *subtree* merge strategy can be used for subsequent merges from an independent project into a subdirectory (subtree) in a main project. It automatically figures out where the subproject was put. This topic, and the concept of subtrees, will be described in more detail in *Chapter 11, Managing Subprojects*.

The default **ort** (**Ostensibly Recursive's Twin**) merge strategy, and its predecessor, the *recursive* merge strategy, are named after how such a strategy deals with multiple merge bases and criss-cross merges. In the case of more than one merge base (which means that there is more than one common ancestor that can be used for a three-way merge), such a strategy creates a merge tree (conflicts and all) from the ancestors as a merge base – that is, it merges recursively. Of course, again, these common ancestors being merged can have more than one merge base.

Some strategies are customizable and have their own options. You can pass an option to a merge algorithm with `-X<option>` (or `--strategy-option=<option>`) on the command line, or set it with the appropriate configuration variables. You will discover more about merge options in the section *Resolving merge conflicts*, when we will discuss solving merge conflicts.

A reminder – merge drivers

Chapter 3, Managing Your Worktrees, introduced git attributes – among others, **merge drivers**. These *drivers* are user-defined and deal with merging file contents if there is a conflict, replacing the default three-way file-level merge. Merge *strategies*, in contrast, deal with DAG-level merging (and tree-level – that is, merging directories), and you can only choose from the built-in options.

A reminder – signing merges and merging tags

In *Chapter 6, Collaborative Development with Git*, you learned about signing your work. While using merge to join two lines of development, you can either merge a signed tag, sign a merge commit, or both. Signing a merge commit is done with the `-S` / `--gpg-sign` option to use the `git merge` or `git commit` command; the latter is used if there are conflicts, or if the `--no-commit` option was used while merging.

Copying and applying a changeset

The merging operation is about joining two lines of development (two branches), including all the changes since their divergence. This means, as described in *Chapter 8, Advanced Branching Techniques*, that if there is one commit on the less stable branch (for example, `master`) that you want to have also in a more stable branch (for example, `maint`), you cannot use the merge operation. You need to create a copy of such a commit. A situation such as this should be avoided (using topic branches), but it can happen, and handling it is sometimes necessary.

Sometimes, the changes that need to be applied come not from the repository (as a revision in the DAG to be copied) but in the form of a patch – that is, a unified diff or an email generated with `git format-patch` (with a patch, plus a commit message).

Git includes the `git am` tool to handle the mass application of commit-containing patches.

Both of these commands are useful on their own, but understanding these methods of getting changes is also useful to understand how cherry-picking and rebasing work.

Cherry-pick – creating a copy of a changeset

You can create a copy of a commit (or a series of commits) with the `cherry-pick` command. Given a series of commits (usually just a single commit), it applies the changes each one introduces, recording a new commit for each change.

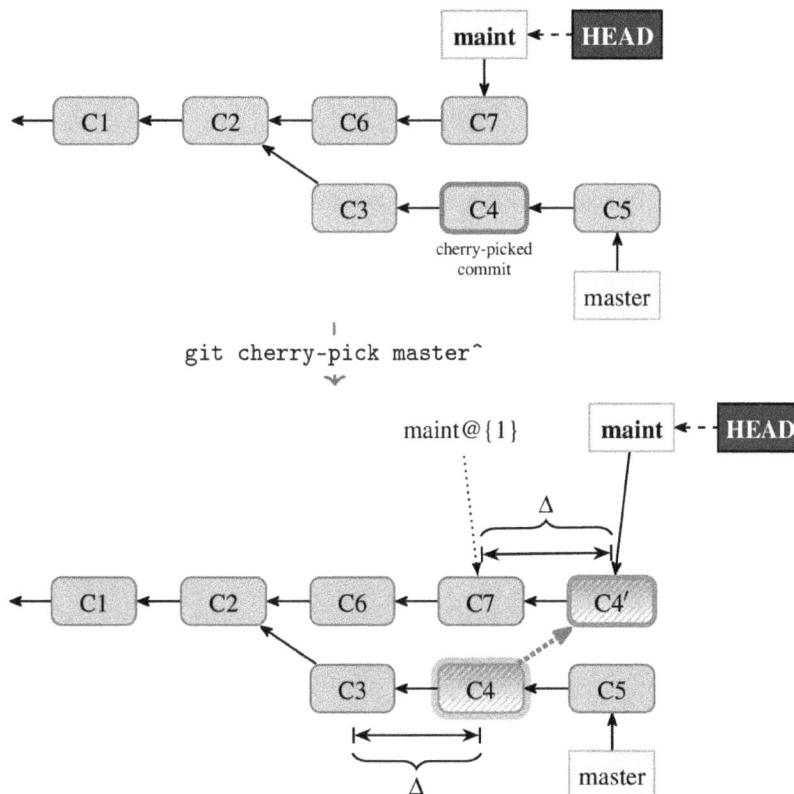

Figure 9.4 – Cherry-picking the C4 commit from master to maint

An example of a cherry-pick operation is shown in *Figure 9.4*. (Note that here the thick dotted arrow from **C4** to **C4'** denotes a copy; it is not a reference.)

The copying of changes does not mean that the snapshot (that is, the state of a project) is the same in the original (**C4** in *Figure 9.4*) and in the copy (**C4'** in *Figure 9.4*); the latter will include other changes while missing others. Also, while the changes will usually be the same (as they are in *Figure 9.4*, where the difference between **C3** and **C4** and the diff between **C7** and **C4'** is the same), they can also be different – for example, if part of the changes was already present in the earlier commits.

Note that, by default, Git does not save information about where the cherry-picked commit came from. You can append this information to an original commit message, as a **(cherry-picked from the commit <sha-1>)** line with `git cherry-pick -x <commit>`. This is only done for cherry-picks without conflicts. Remember that this information is only useful if you have access to the copied commit. Do not use it if you are copying commits from the private branch, as other developers won't be able to make use of that information.

Revert – undoing the effect of a commit

Sometimes, it will turn out that, even with a code review, there will be some bad commits that you need to reverse (perhaps one turned out to be a not-so-good idea, or it contains bugs). If the commit is already made public, you cannot simply remove it; you need to undo its effects. This issue will be explained in detail in *Chapter 10, Keeping History Clean*.

This "undoing of a commit" can be done by creating a commit with a reversal of changes, something like cherry-picking but applying the reversal of changes. This is done with the `revert` command (see *Figure 9.5*).

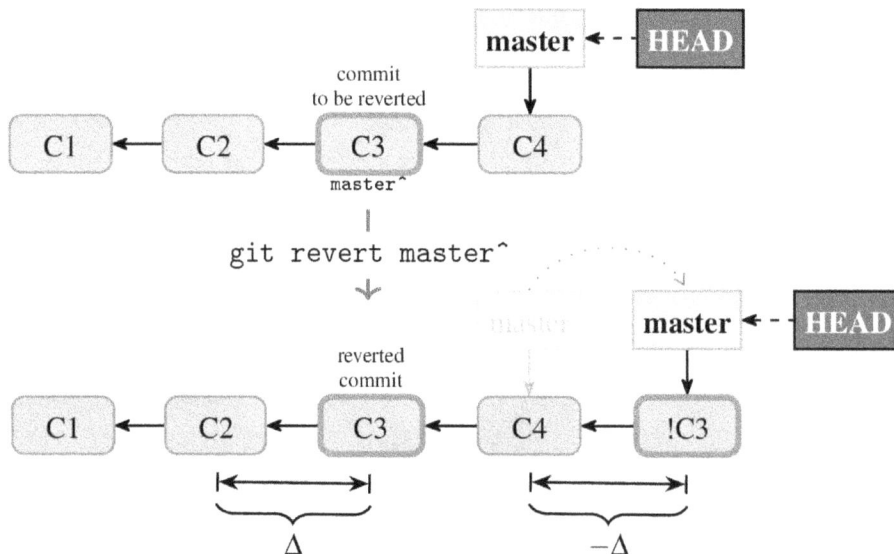

Figure 9.5 – The effect of using 'git revert master^' on a 'master' branch – creating a new commit, denoted !C3, that undoes changes in the C3 commit

The name of this operation might be misleading. If you want to revert all the changes made to the whole working area, you can use `git reset` (in particular, with the `--hard` option). If you want to revert changes made to a single file, use `git checkout <file>` or `git restore <file>`. Both of these are explained in detail in *Chapter 3, Managing Your Worktrees*. The `git revert` command records a new commit to reverse the effect of the earlier commit (often, a faulty one).

Applying a series of commits from patches

Some collaborative workflows include exchanging the changes as patches via email (or another communication medium). This workflow is often encountered in open source projects; it is often easier for a new or a sporadic contributor to create a specially crafted email (for example, with `git format-patch`) and send it to a maintainer or a mailing list, rather than setting up a public repository and sending a pull request.

You can apply a series of patches from a mailbox (in the `mbox` or `maildir` format; the latter is just a series of files) with the `git am` command. If these emails (or files) were created from the `git format-patch` output, you can use `git am --3way` to use the three-way file merge if there are conflicts. Resolving conflicts will be discussed in in the section, *Resolving merge conflicts*.

You can find both tools to help use the patch submission process by sending a series of patches – for example, from the pull request on GitHub (e.g., the *GitGitGadget* GitHub app, or the older *submitGit* web app, to submit patches from GitHub's pull request to the Git project mailing list) – and tools that track web page patches sent to a mailing list (for example, the *patchwork* tool).

Cherry-picking and reverting a merge

This is all good, but what happens if you want to cherry-pick or revert a merge commit? Such commits have more than one parent; thus, they have more than one change associated with them.

In this case, you have to tell Git which change you want to pick up (in the case of cherry-pick), or back out (in the case of revert) with the `-m <parent number>` option – for example, `-m1`.

Note that reverting a merge undoes the changes, but it does not remove the merge from the history of the project. See the section on reverting merges in *Chapter 10, Keeping History Clean*.

Rebasing a branch

Besides merging, Git supports an additional way to integrate changes from one branch into another – namely, the **rebase operation**.

Like a merge, it deals with the changes since the point of divergence (at least, by default). However, while a merge creates a new commit by joining two branches, rebase takes the new commits from one branch (i.e., takes the commits since the divergence) and reapplies them on top of the other branch – see *Figure 9.6* for an example.

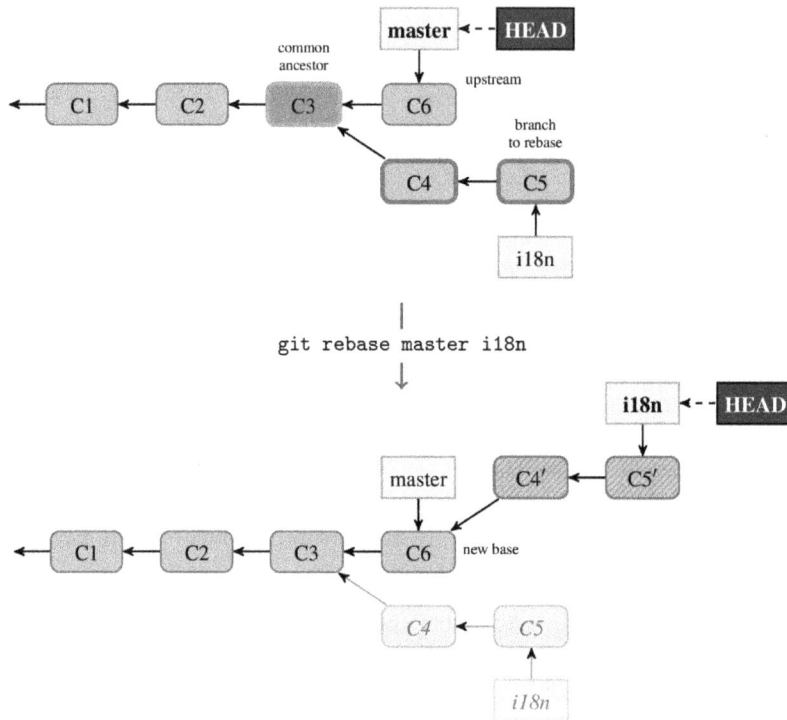

Figure 9.6 – The effects of the rebase operation

With merge, you first switch to the branch to be merged and then use the merge command to select a branch to merge in. With rebase, it is a bit different. First, you select a branch to rebase (i.e., the changes to reapply) and then use the rebase command to select where to put it. In both cases, you first check out the branch to be modified, where a new commit or commits would be (a merge commit in the case of merging, and a replay of commits in the case of rebasing):

```
$ git switch i18n
Switched to branch 'i18n'
$ git rebase master
Successfully rebased and updated refs/heads/master.
```

Alternatively, you can use git rebase master i18n as a shortcut. In this form, you can easily see that the rebase operation takes the master..i18n range of revisions (this notation is explained in *Chapter 4, Exploring Project History*), replays it on top of master, and finally, points i18n to the replayed commits.

Note that old versions of commits don't vanish, at least not immediately. They will be accessible via a reflog (and ORIG_HEAD) for a grace period. This means that it is not that hard to check how replaying changed the snapshots of a project and, with a bit more effort, how changesets themselves have changed.

Merge versus rebase

We have these two ways of integrating changes – merge and rebase. How do they differ, and what are their advantages and disadvantages? We can see by comparing *Figure 9.2* in the *Creating a merge commit* section with *Figure 9.5* in the *Rebasing a branch* section.

First, merge doesn't change history (see *Chapter 10, Keeping History Clean*). It creates and adds a new commit (unless it is a fast-forward merge; then, it just advances the branch head), but the commits that were reachable from the branch remain reachable. This is not the case with rebase. Commits get rewritten, old versions are forgotten, and the DAG of revisions changes. What was once reachable might no longer be reachable. This means that you should not rebase published branches.

Secondly, merge is a one-step operation, with one place to resolve merge conflicts. The rebase operation is multi-step; the steps are smaller (if you follow the recommended practices and keep changes small – see *Chapter 15, Git Best Practices*), but there are more of them.

Linked to this is the fact that the merge result is based (usually) on three commits only, and that it does not take into account what happens on either of the branches that are integrated step by step; only the endpoints matter. Conversely, rebase reapplies each commit individually, so the road to the final result matters here.

Third, the history looks different; you get a simple linear history with rebase, while using the merge operation leads to a complex history, with the lines of development forking and joining. The history is simpler for rebase, but you lose the information that the changes were developed on a separate branch and that they were grouped together, which you get with merge (at least with `--no-ff`). There is even the `git-resurrect` script in the Git contrib tools that uses the information stored in the commit messages of the merge commits to resurrect the old, long-deleted feature branches.

The last difference is that, because of the underlying mechanism, rebase does not, by default, preserve merge commits while reapplying them. You need to explicitly use the `--rebase-merges` option. The merge operation does not change the history, so merge commits are left as they are.

Rebase backends

The previous section described two mechanisms to copy or apply changes – the `git cherry-pick` command and the pipeline from `git format-patch` to `git am --3way`. Either of them can be used by `git rebase` to reapply commits.

The default is to use the merge-based workflow, as if `git rebase` was called with the `--merge` option. The default `'ort'` merge strategy allows rebase to be aware of the renames on the upstream side (where we put the replayed commits). With this option, you can also select a specific merge strategy and pass options to it.

To switch to a patch-based strategy, use `git rebase --apply`. In this case, you can pass some options to `git am` that does the actual replaying of changesets.

These options will be described later when we discuss conflicts.

There is also an interactive rebase with its own set of options. This is one of the main tools in *Chapter 10, Keeping History Clean*. It can be used to execute tests after each replayed commit to check that the replay is correct.

Advanced rebasing techniques

You can also have your rebase operation replay on something other than the target branch of the rebase with --onto <newbase>.

Let's assume that your featureA topic branch is based on the unstable development branch named next, as it is dependent on some feature that is not yet ready and not yet present in the stable branch (master). If the functionality on which featureA depends is deemed stable and merged into master, you would want to move this branch from being forked from next to being forked from master. Alternatively, perhaps you started the server branch from the related client branch, but you want to make it more obvious that they are independent.

You can do this with git rebase --onto master next featureA in the first case, and git rebase --onto master server client in the second one (which is shown in *Figure 9.7*).

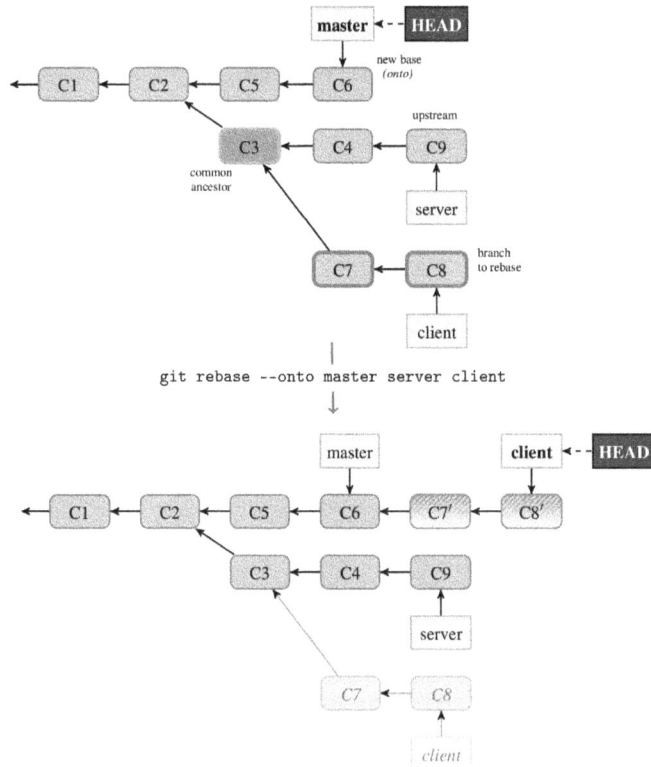

Figure 9.7 – The rebasing branch, moving it from one branch to another

Alternatively, perhaps you want to rebase only a part of the branch. You can do this with `git rebase --interactive`, but you can also use `git rebase --onto <new base> <starting point> <branch>`. You can even choose to rebase the whole branch (usually, an orphan branch) with the `--root` option. In this case, you would replay the whole branch and not just a selected subset of it.

You can also keep the base commit as is with `--keep-base`, instead of following the upstream. With the `--fork-point` option, you can make Git find a better common ancestor using reflog (to find where the branch was created) if it is possible.

Squash merge

If the changes made on a branch are not worth preserving in detail and only their result is, you can use **squash merge** as a way to integrate them as a single commit. This can happen if the branch you want to integrate is full of temporary, work-in-progress commits.

With `git merge --squash`, Git will produce the same result with respect to the working tree (and to the staging area) as if a real merge happened, but it will not perform the commit (the `--commit` option to `git merge` is incompatible with `--squash`). This is done in such a way that the next git commit will create an ordinary commit, not a merge commit. See *Figure 9.8* for a comparison of the merge types.

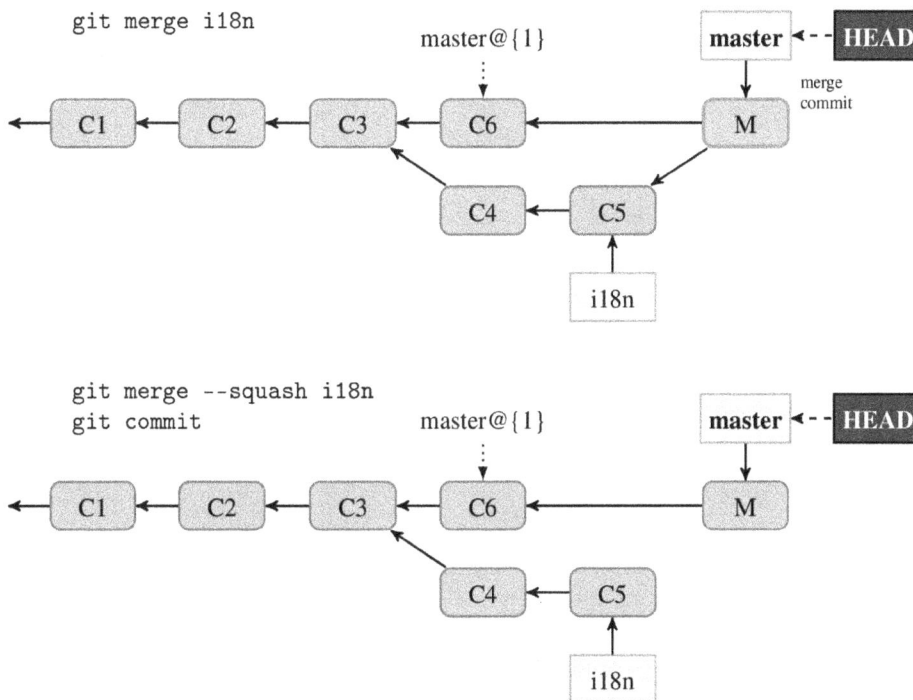

Figure 9.8 – An ordinary merge versus a squash merge for the same set of branches

By default, the commit message of a squashing commit begins with **Squashed commit of the following:**, followed by (as shown in the example in *Figure 9.8*) the result of `git log master..i18n`. However, note that this technique should be used only if we intend to drop (delete) the "merged" branch. This is because Git might have trouble merging any further development on the squash-merged branch, as the graph of revisions does not indicate that the commit was the result of a merge.

The alternative is to use the `squash` command of an interactive rebase.

Resolving merge conflicts

Merging in Git is typically fairly easy. Since Git stores and has access to a full graph of revisions, it can automatically find where the branches diverged and merge only those divergent parts. This works even in the case of repeated merges, so you can keep a very long-lived branch up to date by repeatedly merging into it or rebasing it on top of new changes.

However, it is not always possible to automatically combine changes. There are problems that Git cannot solve because, for example, there were different changes to the same area of a file on different branches. These problems are called **merge conflicts**. Similarly, there can be problems while reapplying changes, although you would still get merge conflicts in case of problems.

The three-way merge

Unlike some other version control systems, Git does not try to be overly clever about merge conflict resolutions and try to solve them all automatically. Git's philosophy is to be smart about determining the cases when a merge can be easily done automatically (for example, taking renames into account) and, if automatic resolution is not possible, to not be overly clever about trying to resolve it. It is better to bail out and ask users to resolve a merge, perhaps unnecessary with a smart algorithm, than to automatically create an incorrect one.

Git uses the **three-way merge algorithm** to come up with the result of the merge, comparing the common ancestors (*base*), the side merged in (*theirs*), and the side merged into (*ours*). This algorithm is very simple, at least at the tree level – that is, the granularity level of files. The following table explains the rules of the algorithm:

ancestor (base)	HEAD (ours)	branch (theirs)	result
A	A	A	A
A	A	B	B
A	B	A	B
A	B	B	B
A	B	C	merge

Table 9.1 – How a three-way merge algorithm works

As shown in the preceding table, the rules for the trivial tree-level three-way merges are as follows:

- If only one side changes a file, take the changed version

- If both sides have the same changes, take the changed version

- If one side has a different change from the other, there is a merge conflict at the content level

It is a bit more complicated if there is more than one ancestor, or if a file is not present in all the versions, but usually, it is enough to just know and understand these rules.

If one side changes a file differently from the other (where the type of the change counts – for example, renaming a file on one branch doesn't conflict with the changing contents of the file on the other branch), Git tries to merge the files at the content level, using the provided **merge driver** if it is defined, and the content-level three-way merge otherwise (for text files).

The three-way file merge examines whether the changes touch different parts of a file (different lines are changed, and these changes are well separated by more than three lines (the context size) away from each other). If these changes are present in different parts of the file, Git resolves the merge automatically (and tells us which files are **auto-merged**).

However, if you change the same part of the same file differently in the two branches that you're merging together, Git won't be able to merge them cleanly:

```
$ git merge i18n
Auto-merging src/rand.c
CONFLICT (content): Merge conflict in src/rand.c
Automatic merge failed; fix conflicts and then commit the result.
```

This problem (a **merge conflict**) is then left for the user to resolve.

Examining failed merges

If Git is unable to automatically resolve a merge (or if you have passed the --no-commit option to the git merge command), it will not create a merge commit. It will pause the process, waiting for you to resolve the conflict.

You can then always abort the merging process with git merge --abort.

Conflict markers in the working tree

If you want to see which files are still unmerged at any point after a merge conflict, you can run git status:

```
$ git status
On branch master
You have unmerged paths.
```

```
   (fix conflicts and run "git commit")
   (use "git merge --abort" to abort the merge)

Unmerged paths:
   (use "git add <file>..." to mark resolution)
          both modified:
 src/rand.c
```

Anything that has not been resolved is listed as unmerged. In the case of content conflicts, Git uses standard conflict markers, putting them around the place of conflict, with the *ours* and *theirs* versions of the conflicted area in question. Your file will contain a section that will look somewhat like the following:

```
<<<<<<< HEAD:src/rand.c
fprintf(stderr, "Usage: %s <number> [<count>]\n", argv[0]);
=======
fprintf(stderr, _("Usage: %s <number> [<count>\n"), argv[0]);
>>>>>>> i18n:src/rand.c
```

This means that the *ours* version on the current branch (HEAD) in the `src/rand.c` file is there at the top of this block, between the `<<<<<<<` and `=======` markers, while the *theirs* version on the `i18n` branch being merged (also from `src/rand.c`) is there at the bottom, between the `=======` and `>>>>>>>` markers.

You need to replace this whole block by the resolution of the merge, either by choosing one side (and deleting the rest) or combining both changes, for example:

```
  fprintf(stderr, _("Usage: %s <number> [<count>]\n"), argv[0]);
```

To help you avoid committing unresolved changes by mistake, Git by default checks whether committed changes include something that looks like conflict markers, refusing to create a merge commit if it finds them. You can force this check to be skipped with the `--no-verify` option.

If you need to examine a common ancestor version to resolve a conflict, you can switch to `diff3-` or `zdiff3-`like conflict markers, which have an additional block, separated by ‖‖‖‖‖‖. This new block shows the common ancestor (*ours*) version:

```
<<<<<<< HEAD:src/rand.c
fprintf(stderr, "Usage: %s <number> [<count>]\n", argv[0]);
|||||||
fprintf(stderr, "Usage: %s <number> [<count>\n", argv[0]);
=======
fprintf(stderr, _("Usage: %s <number> [<count>\n"), argv[0]);
>>>>>>> i18n:src/rand.c
```

You can replace merge conflict markers individually on a file-per-file basis by rechecking the file again, using the following command:

```
$ git checkout --conflict=diff3 src/rand.c
```

If you prefer to use this format all the time, you can set it as the default for future merge conflicts by setting merge.conflictStyle to diff3 or zdiff3 (from the default of merge).

Three stages in the index

How does Git keep track of which files are merged and which are not? Conflict markers in the working directory files would not be enough. Sometimes, there are legitimate contents that look like commit markers (for example, a file that contains an example of a merge conflict or files in the AsciiDoc format), and there are more conflict types than **CONFLICT(content)**. How does Git, for example, represent the case where both sides renamed the file but in a different way, or where one side changed the file and the other side removed it?

It turns out that it is another use for the staging area of the commit (a merge commit in this case), which is also known as the index. In the case of conflicts, Git stores all of the conflicted file versions in the index under stages; each stage has a number associated with it.

- Stage 1 is the common ancestor (*base*)
- Stage 2 is the merged-into version from HEAD – that is, the current branch (*ours*)
- Stage 3 is from MERGE_HEAD, the version you're merging in (*theirs*)

You can see these stages for the unmerged files with the low-level (plumbing) command, git ls-files --unmerged (or for all the files with git ls-files --stage):

```
$ git ls-files --unmerged
100755 ac51efdc3df4f4fd318d1a02ad05331d8e2c9111 1

src/rand.c
100755 36c06c8752c78d2aaf89571132f3bf7841a7b5c3 2

src/rand.c
100755 e85207e04dfdd50b0a1e9febbc67fd837c44a1cd 3

src/rand.c
```

You can refer to each version with the :<stage number>:<pathname> specifier. For example, if you want to view a common ancestor version of src/rand.c, you can use the following:

```
$ git show :1:src/rand.c
```

If there is no conflict, the file is in stage 0 of the index.

Examining differences – the combined diff format

You can use the status command to find which files are unmerged, and conflict markers do a good job of showing conflicts. But how do we see only conflicts before we work on them, and how do we see how they were resolved? The answer is git diff.

One thing to remember is that for merges, even merges in progress, Git will show the so-called **combined diff** format. It looks like the following (for a conflicted file during a merge):

```
$ git diff
diff --cc src/rand.c
index 293c8fc,4b87d29..0000000
--- a/src/rand.c
+++ b/src/rand.c
@@@ -14,16 -14,13 +14,26 @@@ int main(int argc, char *argv[]


   return EXIT_FAILURE;

 }

++<<<<<<< HEAD:src/rand.c
 +fprintf(stderr, "Usage: %s <number> [<count>]\n", argv[0]);
++=======
+ fprintf(stderr, _("Usage: %s <number> [<count>\n"), argv[0]);
++>>>>>>> i18n:src/rand.c
```

You can see a few differences from the ordinary unified diff format described in *Chapter 2, Developing with Git*. First, this uses diff --cc in the header to denote that it uses the compact combined format (it would use diff --combined instead if you used the git diff -c command). The extended header lines, such as index 293c8fc,4b87d29..0000000, take into account that there is more than one source version. The chunk header, @@@ -14,16 -14,13 +14,26 @@@, is modified (and is different from the one for the ordinary patch) to prevent people from trying to apply a combined diff as a unified diff – for example, with the patch -p1 command.

Each line of the diff command is prefixed by two or more characters (two in the most common cases of merging two branches); the first character informs us about the state of the line in the first preimage (*ours*) as compared to the result, the second character informs us about the other preimage (*theirs*), and so on. For example, ++ means that the line was not present in either of the versions being merged (here, in this example, you can find it on the line with the conflict marker).

Examining differences is even more useful for checking the resolution of a merge conflict.

To compare the result (i.e., the current state of the working directory) with the version from the current branch (i.e., merged into) – that is, the *ours* version – you can use `git diff -ours`. This also applies to the version being merged (*theirs*) and the common ancestor version (*base*).

How did we get there – git log --merge

Sometimes, we need more context to decide which version to choose or to resolve a conflict. One such technique is reviewing a little bit of history, recalling why the two lines of development that are merged touched the same area of code.

To get the full list of divergent commits that were included in either branch, we can use the triple-dot syntax that you learned about in *Chapter 4*, *Exploring Project History*, adding the `--left-right` option to make Git show which side the given commit belongs to:

```
$ git log --oneline --left-right HEAD...MERGE_HEAD
```

We can further simplify this and limit the output to only those commits that touched at least one of the conflicted files, with a `--merge` option to `git log`, for example:

```
$ git log --oneline --left-right --merge
```

This can be helpful in quickly giving you the context you need to understand why something conflicts and how to intelligently resolve it.

Avoiding merge conflicts

While Git prefers to fail to auto-merge clearly, rather than trying elaborate merge algorithms, there are a few tools and options that you can use to help Git avoid merge conflicts.

Useful merge options

One of the problems while merging branches might be that they use different end-of-line normalization or clean/smudge filters (see *Chapter 3*, *Managing Your Worktrees*). This might happen when one branch added such a configuration (e.g., changing a git attributes file) and the other did not. In the case of end-of-line character configuration changes, you would get a lot of spurious changes, where lines differ only in the EOL (end-of-line) characters. In both cases, while resolving a three-way merge, you can make Git run a virtual checkout and check-in of all three stages of a file. This is done by passing the `renormalize` option to the `'ort'` merge strategy (`git merge -Xrenormalize`). This would, as the name suggests, normalize end-of-line characters, making them the same for all stages.

Changing how end of line is defined can contribute to whitespace-related conflicts. It's pretty easy to tell that this is the case when looking at the conflict, as every line is removed on one side and added again on the other, and `git diff --ignore-whitespace` shows a more manageable conflict (or even a conflict that is resolved). If you see that you have a lot of whitespace issues in a merge, you can abort and redo it, this time with `-Xignore-all-space`, `-Xignore-space-change`, `-Xignore-space-at-eol`, or `-Xignore-cr-at-eol`.

Note that whitespace changes mixed with other changes to a line are not ignored.

Sometimes, mis-merges occur due to unimportant matching lines (for example, braces from distinct functions). You can make Git spend more time minimizing differences by selecting `patience`, a `histogram`, or a `minimal` diff algorithm with, `-Xdiff-algorithm=patience`, and so on.

If the problem is mis-detected renamed files, you can adjust the rename threshold with `-Xfind-renames=<n>`.

Rerere – reuse recorded resolutions

The **rerere** (**reuse recorded resolutions**) functionality is a bit of a hidden feature. As the name of the feature implies, it makes Git remember how each conflict was resolved chunk by chunk, so that the next time Git sees the same conflict, it will be able to resolve it automatically. However, note that Git will stop at resolving conflicts and does not automatically commit the said rerere-based resolution, even if it resolves it cleanly (if it is superficially correct).

Such a functionality is useful in many scenarios. One example is a situation when you want a long-lived (i.e., long development) branch to merge cleanly at the end of its cycle, but you do not want to create intermediate merge commits. In this situation, you can do **trial merges** (merge, and then delete merge), saving information about how merge conflicts were resolved to the rerere cache. With this technique, the final merge should be easy, as most of it will be cleanly resolved from the resolutions recorded earlier.

Another situation in which you can make use of the rerere cache is when you merge a bunch of topic branches into a testable permanent branch. If the integration test for a branch fails, you want to be able to rewind the failed branch but don't want to lose the work spent on resolving a merge.

Alternatively, perhaps you have decided that you would rather use rebase than merge. The rerere mechanism allows us to translate the merge resolution into the rebase resolution.

To enable this functionality, simply set `rerere.enabled` to `true`, or create the `.git/ rr-cache` file.

Dealing with merge conflicts

Let's assume that Git was not able to auto-merge cleanly and that there are merge conflicts that you need to resolve to be able to create a new merge commit. What are your options?

Aborting a merge

First, let's cover how to get out of this situation. If you weren't perhaps prepared for conflicts or you don't know enough about how to resolve them, you can simply back out from the merge you started with `git merge --abort`.

This command tries to reset to the state before you started a merge. It might be not able to do this if you did not start from a clean state. Therefore, it is better to stash away changes, if there are any, before performing a merge operation (which you can do with `--autostash`, or the `merge.autoStash/rebase.autoStash` configuration options).

Selecting the ours or theirs version

Sometimes, it is enough to choose one version in the case of conflicts. If you want to resolve all the conflicts this way, forcing all the chunks to resolve in favor of *ours* or *theirs* version, you can use the `-Xours` or `-Xtheirs` merge strategy option, respectively. Note that `-Xours` (the **merge option**) is different from `--strategy=ours` (the **merge strategy**); the latter creates a merge commit where the project state is the same as the *ours* version, instead of taking the *ours* version only for conflicted files.

If you want to do this only for selected files, you can again check out the file with the *ours* or *theirs* version with `git checkout --ours` or `git checkout --theirs`, respectively. Note that during the rebase, the *ours* and *theirs* version may appear to be swapped.

You can examine the *base*, *ours*, or *theirs* version with `git show :1:file`, `git show:2:file`, or `git show:3:file`, respectively, as described earlier.

Scriptable fixes – manual file remerging

There are types of changes that Git can't handle automatically, but they are scriptable fixes. The merge can be done automatically, or at least is much easier, if we transform the *ours*, *theirs*, or *base* version first. Renormalization after changing how the file is checked out and stored in the repository (i.e., eol and clean/smudge filters) and handling the whitespace change are built-in options. Another example, but without built-in support, is changing the encoding of a file or another scriptable set of changes, such as renaming variables.

To perform a scripted merge, you first need to extract a copy of each of these versions of the conflicted file, which can be done with the `git show` command and with `:<stage>:<file>`:

```
$ git show :1:src/rand.c >src/rand.common.c
$ git show :2:src/rand.c >src/rand.ours.c
$ git show :3:src/rand.c >src/rand.theirs.c
```

Now that you have in the working area the contents of all three stages of the files, you can fix each version individually – for example, with `dos2unix` or `iconv`. You can then remerge the contents of the file with the following command:

```
$ git merge-file -p \
  rand.ours.c rand.common.c rand.theirs.c >rand.c
```

Using graphical merge tools

If you want to use a graphical tool to help you resolve merge conflicts, you can run `git mergetool`, which fires up a visual merge tool and guides the invoked tool through all the merge conflicts.

It has a wide set of preconfigured support for various graphical merge helpers. You can configure which tool you want to use with `merge.tool`. If you don't do this, Git will try all the possible tools in the sequence, which depends on the operating system and the desktop environment.

You can also configure a setup for your own tool.

Marking files as resolved and finalizing merges

As described earlier, if there is a merge conflict for a file, it will have three stages in the index. To mark a file as resolved, you need to put the contents of a file into stage 0. This can be done by simply running `git add <file>` (running `git status` will give you this hint).

When all the conflicts are resolved, you need to simply run `git commit` to finalize the merge commit (or you can skip marking each file individually as resolved and just run `git commit -a`). The default commit message for merge summarizes what we merge, including a list of the conflicts, if any were present. You can make Git add a shortlog of the merged-in branches with the `--log` option for a single merge, or set it up permanently with the `merge.log` configuration variable.

Resolving rebase conflicts

When there is a problem with applying a patch or a patch series, cherry-picking or reverting a commit, or rebasing a branch, Git will fall back to using the three-way merge algorithm. How to resolve such conflicts is described in the earlier sections.

> **Important note**
> Note that when using merging strategies (the default), for technical reasons, *ours* is the so-far rebased series – that is, the branch being integrated – while *theirs* is the working branch (the branch rebased onto).

However, for some of these methods, such as rebase, applying a mailbox (`git am`), or cherry-picking a series of commits, that are done stage by stage (a sequencer operation), there are other issues – namely, what to do if there is a conflict during such a stage.

You have three options:

- You can resolve the conflict and continue the operation with the `--continue` parameter (or, in the case of `git am`, also `--resolved`)

- You can abort the whole operation and reset HEAD to the original branch with `--abort`

- You can use `--skip` to drop a revision, perhaps because the commit is already present in the upstream and we can drop it during replaying

git-imerge – an incremental merge and rebase for git

Both rebase and merge have their disadvantages. With merge, you need to resolve one big conflict (although using test merges and rerere to keep up-to-date proposed resolutions could help with this) in an all-or-nothing fashion. There is almost no way to save partially a done merge or to test it; `git stash` can help, but it might be an inadequate solution.

Rebase, conversely, is done in a step-by-step fashion, but it is not ideal for collaboration; you should not rebase published parts of the history of the project. You can interrupt a rebase, but it leaves you in a strange state (on an anonymous branch).

That's why the `git imerge` third-party tool was created. It presents conflicts pair-wise in small steps. It records all the intermediate merges in such a way that they can be shared, so one person can start merging and the other can finish it. The final resolution can be stored as an ordinary merge, an ordinary rebase, or a rebase with history.

Summary

This chapter has shown us how to effectively join two lines of development together, combining commits they gathered since their divergence.

First, we got to know various methods of combining changes – merge, cherry-pick, and rebase. This part focused on explaining how these functionalities work at higher levels – at the level of the DAG of revisions. You learned how merge and rebase work and what the difference is between them. Some of the more interesting uses of rebase, such as transplanting a topic branch from one long-lived branch to another, were also shown.

Then, you learned what to do if Git is not able to automatically combine changes – that is, what can be done in the presence of a merge conflict. The important part of this process is to understand how the three-way merge algorithm works, as well as how the index and the working area are affected if there are conflicts. You now know how to examine failed merges, examine proposed resolutions, avoid conflicts, and finally, resolve them and mark them as resolved.

The next chapter, *Keeping History Clean*, will explain why we might want to rewrite history to keep it clean (and what that means). One of the tools to rewrite history is an interactive rebase, a close cousin of an ordinary rebase operation described in this chapter. You will learn various methods to rewrite commits: how to reorder them, how to split them if they are too large, how to squash the fixing commit with the commit it corrects, and how to remove a file from history. You will discover what you can do if you cannot rewrite history (understanding why rewriting published history is bad) but you still need to correct it, with `git replace` and `git notes` commands. We will also discuss other applications of these mechanisms.

Questions

Answer the following questions to test your knowledge of this chapter:

1. What are the advantages and disadvantages of using merge to integrate changes?
2. What are the advantages and disadvantages of using rebase to integrate changes?
3. How do we avoid resolving similar conflicts again and again during a merge or rebase?
4. How can we discover whether we are in the middle of merge or rebase, and remind ourselves on how to resolve conflict or abort an operation?

Answers

Here are the answers to the questions given above:

1. With merge, you do the integration in a single step (which can be an advantage or disadvantage), and you need to test only a single commit – the result of the merge. You can easily see where the branch began and where it ended. The first-parent view can serve as a summary of the integrated branches.

2. With rebase, you do the integration step by step (which can be a disadvantage or an advantage). Each of the rebased commits might need testing. The resulting history is much simpler, more linear, and easier to see. Using `bisection` to find regression bugs should be faster with linear history.

3. You can use the rerere mechanism, which automatically reapplies recorded conflict resolutions.

4. Use the `git status` command.

Further reading

To learn more about the topics that were covered in this chapter, take a look at the following resources:

- Scott Chacon, Ben Straub: *Pro Git*, 2nd Edition (2014): `https://git-scm.com/book/en/v2`:

  - *Chapter 3.6*, *Git Branching – Rebasing*

  - *Chapter 7.8*, *Git Tools – Advanced Merging*

  - *Chapter 7.9*, *Git Tools – Rerere*

- Julia Evans, *git rebase: what can go wrong?* (2023): `https://jvns.ca/blog/2023/11/06/rebasing-what-can-go-wrong-/`

- Julia Evans, *How git cherry-pick and revert use 3-way merge* (2023): `https://jvns.ca/blog/2023/11/10/how-cherry-pick-and-revert-work/`

- Junio C Hamano, *Where do evil merges come from?* (2013): `https://git-blame.blogspot.com/2013/04/where-do-evil-merges-come-from.html`

- Nick Quaranto, *git ready – keep either file in merge conflicts* (2009): `https://gitready.com/advanced/2009/02/25/keep-either-file-in-merge-conflicts.html`

- *Learn to use email with git!*: `https://git-send-email.io/`

Keeping History Clean

The previous chapter, *Merging Changes Together*, described how to join changes developed by different people (as described in *Chapter 6, Collaborative Development with Git*), or just developed in a separate feature branch (as shown in *Chapter 8, Advanced Branching Techniques*). One of the techniques was rebase, which can help bring a branch to be merged to a better state. However, if we are rewriting history, perhaps it would be possible to also modify the rebased commits to be easier for review, making the development steps of a feature clearer. If rewriting is forbidden, can one make history cleaner without it? How do we fix mistakes if we cannot rewrite project history?

This chapter will answer all those questions. It will explain why one might want to keep a clean history, when it can and should be done, and how it can be done. Here you will find step-by-step instructions on how to reorder, squash, and split commits. This chapter will also describe how to do large-scale history rewriting (for example, the clean-up after imports from other VCS) and what to do if one cannot rewrite history: in other words, using **reverts**, replacements, and notes.

To really understand some of the topics presented here, and to truly master their use, you need some basics of Git internals. These are presented at the beginning of this chapter.

In this chapter, we will cover the following topics:

- The basics of the object model of Git repositories
- Why you shouldn't rewrite published history, and how to recover from doing so
- The interactive rebase: reordering, squashing, splitting, and testing commits
- Large-scale scripted history rewriting
- Reverting a revision, reverting a merge, and re-merging after a reverted merge
- Amending history without rewriting with replacements
- Appending additional information to objects with notes

An introduction to Git internals

To really understand and make good use of at least some of the methods described in this chapter, you will need to understand at least the very basics of Git internals. Among other things, you will need to know how Git stores the information about revisions.

When describing Git internals, it will be useful to create different types of data to later examine. This can be achieved with a set of low-level commands that Git provides as a supplement to user-facing high-level commands. These low-level commands operate on the level of the internal representation instead of using friendly abstractions. That makes those commands very flexible and powerful, though perhaps not user-friendly.

Git objects

In *Chapter 4*, *Exploring Project History*, you learned that Git represents history as the **Directed Acyclic Graph** (**DAG**) of revisions, where each revision is a graph node represented as a **commit object**. Each commit is identified by the SHA-1 identifier. We can use this identifier (in its full or ambiguous shortened form) to refer to any given version.

The commit object consists of revision metadata, links to zero or more parent commits, and the snapshot of the project's files at the revision that it represents. The revision metadata includes information about who made changes and when, who created the commit object (who entered changes into the repository) and when, and of course the commit message.

Beyond this fact, it is also useful, in some cases, to know how Git internally represents the snapshot of a project's files at the given revision. Git uses **tree objects** to represent directories, and **Binary Large Objects** (**blobs**) to represent the contents of a file.

Aside from the commit, tree, and blobs, there might also be **tag objects** representing annotated and signed tags.

Each object is identified by the SHA-1 hash function over its contents, or to be more exact, over the type and the size of the object plus its contents. Such a content-based identifier does not require a central naming service. Thanks to this fact, each and every distributed repository of the same project will use the same identifiers, and we do not have to worry about name collisions:

```
# calculate SHA-1 identifier of blob object with Git
$ printf "foo" | git hash-object -t blob --stdin
19102815663d23f8b75a47e7a01965dcdc96468c
# calculate SHA-1 identifier of blob object by hand
$ printf "blob 3\0foo" | sha1sum
19102815663d23f8b75a47e7a01965dcdc96468c
```

> **Object identifier – SHA-1 to SHA-256 transition**
>
> Over time, flaws in the SHA-1 hash function have been discovered. Therefore, Git will transition to using SHA-256 while providing interoperability. At the time of writing this, Git was still using SHA-1 by default.

We can say that the Git repository is the content-addressed object database. That is, of course, not all there is; there are also references (branches and tags), various configuration files, and other things.

Let's describe Git objects in more detail, starting from the bottom up. We can examine objects with the low-level `git cat-file` command:

- **Blob**: These objects store the contents of the file at the given revision. Such an object can be created using the low-level `git hash-object -w` command. Note that if different revisions have the same contents of a file, it is stored only once thanks to content-based addressing:

  ```
  $ git cat-file blob HEAD:COPYRIGHT
  Copyright (c) 2014 Company
  All Rights Reserved
  ```

- **Tree object**: These objects represent directories. Each tree object is a list of entries, sorted by filename. Each entry is composed of combined permissions and type, the name of the file or directory, and the link (that is, SHA-1 identifier) of an object connected with the given path, either the tree object (representing the subdirectory), the blob (representing the file contents), or rarely the commit object (representing the submodule; see *Chapter 11, Managing Subprojects*). Note that if different revisions have the same contents of a subdirectory, it will be stored only once thanks to content-based addressing:

  ```
  $ git cat-file -p HEAD^{tree}
  100644 blob 862aafd...
  COPYRIGHT
  100644 blob 25c3d1b...
  Makefile
  100644 blob bdf2c76...
  README
  040000 tree 7e44d2e...
  src
  ```

Note that the real output includes full 40-character SHA-1 identifiers, not shortened ones as in the preceding example. You can create tree objects out of the index (which you can create using the `git update-index` command) with `git write-tree`.

- **Commit object**: These objects represent revisions. Each commit is composed of a set of headers (key-value data) that includes zero or more `parent` lines and exactly one tree line with the link to the `tree` object representing a snapshot of the repository contents (the top directory of a project). You can create a commit with a given tree object as a revision snapshot by using the low-level `git commit-tree` command, or by simply using `git commit`:

```
$ git cat-file -p HEAD
tree 752f12f08996b3c0352a189c5eed7cd7b32f42c7
parent cbb91914f7799cc8aed00baf2983449f2d806686
parent bb71a804f9686c4bada861b3fcd3cfb5600d2a47
author Joe Hacker <joe@example.com> 1401584917 +0200
committer Bob Developer <bob@example.com> 1401584917 +0200

Merge remote branch 'origin/multiple'
```

- **Tag object**: These objects represent annotated tags, of which signed tags are a special case. Tags (lightweight and annotated) give a permanent name to a commit (such as `v0.2`) or any object. Tag objects also consist of a series of headers (including links to the tagged object) and the tag message. You can create the tag object with a low-level `git mktag` command, or simply with `git tag`:

```
$ git cat-file tag v0.2
object 5d2584867fe4e94ab7d211a206bc0bc3804d37a9
type commit
tag v0.2
tagger John Tagger <john@example.com> 1401585007 +0200

random v0.2
```

Internal datetime format

The Git internal format for the author, the committer, and the tagger dates is **\<unix timestamp\> \<timezone offset\>**. The Unix timestamp (POSIX time) is the number of seconds since the Unix epoch, which is 00:00:00 **Coordinated Universal Time** (UTC), Thursday, January 1st, 1970 (1970-01-01T00:00:00Z), not counting leap seconds. This denotes when the event took place. You can print the Unix timestamp with `date "%s"` and convert it into other formats with `date --date="@<timestamp>"`.

The timezone offset is a positive or negative offset from UTC, in the **HHMM** (hours, minutes) format. For example, CET (the timezone that is 2 hours ahead of UTC) is +0200. This can be used to find the local time for an event.

The relationship between different types of Git objects mentioned here is shown in *Figure 10.1*. It represents a typical case, with a tag pointing to a commit and with commits sharing the same contents of at least some files.

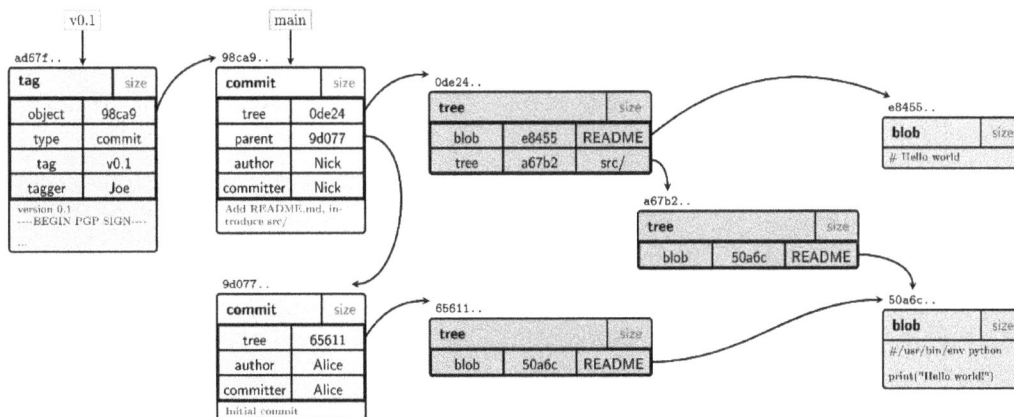

Figure 10.1 – The Git repository object model

Some Git commands work on any type of object. For example, you can tag any type of object, not only commits. You can, among other things, tag a blob to keep some unrelated piece of data in the repository and have it available in each clone. Public keys can be such data.

Notes and replacements, which will be described later in this chapter, also work on any type of object.

Plumbing and porcelain Git commands

Git was developed in a bottom-up fashion. This means that its development started from basic blocks and built upward. Many of the user-facing commands were once built as shell scripts utilizing these basic low-level blocks to do their work. Thanks to this, we can distinguish between the two types of Git commands.

The better-known type is **porcelain** commands, which are high-level user-facing commands (*porcelain* is a play on words on calling engine-level commands *plumbing*). The output of these commands is intended for the end user. This means that their output can be changed to be more user-friendly. Therefore, their output can be different in different Git versions. The user is smart enough to understand what happens if they are presented with additional information, changed wording, or changed formatting (for example).

This is not the case for the scripts you may write in this chapter, such as those used for rewriting with `git filter-repo`. Here, you need unchanging output – well, at least for the scripts that are used more than once (as hooks, as `.gitattribute` drivers, and as helpers). You can often find a switch, usually named `--porcelain`, that ensures that the command output is immutable. For other commands, the solution is to specify the format fully. Alternatively, you can use low-level commands intended for scripting: **plumbing** commands. These commands usually do not have user-friendly defaults, not to mention a "do what I mean" quality. Their output also does not depend on the Git configuration; not many of them can be configured via the Git configuration file.

The git (1) manpage includes a list of all the Git commands, separated into porcelain and plumbing. The distinction between plumbing and porcelain commands was mentioned as a tip in *Chapter 3*, *Managing Your Worktrees*, when we encountered the first low-level plumbing command without a user-facing and user-friendly porcelain equivalent.

Rewriting history

Many times, while working on a project, you may want to revise your commit history. One reason for this could be to make it easier to review before submitting the changes upstream. Another reason would be to take reviewer comments into account in the next improved version of changes. Or perhaps you'd like to have a clear history while finding regressions using bisection, as described in *Chapter 4*, *Exploring Project History*.

One of the great things about Git is that it makes revising and rewriting history possible while providing a wide set of tools to revise history and make it clean.

> **Views on rewriting history**
>
> There are two conflicting views among users of the version control system. One states that history is sacred and that you should show the true history of the development, warts, and all. The other states that you should clean up the new history for better readability before publishing it.

An important issue to note is that even though we talk about "rewriting" the history, objects in Git (including commits) are **immutable**. This means that "rewriting" really means creating a modified copy of commits, a new path in the DAG of revisions. Then, the appropriate branch reference is switched to point to the just-created new path, to the changed copy of the history. The original, pre-rewrite commits are still there in the repository, referenced and available from the reflog (and also from ORIG_HEAD). Well, at least they will be there until they get pruned (that is, deleted) as unreferenced and unreachable objects during garbage collection, though this only happens after the reflog expires.

Amending the last commit

The simplest case of history rewriting is correcting the latest commit on a branch.

Sometimes you might notice a typo (an error) in a commit message, or that you have committed an incomplete change in the last revision. If you have not pushed (published) your changes, you can **amend** the last commit. This is done with the --amend option to the git commit command.

The result of amending a commit is shown in *Figure 6* in *Chapter 2, Developing with Git*. Note that there is no functional difference between amending the last commit and changing some commits deeper in the history. In both cases, you are creating a new commit, leaving the old version referenced by the reflog. The difference is in what happens to other commits.

Here, the index (that is, the explicit staging area for commits) shows its usefulness again. For example, if you want to simply fix only the commit message and you do not want to make any other changes, you can use `git commit --amend` (note the lack of an `-a` or `--all` option). This works even if you have started work on a new commit – at least, assuming that you didn't add any changes to the index. If you did, you can put them away temporarily with `git stash`, fix the commit message of the last commit, and then pop stashed changes and restore the index with `git stash pop --index`.

If, on the other hand, you realize that you have forgotten some changes, you can just edit the files and use `git commit --amend --all`. If the changes are interleaved, you can use `git add` or its interactive version (utilizing knowledge from *Chapter 3, Managing Your Worktrees*) to create the contents you want to have, finalizing it with `git commit --amend`.

The interactive rebase

Sometimes you might want to edit commits deeper in the history or reorganize commits into a logical sequence of steps. One of the built-in tools in Git that you can use for this purpose is `git rebase --interactive`.

Here, we will assume that you are working on a feature using a separate topic branch, as well as a topic branch workflow described and recommended in *Chapter 8, Advanced Branching Techniques*. We will also assume that you are doing the work in a series of logical steps rather than in one large commit.

When implementing a new feature, you usually won't do it perfectly from the very beginning. You would want to introduce it in a series of small self-contained steps (see *Chapter 15, Git Best Practices*) to make code review, code audit, and bisection (finding the cause of regressions bugs) easier. Often, you will only see how to split it better after finishing the work. It is also unreasonable to expect that you would not make mistakes while implementing a new feature.

Before submitting the changes (by either pushing them to the central repository, pushing them to your own public repository and sending pull requests, or using some other workflow described in *Chapter 6, Collaborative Development with Git*), you will often want to update your branch to the up-to-date state of the project to make it easier to merge. By rebasing your changes on top of the current state and having them up-to-date, you will make it easier for the maintainer (the integration manager) to ultimately merge your changes when they are accepted for inclusion into the mainline. **Interactive rebase** allows you to clean up history, as described earlier, while doing this work.

Aside from tidying up changes before publishing them, there is also additional use for tools such as interactive rebases. While working on a more involved feature, the very first submission is not always accepted into an upstream and added to the project. Often, the process of patch review finds problems with the code, or with the explanation of the changes. Perhaps something is missing (for example, the feature might lack documentation or tests), some commit needs to be fixed, or the submitted series of patches (or the branch submitted in the pull request) should be split into smaller commits for easy review. In this case, you would also use an interactive rebase (or an equivalent tool) to prepare a new version to submit, taking into account the results of the code inspection.

Reordering, removing, and fixing commits

Rebase, as described in *Chapter 9, Merging Changes Together*, consists of taking a series of changes of the commits being rebased and reapplying them on top of a new base (a new commit). In other words, rebase moves changesets, not snapshots. Git starts the interactive rebase by opening the instructions sheet corresponding to those operations of reapplying changes in the editor.

> **Tip**
>
> You can configure the text editor used for editing the rebase instruction file separately from the default editor (which is used, for example, to edit commit messages) with the sequence. editor configuration variable, which can in turn be overridden by the GIT_SEQUENCE_ EDITOR environment variable.

Like in the case of the template for editing commits, the instruction sheet is accompanied by the comments explaining what you can do with it (note that if you are using older Git, some interactive rebase commands might be missing from this sheet):

```
pick 89579c9 first commit in a branch
pick d996b71 second commit in a branch
pick 6c89dee third commit in a branch

# Rebase 89579c9..6c89dee onto b8fffe1 (3 commands)
#
# Commands:
#  p, pick = use commit
#  r, reword = use commit, but edit the commit message
#  e, edit = use commit, but stop for amending
#  s, squash = use commit, but meld into previous commit
#  f, fixup = like "squash", but discard this commit's log message
#  x, exec = run command (the rest of the line) using shell
#  d, drop = remove commit
#
# These lines can be re-ordered; they are executed from top to bottom.
#
# If you remove a line here THAT COMMIT WILL BE LOST.
#
# However, if you remove everything, the rebase will be aborted.
```

Note that empty commits will be marked with # empty at the end of the line. Depending on your version of Git and your configuration, the instruction sheet may include more commands.

As explained in the comments, the instructions are in the order of execution, starting from the instruction on the top to create the first commit (with the new base as its parent) and ending at the bottom with the instruction copying the commit at the tip of the branch being rebased. This means that revisions are listed in an increasing chronological order, with older commits first. This is the reverse order from the git log output, with the most recent commit first (unless you are using git log --reverse). This is quite understandable; the rebase reapplies changesets in the order in which they were added to the branch, while the log operation shows commits in the order of reachability from the tips.

Each line of the instruction sheet consists of three elements, separated by spaces:

- First, there is a one-word command. By default, the interactive rebase starts with pick. Each command has a one-letter shortcut that you can use instead of the long form, as shown in the comments (for example, you can use p in place of pick).

- Next, there is a uniquely shortened SHA-1 identifier of a commit to be used with the command. Strictly speaking, it is the identifier of a commit being rebased, which it had before the rebase process started. This shortened SHA-1 identifier is used to pick the appropriate commit (for example, when reordering lines of the interactive rebase instruction sheet, which effectively means **reordering commits**).

- Lastly, there is the description (subject) of a commit. It is taken from the first line of the commit message. More specifically, it is the first paragraph of the commit message with the line breaks removed, where a paragraph is defined as the set of subsequent lines of text, separated from other paragraphs by at least one empty line – that is, two or more end-of-line characters. This is one of the reasons why the first line of the commit message should be a short description of changes (see *Chapter 15*, *Git Best Practices*). This description is for you to help decide what to do with the commit; Git uses its SHA-1 identifier and ignores the rest of the line.

Reordering commits with the interactive rebase is as simple as reordering lines in the instruction sheet. Note, however, that if the changes are not independent, you might need to resolve conflicts, even if there would be no merge conflicts without reordering. In such cases, as instructed by Git, you will need to fix conflicts, mark conflicts as resolved (for example, with git add), and then run git rebase --continue. Git will remember that you are in the middle of an interactive rebase, so you don't need to repeat the --interactive option.

The other possibility of dealing with a conflict is skipping a commit, rather than resolving a conflict, by running git rebase --skip. By default, rebase removes changes that are already present in upstream; you might want to use this command in case the rebase didn't detect correctly that the commit in question is already there in the branch we are transplanting revisions onto. In other words, do skip a commit if you know that the correct resolution of a conflict is an empty changeset.

> **Tip**
>
> You can also make Git present you with the instruction sheet again at any time when rebase stops for some reason (including an error in the instruction sheet, such as using the `squash` command with the first commit). You can do this with the `git rebase --edit-todo` `command`. After editing it, you can continue the rebase.

To **remove changes** you simply need to remove the relevant line from the instruction sheet or comment it out, or better yet, use the `drop` command. You can use it to drop failed experiments or to make it easier on the rebase by deleting changesets that you know are already present in the upstream being rebased onto (though perhaps in a different form). Note though that removing the instruction sheet altogether aborts the rebase.

To **fix a commit**, change the `pick` command preceding the relevant commit in the instruction sheet to `edit` (or just `e`). This would make the rebase stop at this commit, that is, at this step of reapplying changes, similar to what happens in the case of a conflict. To be precise, the interactive rebase applies the commit in question, making it the `HEAD` commit, and then stops the process, giving control to the user . You can then fix this commit as if it were the current one with `git commit --amend`, as described in *Amending the last commit*. After changing it to your liking, run `git rebase --continue`, as explained in the instruction that Git prints.

> **Tip**
>
> A proper Git-aware command line prompt, such as the one from the Git `` `contrib/` `` directory in the Git source code, would tell you when you are in the middle of the rebase (see *Chapter 13*, *Customizing and Extending Git*). If you are not using such a prompt, you can always check what's happening with `git status`, which says that there is a rebase in progress in such cases. You can also find instructions on what you can do next there.

Alternatively, you can always go to the state before starting the rebase with the `git rebase --abort` command.

If you only want to change the commit message (for example, to fix a spelling error or to include additional information), you can skip the need to run `git commit --amend` and then `git rebase --continue` by using `reword` (or `r`) instead of `edit`. Git will then automatically open the editor with the commit message. Saving changes and exiting the editor will commit the changes, amend the commit, and continue the rebase.

Squashing commits

Sometimes you might need to make one commit out of two or more, squashing them together. Maybe you decided that it didn't make sense to split the changes and that they are better together.

With the interactive rebase you can reorder these commits as needed, so that they are next to each other. Then, keep the `pick` command for the first of the commits to be concatenated together (or change it to the `edit` command). For the rest of the commits, replace the `pick` command with either the `squash` or the `fixup` command. Git will then accumulate the changes and create the commit with all of them together. The suggested commit message for the folded commit is the commit message of the first commit, with the messages of the commits with the `squash` command appended. Commit messages with the `fixup` command are omitted. This means that the `squash` command is useful for squashing changes, while `fixup` is useful for adding fixes. If commits had different authors, the folded commit will be attributed to the author of the first commit. The committer will be you, the person performing the rebase.

Let's assume that you noticed that you forgot to add some parts of the changes to the commit. Perhaps it is missing tests (or just negative tests) or documentation. The commit is in the past, so you cannot simply add to it by amending it. You could use the interactive rebase or the patch management interface to fix it, but it is often more effective to create the commit with forgotten changes and squash it later.

Similarly, when you notice that the commit you created a while ago has a bug, instead of trying to edit it immediately, you can create a `fixup` commit with a bug fix to be squashed later.

If you use this technique, there might be some delay between noticing the need to make changes or fix a bug and creating the appropriate commit. This gap includes the time taken for the rebase operation.

How do you then mark the commit being created for squashing or fixup? If you use the commit message beginning with the magic `squash!` ... or `fixup!` ... strings, respectively, preceding the description (the first line of the commit message, sometimes called the **subject**) of a commit to be squashed into, you can then later ask Git to **autosquash** them, thus automatically modifying the to-do list of `rebase -i`. You can request this on an individual basis with the `--autosquash` option, or you can enable this behavior by default with the `rebase.autoSquash` configuration variable. To create an appropriate "magic" commit message, you can use `git commit --squash/--fixup` (when creating the commit to be squashed into or the bugfix commit).

Splitting commits

Sometimes, you might want to make two commits or more out of one commit, splitting it into two or more parts. You may have noticed that the commit is too large, perhaps because it tries to do too much, and should be split into smaller pieces. Or perhaps you have decided that some part of a changeset should be moved from one commit to another, and extracting it into a separate commit is a first step toward accomplishing that.

Git does not provide a one-step built-in command for this operation. Nevertheless, splitting commits is possible with the clever use of the interactive rebase.

To split a given commit, first mark it with the `edit` action. As described earlier, Git will stop at the specified commit and give the control back to the user. In the case of splitting a commit, when returning control to Git with `git rebase --continue`, you would want to have two commits in place of one.

The problem of splitting a commit is comparable to the problem of having different changes tangled together in the working directory from *Chapter 2, Developing With Git* (the section about the interactive commit), and *Chapter 3, Managing Your Worktrees*. The difference is that in the case of splitting a commit with the interactive rebase, when the rebase stops for editing, the commit is already created and copied from the branch being rebased. This is simple to fix with `git reset HEAD^`; as described in *Chapter 3, Managing Your Worktrees*, this command will keep the working area at the (entangled) state of the commit to be split while moving the `HEAD` pointer and the staging area for the commit to the state before this revision. Then you can interactively add those changes that you want to have in the first commit to the index, by composing the intermediate step in the staging area. Next, you should check whether you have what you want in the index, then create a commit from it using `git commit` without the `-a` or `--all` option. Repeat these last two steps as often as necessary.

For the last commit in the series (which would be the second one, if you are splitting the commit in two), you can do one of two things. The first option is to add everything to the index, making the working copy clean, and create a commit from the index. The other option is to create a commit from the state of the working area (`git commit --all`). If you want to keep or start from the commit message of the original commit to be split, you can provide it with the `--reuse-message=<commit>` or the `--reedit-message=<commit>` option while creating a commit. I think the simplest way of naming a commit that was split (or that is being split) is to use reflog – it will be the `HEAD@{n}` entry just before `reset: moving to HEAD^` in the `git reflog` output.

Instead of crafting the commit in the staging area (the index) starting from the parent of the commit to be split and adding changes, perhaps interactively, you could start directly from the final state— the commit to be split—and remove the changes intended for the second step. This can be done, for example, with `git reset --patch HEAD^`. Frankly, you can use any combination of techniques from *Chapter 3, Managing Your Worktrees*. I find graphical commit tools such as `git gui` quite useful for this purpose (you can find out about graphical commit tools, including some examples, in *Chapter 13, Customizing and Extending Git*).

If you are not absolutely sure that the intermediate revisions you are creating in the index are consistent (they compile, pass the test suite, and so on), you should use `git stash save --keep-index` to stash away the not-yet-committed changes, bringing the working area to the state composed in the index. You can then test the changes and amend the staging area if fixes are necessary.

Alternatively, you can create the commit from the index and use a plain `git stash` command to save the state of the working area after each commit. You can then test and amend the created intermediate commit if fixes are necessary. In both cases, you need to restore the changes with `git stash pop` before working on a new commit in the split.

Testing each rebased commit

A good software development practice is to test each change before committing it. However, this practice is not always followed. Let's assume that you forgot to test some commit or skipped it because the change seemed trivial and you were pressed for time. The interactive rebase allows you to **execute** tests (to be precise: any command) during the rebase process with the `exec` (or `x`) action. It is run between steps of rebasing commits. The `exec` command itself is formatted in a different way from the commands described earlier in this chapter: instead of SHA-1 and a summary of a commit, you provide the command to run.

The `exec` command launches the provided command (given by the rest of the line) in a shell: the one specified in the `SHELL` environment variable, or the default shell if `SHELL` is not set. This means that you can use shell feature. For the POSIX shell, this would mean using `cd` to change directories, `>` to redirect command output, `;` and `&&` to sequence multiple commands, and so on. It is important to remember that the command to be executed is run from the root of the working tree, not from the current directory (i.e., not from the subdirectory you were in when starting the interactive rebase).

If you are strict about not publishing untested changes, you might have worried about the fact that rewritten commits rebased on top of the new changes might not pass tests, even if the original commits did. You can, however, make the interactive rebase test each commit with the `--exec` option. Here is an example:

```
$ git rebase --interactive --exec "make test"
```

This would modify the starting instruction sheet, inserting `exec make test` after each entry:

```
pick 89579c9 first commit in a branch
exec make test
pick d996b71 second commit in a branch
exec make test
pick 6c89dee third commit in a branch
exec make test
```

External tools – patching management interfaces

You might prefer to fix the old commit immediately at the time when you notice the bug, instead of postponing it until the time when the branch is rebased. The latter is usually done just before the branch is sent for review (to publish it). This might be quite some time after realizing the need to edit the past commit.

Git itself doesn't make it easy to fix the found bug straight away, or at least, not with built-in tools. You can, however, find third-party external tools that implement the patch management interface on the top of Git. Examples of such tools include **Stacked Git (StGit)** and **Git Quilt (Guilt)** – the latter is unmaintained, but still usable.

These tools provide similar functionality to **Quilt** (that is, pushing or popping patches to and from a stack). With such tools, you have a set of work-in-progress "floating" patches in the Quilt-like stack. You also have accepted changes in the form of proper Git commits. You can convert between **patch** and commit and vice-versa, move and edit patches around, move and edit commits (which is done by turning the commit and its children into patches, reordering or editing patches, and then turning patches back into commits again), squash patches, and so on.

This is, however, an additional tool to install, an additional set of operations to learn (even if they make your work easier), and an additional set of complications coming from the boundary between Git and the tool in question. An interactive rebase is powerful enough nowadays, and, with autosquash, the need for another layer on top of Git is lessened.

Rewriting project history with Git filter-repo

In some use cases, you might need to use a more powerful tool for rewriting and cleaning up history than the interactive rebase. You might want something that would rewrite full history non-interactively when given some specified algorithm for doing the rewrite. Such a situation is a task for the `git filter-repo` command.

This is an external project that needs to be installed in addition to Git. However, as it is a single-file Python script, installing it is trivial in most cases. It is now recommended by the Git project to use `git filter-repo` project instead of the built-in `git filter-branch` command (which is now deprecated).

The calling convention of this command is rather different than the convention for the interactive rebase. By default, it operates on the whole history of the project, changing the full graph of revisions, though you can limit the operation to a selected branch or set of branches with the `--refs` option.

This command rewrites the Git revision history by applying custom filters on each revision to be rewritten. That's another difference: rebase works by reapplying changesets, while `filter-branch` works with snapshots. One of the consequences of this is that for `git filter-repo`, a merge is simply a kind of a commit object, while the rebase drops merges and puts commits into a line, at least unless you use the `--rebase-merges` option.

Of course, with `git filter-repo`, you describe how to do the rewrite with appropriate options instead of doing the rewriting interactively. This means that the speed of the operation is not limited by the speed of user interaction but by I/O.

> **Safety check**
>
> Since `git filter-repo` is usually used for massive rewrites and does irreversible rewriting of the project's history, it needs to be run from the fresh clone. This means that the user would always have a good backup in the form of a separate clone. If anything goes wrong, you can simply delete your clone and restart.
>
> You can make `git filter-repo` ignore the fresh clone check with the `--force` option.

Running filter-repo without filters

If you specify no filters, `filter-repo` will error out unless you specify `--force`. In this case, the commits will be recommitted without any changes. Such usage would normally have no effect, but it is permitted to allow you to compensate for some Git bugs in the future.

This means that `git filter-repo --force`, without other options, can be used to make effects implemented by replacement refs permanent. This way, you can use the following technique: use `git replace` on specified commits to alter history, ensure that it looks correct, and then make the modification permanent. This is the simplest way to do commit parent rewriting.

> **Important Note**
>
> The `git filter-repo` command command respects **replacements** (refs in the `refs/replace/` namespace). Replacements is a technique to affect the history (or rather, a view of it) without rewriting any revisions. It will be explained later in the *Replacements mechanism* section.

Available filter types for filter-repo

There is a large set of different filtering options to specify how to rewrite history. You can specify more than one option; they are applied in the order in which they are presented.

You can run the command multiple times to achieve your desired results. The `--analyze` option can be used to analyze repository history, creating a directory of reports, which (among other things) mention renames and list object sizes. This information may be useful in choosing how to filter your repo and to verify the changes.

The `git filter-repo` command supports the following types of filters:

- Filtering based on paths, which specifies the paths to select or exclude. Note that renames are not followed, so you may need to specify both the old and new names of the path.

- Renaming paths, which may be combined with path filtering.

- Content editing filters, which involve replacing text in a project's files, removing large blobs (files), or removing specified blobs (versions of file contents).

- Filtering commit messages with special support for filtering author names and emails with the help of `.mailmap` or a mailmap-like file.

- Renaming tags, which involves replacing one tag prefix with another.

For flexibility, `filter-repo` also allows you to specify functions in Python to further filter all changes using custom API, with various `--<something>-callback` options, such as (for example) `--filename-callback` or `--commit-callback`.

You can also configure how commits are rewritten and pruned. For example, you can decide whether to re-encode commit message into UTF-8, or whether to prune commits that have become empty (that is, ones that bring no changes to the project).

Examples of using filter-repo

Let's assume that you committed the wrong file to the repository by mistake, and you want to **remove that file from history**. Perhaps this was a site-specific configuration file with passwords or their equivalent. Or perhaps, during git add ., you included a generated file that was not properly ignored (such as perhaps a large binary file). Alternatively, it might have turned out that you don't have the distribution rights to a file and you need to have it removed to avoid copyright violations. Using git rm --cached would only remove it from future commits. You can also quite easily remove the file from the latest version by amending the commit (as described earlier in this chapter).

Let's assume that the file is called passwords.txt. To excise it from the entire history, you can use the following command:

```
$ git filter-repo --path 'passwords.txt' --invert-paths
```

If you want to delete all .DS_Store files in any directory (and not only from the top directory of the project), you can use one of two commands. Here is the first option:

```
$ git filter-repo --invert-paths --path '.DS_Store' --use-base-name
```

You can also use the following option:

```
$ git filter-repo --invert-paths --path-glob '*/.DS_Store' --path
'.DS_Store'
```

You can use filter-repo to **remove all API keys or passwords** from the history by specifying the text to replace in files. You can match literal text, shell globs, or regular expressions, and specify the replacement (it will be ***REMOVED*** if the replacement is not specified). For example, to remove accidentally committed GitHub Personal Access Tokens, you can use the file specifying the list of expressions, one per line. Let's say that you create an expressions.txt file with the following contents:

```
regex:ghp_ua[A-Za-z0-9]{20,}==><access_token>
```

Then you need to run the following command:

```
$ git filter-repo --replace-text expressions.txt
```

You can use filter-repo to permanently **join two repositories**, connecting histories. You can also use it to **split history** in two. The simplest solution is to use replacements, check that the joined or split history has rendered correctly, and then make the replacements permanent. For example, to split history and remove everything later than v1.0 tag, you can use the following:

```
$ git replace --graft v1.0^{commit}
$ git filter-repo --force
```

Another common case is to **fix erroneous names or email addresses in commits**. Perhaps you forgot to run `git config` to set your name and email address before you started working and Git guessed them incorrectly (if it couldn't guess, it would ask before allowing a commit). Maybe you want to open the sources of a formerly proprietary closed source program and need to change your internal corporate email to your personal address. We'll say that you want the change to be permanent instead of relying on the `.mailmap` file.

In any case, you can change email addresses in a whole history with `filter-repo`:

```
$ git filter-repo --use-mailmap
```

If you are open-sourcing a project, you could also want to add the `Signed-off-by:` lines for the Digital Certificate of Origin (see *Chapter 15, Git Best Practices*), and add the trailer to the commit message if one is not already present:

```
$ git filter-repo --message-callback '
  if b"Signed-off-by:" not in message:
    message += "\n\nSigned-off-by: Joe Hacker <joe@h.com>"
  return message
```

Suppose that you have noticed a typo in the name of a subdirectory, such as `inlude/` instead of `include/`. This can be fixed simply by running the following:

```
$ git filter-repo --path-rename inlude/:include/
```

Often, some part of a larger project will take on a life on its own. In those instances, it begins to make sense to separate the part from the project it started in. We would want to extract the history of this part to make its **subdirectory the new root**. To rewrite history in this way and discard all other history, you can run the following:

```
$ git filter-repo --subdirectory-filter lib/foo
```

However, perhaps a better solution would be to use a specialized third-party tool, namely `git subtree`. This tool (and its alternatives) will be discussed in *Chapter 11, Managing Subprojects*.

External tools for large-scale history rewriting

The `git filter-repo` project is not the only solution for a large-scale rewriting of the project's history. There are other tools that are more specialized, perhaps including lots of predefined clean-up operations or providing some level of interactivity with the ability for scripted rewrites (with a **Read–Evaluate–Print Loop** (**REPL**), similar to interactive shells in some interpreted programming languages).

Removing files from the history with the BFG Repo Cleaner

The BFG Repo Cleaner is a specialized alternative to using `git filter-repo`. It is specialized for the purpose of cleaning bad data out of your Git repository history by removing files and directories and replacing text in files (for example, accidentally committed passwords or API keys with their placeholders). It can use multiple cores with parallel processing – BFG is written in Scala and uses JGit as a Git implementation.

BFG provides a set of command-line parameters that are specialized for removing files and fixing them, such as `--delete-files` or `--replace-text`, a "query language" of sorts. It lacks the flexibility of other tools. Nowadays, `filter-repo` can do everything it can. There is even **bfg-ish**, a reimplementation of BFG based on `filter-repo`.

One issue you need to remember is that BFG assumes that you have fixed the contents of your current commit.

Editing the repository history with reposurgeon

The **reposurgeon** project was originally created to help clean up artifacts created by the repository conversion (migrating from one version control system to another). It relies on being able to parse, modify, and emit the command stream in the `git fast-import` format, which is a common export and import format among source control systems nowadays thanks to it being version control-agnostic. The `git filter-repo` tool, which was described earlier in this chapter, is also based on processing fast-import streams.

It can be used for history rewriting, including editing past commits and metadata, excising commits, squashing (coalescing) and splitting commits, removing files and directories from history, and splitting and joining history.

The advantage that `reposurgeon` has over `git filter-repo` is that it can be run in two modes: either as an interactive interpreter, a kind of debugger or editor for history, with command history and tab completion; or in a batch mode to execute commands given as arguments. This allows users to interactively inspect history and test changes, and then batch run them for all the revisions.

The disadvantage is in having to install and then learn to use a separate tool.

The perils of rewriting published history

There is, however, a very important principle to know about: you should never (or at least not without a very, very good reason) rewrite *published* history, especially when it comes to those commits that were pushed to the public repository or were otherwise made public. What you can do is change the parts of the graph of revisions that are private.

The reason behind this rule is that rewriting published history could cause trouble for downstream developers if they have based their changes on revisions that were rewritten.

This means that it is safe to rewrite and rebuild those public branches that are explicitly stated and documented to be in flux, for example, as a way of showing work in progress (such as 'proposed-updates' type of branch, that is used to test merge all feature branches – see the *Visibility without integration* and *Progressive-stability branches* sections in *Chapter 8, Advanced Branching Techniques*). Another possibility for the safe rewriting of a public branch is to do it at specific stages of the project's life, namely after creating a new release; again, this needs to be documented.

The consequences of upstream rewrites

Now you will see, in a simple example, the perils of rewriting published history (for example, rebasing) and how it causes trouble. Let's assume that there are two public branches that are of interest: master and subsys. The latter is based on (forked from) the former. Let's also assume that a downstream developer (who might be you) created a new topic branch based on the subsys branch for their own work, but did not publish it yet; it is only present in their local repository. This situation is shown in *Figure 10.2* (the revisions below the dashed lines, denoted by darker color, are present only in the local repository of the downstream developer).

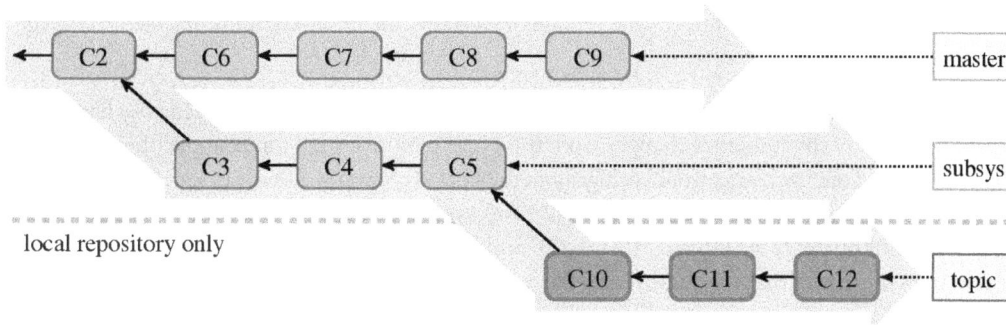

Figure 10.2 – The state of the local repository of a downstream developer before the rewrite of the published history, with the new local work that was put on a topic branch

Then, the upstream developer rewrites the subsys branch to start from the current (topmost) revision in the master branch. This operation is called rebase and was described in *Chapter 9, Merging Changes Together* (the previous chapter). Let's assume that during the rewrite, one of the commits was dropped; perhaps the same change was already present in master and was skipped, perhaps it was dropped for some other reason, or perhaps it was or squashed into the previous commit with the interactive rebase. The public repository now looks as follows:

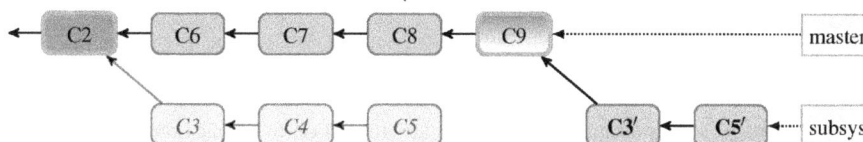

Figure 10.3 – The state of a public upstream repository after rewrite, with an emphasized old base of the rebased branch, plus a new base and rewritten commits (after the rebase)

Note that in the default configuration, Git would refuse to push rewritten history (it would deny a non-fast-forward push). You would need to force the push.

The problem is with merging changes based on the pre-rewrite versions of revisions, such as the `topic` branch in this example.

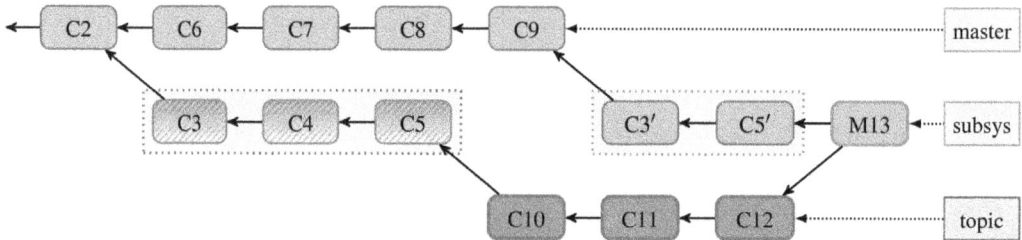

Figure 10.4 – The situation after merging the changes that were based
on pre-rewrite revisions into post-rewrite branches

Notice that the merge brings the pre-rewrite version of revisions, including commits that were dropped during the rebase.

If neither the downstream developer nor the upstream one notices that the published history has been rewritten, and one of them merges changes from the `topic` branch into, for example, the `subsys` branch it was based on, the merge would bring about duplicated commits. As we can see in the example in *Figure 10.3*, after such a merge (denoted by **M13** here), we have both the **C3**, **C4**, and **C5** pre-rewrite commits brought by the topic branch and the **C3'** and **C5'** post-rewrite commits (see *Figure 10.4*). Note that the **C4** commit that was removed in the rewrite is back – it might have been a security bug!

Recovering from an upstream history rewrite

However, what can we do if the upstream has rewritten the published history (for example, rebased it)? Can we avoid bringing the abandoned commits back and merging a duplicate or near-duplicate of the rewritten revisions? After all, if the rewrite is published, changing it would be yet another rewrite.

The solution is to rebase your work to fit with the new version from the upstream, moving it from the pre-rewrite upstream revisions to the post-rewrite ones.

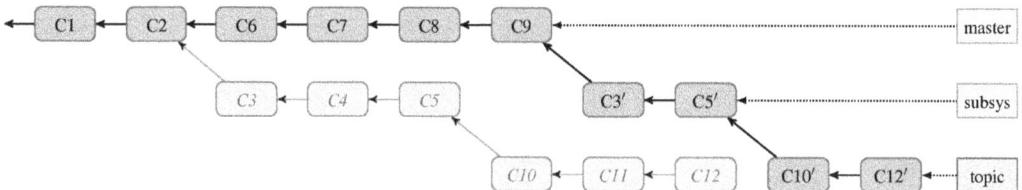

Figure 10.5 – The situation after a downstream rebase of a topic branch

In the case of our example, it would mean rebasing the `topic` branch onto a new (post-rewrite) version of `subsys`, as shown in *Figure 10.5*.

> **Tip**
> You might not have a local copy of the `subsys` branch; in this case, do substitute `subsys` with the respective remote-tracking branch, for example, `origin/subsys`.

Depending on whether the `topic` branch is public or not, it might mean that you are now breaking the promise of unaltered public history for your downstream. Recovering from an upstream rewrite might then result in a ripple of rebases following the rewrite down the river of downstreams (dependent repositories).

An easy case is when `subsys` is simply rebased and the changes remain the same (which means that **C4** vanished because **C6-C9** included it). Then you can simply rebase `topic` on top of its upstream, that is, `subsys`, with the following:

```
$ git rebase subsys topic
```

The `topic` part is not necessary if you are currently on it (if `topic` is the current branch). This rebases everything: the old version of `subsys` and your commits in `topic`. This solution, however, relies on the fact that `git rebase` would skip repeated commits (removing **C3**, **C4**, and **C5**, leaving only **C10'** and **C12'**). It might be better and less error-prone to assume the more difficult case.

The hard case is when rewriting `subsys` involved some changes and was not a pure rebase, or when an interactive rebase was used. In this case, it is better to explicitly move just your changes, namely `subsys@{1}..topic` (assuming that the `subsys@{1}` entry in the `subsys` reflog comes from before rewrite), stating that they are moved on top of the new `subsys`. This can be done with the `--onto` option:

```
$ git rebase --onto subsys subsys@{1} topic
```

You can make Git use the reflog to find a better common ancestor with the `--fork-point` option with the `git rebase` command, such as in the following example:

```
$ git rebase --fork-point subsys topic
```

The rebase would then move the changes to `topic`, starting with the result of the `git merge-base --fork-point subsys topic` command. However, if the reflog of the `subsys` branch does not contain necessary information, Git would fall back upstream, here to `subsys`.

> **Important note**
> You can use the interactive rebase instead of an ordinary rebase like in the narration mentioned earlier for better control at the cost of more work (for example, to drop commits that are already present, but are not detected by the rebase machinery as such).

Amending history without rewriting

What should you do if what you need to fix is in the published part of the history? As described in *The perils of rewriting published history*, changing those parts of the history that were made public can cause problems for downstream developers. You had better not touch this part of the graph of revisions.

There are a few solutions to this problem. The most commonly used one is to put in a new fixup commit with appropriate changes (for example, a typo fix in documentation). If what you need is to remove the changes, deciding that they turned out to be bad to have in the history, you can create a commit to revert the changes.

If you fix a commit or revert one, it would be nice to annotate that commit with the information that it was buggy, as well as which commit fixed (or reverted) it. Even though you cannot (and should not) edit the fixed commit to add this information if the commit is public, Git provides the **notes** mechanism to append extra information to existing commits, which is a bit like publishing an addendum, errata, or amendment. However, remember that notes are not published by default; nonetheless, it is easy to publish them (you just need to remember to do it).

Reverting a commit

If you need to back out an existing commit, undoing the changes it brought, you can use `git revert`. As described in *Chapter 9, Merging Changes Together* (see, for example, *Figure 9.5* in that chapter), the `revert` operation creates a commit with the reverse of any changes. For example, where the original adds a line, reversion removes it; where the original commit removes the line, reversion adds it.

> Trivia
>
> Note that different version control systems use the name revert for different operations. In particular, it is often used to mean resetting the changes to a file back to the latest committed version, throwing away uncommitted changes. It is something that `git reset -- <file>` does in Git.

This is best shown in an example. Let's assume that the last commit on the `multiple` branch has the following summary of its changes:

```
$ git show --stat multiple
commit bb71a804f9686c4bada861b3fcd3cfb5600d2a47
Author: Alice Developer <alice@company.com>
Date:   Sun Jun 1 03:02:09 2014 +0200

    Support optional <count> parameter

 src/rand.c | 26 +++++++++++++++++++++-----
 1 file changed, 21 insertions(+), 5 deletions(-)
```

Reverting this commit (which requires a clean working directory) would create a new revision. This revision undoes the changes that the reverted commit brought:

```
$ git revert bb71a80
[master 76d9e25] Revert "Support optional <count> parameter"
 1 file changed, 5 insertions(+), 21 deletions(-)
```

Git would ask for a commit message, which should explain why you reverted the given revision: how it was faulty, and why it needed to be reverted rather than fixed. The default is to give the SHA-1 of the reverted commit:

```
$ git show --stat
commit 76d9e259db23d67982c50ec3e6f371db3ec9efc2
Author: Alice Developer <alice@example.com>
Date:   Tue Jun 16 02:33:54 2015 +0200

    Revert "Support optional <count> parameter"

    This reverts commit bb71a804f9686c4bada861b3fcd3cfb5600d2a47.

 src/rand.c | 26 +++++----------------------
 1 file changed, 5 insertions(+), 21 deletions(-)
```

Compare the summary of changes for the commit and its revert. In the preceding example, the commit has 21 insertions and 5 deletions, while the revert has 5 insertions and 21 deletions (where line that changed from one version to the other counts as deletion of the old version and insertion of the new).

A common practice is to leave the subject alone (which allows you to easily find reverts) but replace the content with a description of the reasoning behind the revert.

Reverting a faulty merge

Sometimes, you might need to undo the effect of a merge. Suppose that you have merged changes, but it turned out that they were merged prematurely and that the merge brings regressions.

Let's say that the branch that was merged is named `topic`, and that you were merging it into the `master` branch. This situation is shown in *Figure 10.6*.

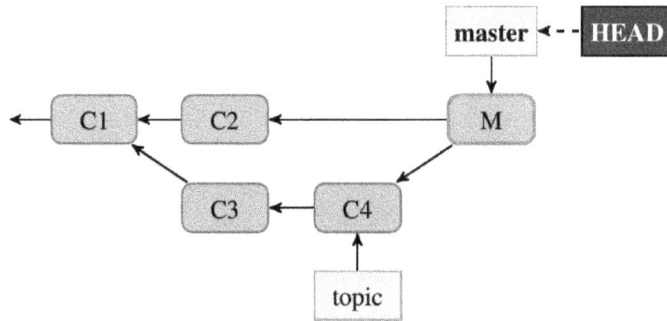

Figure 10.6 – An accidental or premature merge commit, a starting
point for reverting merges and redoing reverted merges.

If you didn't publish this merge commit before you noticed the mistake, and the unwanted merge exists only in your local repository, the easiest solution is to drop this commit with `git reset --hard` HEAD^ (see *Chapter 3, Managing Your Worktrees*, for an explanation of the hard mode of `git reset`).

What do you do if you realize only later that the merge was incorrect, for example after one more commit was created on the `master` branch and published? One possibility is to revert the merge.

However, a merge commit has more than one parent, which means more than one delta (or, more than one changeset). To run `revert` on a merge commit, you need to specify which patch you are reverting, or, in other words, which parent is the mainline. In this particular scenario, assuming that there was one more commit after the merge (and that the merge was two commits back in the history), the command to revert the merge would look like this:

```
$ git revert -m 1 HEAD^^
[master b2d820c] Revert "Merge branch 'topic'"
```

The situation after reverting a merge is shown in *Figure 10.7*.

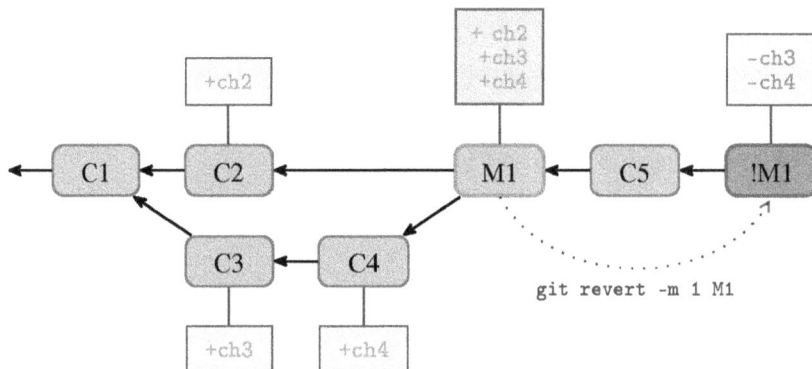

Figure 10.7 The history from the previous figure after reverting merge; the square boxes
attached to selected commits symbolize their changesets in a diff-like format

Starting with the new **!M1** commit (the **!M1** symbol is used to represent negation or reversal of the **M1** commit), it's as if the merge never happened, at least with regards to the changes.

Recovering from a reverted merge

Let's assume that you continued work on a branch whose merge was reverted. Perhaps it was prematurely merged, but it doesn't mean that the development on it has stopped. If you continue to work on the same branch, perhaps by creating commits with fixes, they will get ready in some time, and then you will need to be able to merge them correctly into the mainline again. Or perhaps the mainline will mature enough to be able to accept a merge. Trouble lies ahead if you simply try to merge your branch again the same way as last time.

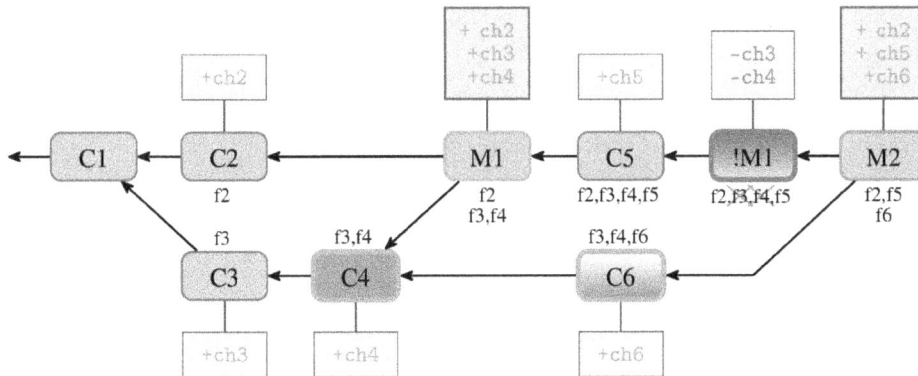

Figure 10.8 – The unexpectedly erroneous result of trying to simply redo a reverted merge

The unexpected result, as shown in *Figure 10.8*, is that Git has only brought the changes since the reverted merge. The changes brought by the commits on a side branch, whose merge got reverted, are not here. In other words, you would get a strange result: the new merge would not include the changes that were created on your branch (the side branch) before the merge that got reverted.

This is caused by the fact that `git revert` undoes changes (the data), but does not undo the history (the DAG of revisions). This means that the new merge sees **C4**, the commit on the side branch just before the reverted merge, as a common ancestor. Since the default three-way merge strategy looks only at the state of the *ours*, *theirs*, and *base* snapshots, it doesn't search through the history to find that there was a revert there. It sees that both the common ancestor **C4** and the merged branch (that is, *theirs*) **C6** do include features brought by the **C3** and **C4** commits, namely **f3** and **f4**, while the branch that we are merging into (that is, *ours*) doesn't have them because of the revert.

For the merge strategy, it looks exactly like the case where one branch deleted something, which means that this change (the removal) is the result of the merge (it looks like the case where there was change on only one side). In particular, it looks like the base and the side branch have the feature, but the current branch doesn't (because of the revert) – so the result doesn't have it either. You can find an explanation of the merging mechanism in *Chapter 9, Merging Changes Together*.

There is more than one option to fix this issue and make Git re-merge the `topic` branch correctly, which means including the **f3** and **f4** features in the result. The option that you should choose depends on the exact circumstances, for example, whether the branch being merged is published or not. You don't usually publish topic branches, and if you do, such as perhaps in the form of the `proposed-updates` branch with all the topic branches merged in, it is with the understanding that they can and probably will be rewritten.

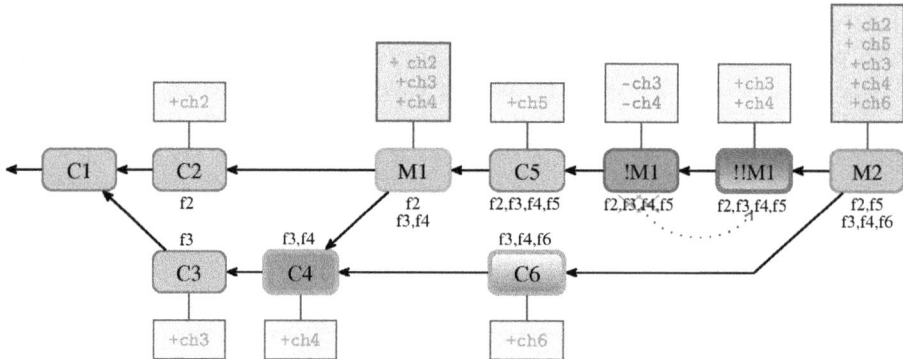

Figure 10.9 The history after re-merging (as M2) a reverted M1 merge, with revering revert !!M1 (replay)

One option is to bring back deleted changes by reverting the revert. The result is shown in *Figure 10.9*. In this case, you have brought changes to match the recorded history.

Another option would be to change the view of the history (perhaps temporarily), for example, by amending it with `git replace`, or by changing the **!M1** merge to a non-merge commit (this will be described later in the chapter). Both of those options are suitable in the situation where at least the parts of the branch being merged, namely `topic`, were published.

If the problem was some bugs in the commits being merged (on the `topic` branch), and the branch being merged was not published, you can fix these commits with the interactive rebase, as described earlier. Rebasing changes the history anyway. Therefore, if you additionally ensure that the new history you are creating with the rebase does not have any revision in common with the old history that includes the failed and reverted merge, re-merging the topic branch would pose no challenges.

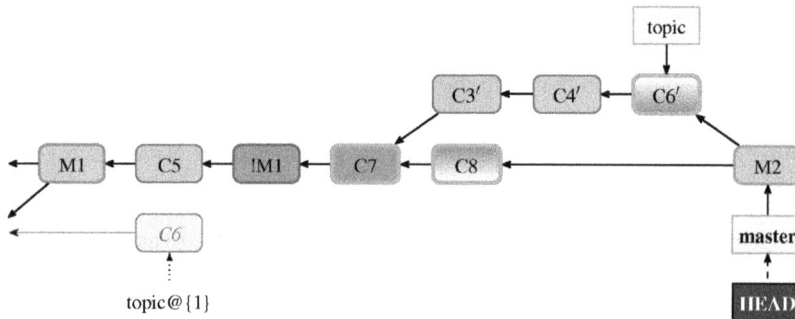

Figure 10.10 – The history after re-merging the rebased branch, which had its merge reverted

Usually you would rebase a topic branch, `topic` in this case, on top of the current state of the branch it was forked from, which is the `master` branch here. This way, your changes are kept up to date with the current work, which makes a later merge easier. Now that the `topic` branch has a new history, merging it into `master` "again", like in *Figure 10.10*, is easy and doesn't give us any surprises or trouble.

A more difficult case would be if the `topic` branch is for some reason required to keep its base (such as being able to merge it into the `maint` branch too). This is not more difficult in the sense that there would be problems with re-merging the `topic` branch after the rebase, but in that we need to ensure that the branch doesn't share history with the reverted merge arc after the rebase. The goal is to have history in the same shape as is shown in *Figure 10*. By default, a rebase tries to fast-forward revisions if they didn't change (for example, leaving **C3** in place if the rebase didn't modify it), so we need to use -f or --force-rebase to force rebasing of unchanged skippable commits (or of --no-ff, which is equivalent) as well. The result is shown in *Figure 10.11*.

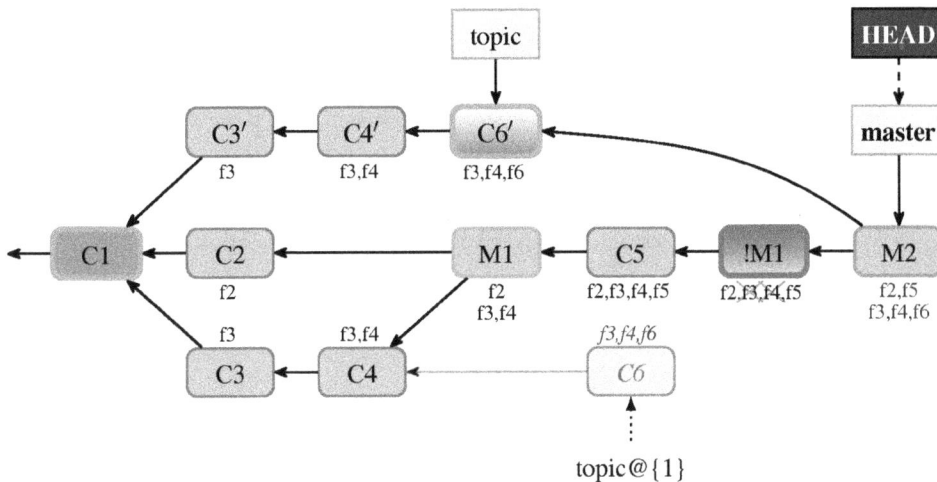

Figure 10.11 The history after re-merging an in-place-rebased topic
branch, where a pre-rebase merge was reverted

So, you should not be blindly reverting the revert of a merge. What to do with the problem of re-merging after a reverted merge depends on how you want to handle the branch being merged. If the branch is being rewritten (for example, using an interactive rebase), then reverting the revert would be an actively wrong thing to do because you could bring back errors that were fixed in the rewrite.

Storing additional information with notes

The notes mechanism is a way to store additional information for an object, usually a commit, without touching the object itself. You can think of it as an attachment, or an appendix, that is "stapled" to an object. Each note belongs to some category of notes so that notes used for different purposes can be kept separate.

Adding notes to a commit

Sometimes you want to add extra information to a commit, particularly information that is available only some time after its creation. It might be, for example, a note that there was a bug found in the commit, and perhaps even that it was fixed in some specified future commit (in case of regression). Perhaps we realized after the commit got published that we forgot to add some important information to the commit message, for example, to explain why it was done. Or maybe we realized after the fact that there is another way of doing it and we want to create a note to ensure that we do not forget about it, and for other developers to share the idea.

Since history is immutable in Git, you cannot do this without rewriting the history (creating a modified copy and forgetting the old version of the history). The immutability of history is important; it allows people to sign revisions and trust that, once inspected, history cannot change. What you can do instead is add the extra message as a note.

Let's assume that codevelopers have switched from `atoi()` to `strtol()` because the former is deprecated. The change was made public since then. However, the commit message didn't include an explanation of why it was deprecated and why it is worth it to switch, even if the code after the change is longer. Let's add the information as a note:

```
$ git notes add \
  -m 'atoi() invokes undefined behaviour upon error' v0.2~3
```

We have added the note directly from the command line without invoking the editor by using the `-m` flag (the same flag as for `git commit`) to simplify the explanation of this example. The note will be visible when running `git log` or `git show`:

```
$ git show --no-patch v0.2~3
commit 8c4ceca59d7402fb24a672c624b7ad816cf04e08
Author: Bob Hacker <bob@company.com>
Date:   Sun Jun 1 01:46:19 2014 +0200

    Use strtol(), atoi() is deprecated

Notes:
    atoi() invokes undefined behaviour upon error
```

As you can see from the preceding output, our note is shown after the commit message in the `Notes:` section. Displaying notes can be disabled with the `--no-notes` option, and (re)enabled with `--show-notes`.

How notes are stored

In Git, notes are stored using extra references in the `refs/notes/` namespace. By default, commit notes are stored using the `refs/notes/commits` ref. This can be changed using the `core.notesRef` configuration variable, which can in turn be overridden with the `GIT_NOTES_REF` environment variable.

If the given ref does not exist, it is not an error, but it means that no notes should be printed. These variables decide both which type of notes are displayed with the commit after the Notes: line and where to write the note created with git notes add.

You can see that the new type of has reference appeared in the repository:

```
$ git show-ref --abbrev commits
fcac4a6 refs/notes/commits
```

If you examine the new reference, you will see that each note is stored in a file named after the SHA-1 identifier of the annotated object. This means that you can have only one note of the given type for one object. You can always edit the note, append to it (with git notes append), or replace its content (with git notes add --force).

In interactive mode, Git opens the editor with the contents of the note, so edit, append, and replace operations work almost the same interactively. As opposed to commits, notes are mutable, or to be more exact, only the latest version of each note is used:

```
$ git show refs/notes/commits
commit fcac4a649d2458ba8417a6bbb845da4000bbfa10
Author: Alice Developer <alice@example.com>
Date:   Tue Jun 16 19:48:37 2015 +0200

    Notes added by 'git notes add'

diff --git a/8c4ceca59d7402fb24a672c624b7ad816cf04e08
b/8c4ceca59d7402fb24a672c624b7ad816cf04e08
new file mode 100644
index 0000000..a033550
--- /dev/null
+++ b/8c4ceca59d7402fb24a672c624b7ad816cf04e08
@@ -0,0 +1 @@
+atoi() invokes undefined behaviour upon error
$ git log -1 --oneline \
   8c4ceca59d7402fb24a672c624b7ad816cf04e08
8c4ceca Use strtol(), atoi() is deprecated
```

Notes for commits are stored in a separate line of (meta) history, but this need not be the case for the other categories of notes. The notes reference can point directly to the tree object instead of the commit object such as for refs/notes/commits.

One important issue that is often overlooked in books and articles is that it is the full path to a file with the note's contents, not the base name of the file, that identifies the object that the note is attached to. If there are many notes, Git can and will use a fan-out directory hierarchy, for example, storing the preceding note at the 8c/4c/eca59d7402fb24a672c624b7ad816cf04e08 path (note the slashes).

Other categories and uses of notes

Notes are usually added to commits. However, even for those notes that are attached to commits, it makes sense, at least in some cases, to store different pieces of information using different categories of notes. This makes it possible to decide which parts of information to display on an individual basis, and which parts to push to the public repository. It also allows us to query for specific parts of information individually.

To create a note in a namespace (category) that is different from the default one (where the default means `notes/commits`, or the value of the configuration variable `core.notesRef` if it is set), you need to specify the category of notes while adding it:

```
$ git notes --ref=issues add -m '#2' v0.2~3
```

Now, by default, Git will only display the `core.notesRef` category of notes after the commit message. To include other types of notes, you must either select the category to display with `git log --notes=<category>` (where `<category>` is either the unqualified or qualified reference name, or a glob; you can therefore use `--notes=*` to show all categories) or configure which notes to display in addition to the default with the `display.notesRef` configuration variable (or the `GIT_NOTES_DISPLAY_REF` environment variable). You can either specify the configuration variable value multiple times, just like for `remote.<remote-name>.push` (or specify a colon-separated list of pathnames if you are using the environment variable), or you can specify a globing pattern:

```
$ git config notes.displayRef 'refs/notes/*'
$ git log -1 v0.2~3
commit 8c4ceca59d7402fb24a672c624b7ad816cf04e08
Author: Bob Hacker <bob@company.com>
Date:   Sun Jun 1 01:46:19 2014 +0200

    Use strtol(), atoi() is deprecated

Notes:
    atoi() invokes undefined behaviour upon error

Notes (issues):
    #2
```

There are many possible uses of notes. You can, for example, use notes to reliably mark which patches (which commits) were **upstreamed** (**forward-ported** to the development branch) or **downstreamed** (**back-ported** to the more stable branch or the stable repository), even if the upstreamed or downstreamed version is not identical, and mark a patch as being **deferred** if it is not ready for either upstream or downstream.

If you require manual input, this is a bit more reliable than relying on the `git patch-id` mechanism to detect when the changeset is already present (which you can do by rebasing, by using `git cherry-pick`, or with the `--cherry`, `--cherry-pick`, or `--cherry-mark` option of `git log`). This is, of course, in case we are not using topic branches from the start, but rather cherry-picking commits.

Notes can also be used to store the results of the post-commit (but pre-merge) **code audit** and to notify other developers of the reason(s) why this version of the patch was used.

Notes can also be used to handle **marking bugs and bug fixes**, as well as **verifying** fixes. You will often find bugs in commits long after they get published; that is why you need notes for this purpose. If you find a bug before publishing, you would rewrite the buggy commit instead.

In this case, when the bug gets reported, and if it was a regression, you would first find which revision introduced the bug (for example, with `git bisect`, as described in *Chapter 4*, *Exploring Project History*). Then you would want to mark this commit, putting the identifier of a bug entry in the issue tracker for the project (which is usually a number, or number preceded by some specific prefix such as **Bug:1385**) in the `bugs`, `defects`, or `issues` category of notes. Perhaps you would also want to include the description of a bug. If the bug affects security, it might be assigned a vulnerability identifier, for example, a **Common Vulnerabilities and Exposures** (**CVE**) number; this information could be put into the note in the `CVE-IDs` category.

Then, after some time, the bug will hopefully get fixed. Just like we marked the commit with the information that it contains the bug, we can additionally annotate it with the information on which commit fixes it, such as in a note in the `fixes` category. Unfortunately, it might happen that the first attempt at fixing it doesn't handle the bug entirely correctly and you have to amend a fix, or perhaps even create a fix for a fix. If you are using bugfix or hotfix branches (topic branches for bugfixes), as described in *Chapter 8*, *Advanced Branching Techniques*, it will be easy to find and apply them together by merging the aforementioned bugfix branch. If you are not using this workflow, then it would be a good idea to use notes to annotate fixes that should be cherry-picked together with a supplementary commit, for example by adding a note in the `alsoCherryPick` or `seeAlso` category, or whatever you want to name this category of notes. Perhaps the original submitter, or a Q&A group, would also get to the fix and test that it works correctly. It would be better if the commit was tested before publishing, but it is not always possible, so `refs/notes/tests` it is.

Third-party tools use (or could use) notes to store additional **per-commit tool-specific information**. For example, **Gerrit**, which is a free web-based team code collaboration tool, stores information about code reviews in `refs/notes/reviews`. This includes the name and email address of the Gerrit user that submitted the change, the time the commit was submitted, the URL to the change review in the Gerrit instance, review labels and scores (including the identity of the reviewer), the name of project and branch, and so on:

```
Notes (review):
    Code-Review+2: John Reviewer <john@company.com>
    Verified+1: Jenkins
    Submitted-by: Bob Developer <bob@company.com>
```

```
Submitted-at: Thu, 20 Oct 2014 20:11:16 +0100
Reviewed-on: http://localhost:9080/7
Project: common/random
Branch: refs/heads/master
```

Notes as cache

Going to a more exotic example, you can use the notes mechanism to store **the result of a build** (either the archive, the installation package, or just the executable), attaching it to a commit or a tag. Theoretically, you could store a build result in a tag, but you would usually expect a tag to contain a **Pretty Good Privacy** (**PGP**) signature and perhaps also the release highlights. Also, you would, in almost all cases, want to fetch all the tags, while not everyone wants to pay for the cost of disk space for the convenience of pre-build executables. You can select whether you want to fetch the given category of notes (for example to skip pre-built binaries) or not from case to case while you autofollow tags. That is why notes are better than tags for this purpose.

Here, the trouble is to correctly generate a binary note. You can binary-safely create a note with the following trick:

```
# store binary note as a blob object in the repository
$ blob_sha=$(git hash-object -w ./a.out)
# take the given blob object as the note message
$ git notes --ref=built add --allow-empty -C "$blob_sha" HEAD
```

You cannot simply use -F ./a.out, as this is not binary-safe – comments (or rather what was misdetected as comments, that is, lines starting with #) would be stripped.

The notes mechanism is also used as a mechanism to enable storing cache for the textconv filter (see the section on gitattributes in *Chapter 3, Managing Your Worktrees*). All you need to do is configure the filter in question, setting its **cachetextconv** to true:

```
[diff "jpeg"]
    textconv = exif
    cachetextconv = true
```

Here, notes in the refs/notes/textconv/jpeg namespace (named after the filter) are used to attach the text of the conversion to a blob.

Notes and rewriting history

Notes are attached to objects they annotate (usually commits) by their SHA-1 identifier. What happens with notes when we are rewriting history then? In the new, rewritten history, SHA-1 identifiers of objects are different in most cases.

It turns out that you can configure this quite extensively. First, you can select which categories of notes should be copied along with the annotated object during the rewrite with the notes.rewriteRef multi-value configuration variable. This setting can be overridden with the GIT_NOTES_REWRITE_ REF environment variable with a colon-separated list of fully qualified notes references and globs (denoting reference patterns to match). There is no default value for this setting; you must configure this variable to enable rewriting.

Second, you can also configure whether to copy a note during rewriting depending on the exact type of the command doing the rewriting (rebase and amend are currently supported as the value of the command). This can be done with the boolean-valued notes.rewrite.<command> configuration variable.

In addition, you can decide what to do if the target commit already has a note while copying notes during a rewrite, for example, while squashing commits using an interactive rebase. You have to decide between overwrite (taking the note from the appended commit), concatenate (which is the default value), cat_sort_uniq (like concatenate, but sorting lines and removing duplicates), and ignore (using the note from the original commit being appended to) for the notes.rewriteMode configuration variable or the GIT_NOTES_REWRITE_MODE environment variable.

Publishing and retrieving notes

So, we have notes in our own local repository. What do we do if we want to share these notes? How do we make them public? How can we, and other developers, get notes from other public repositories?

We can employ our knowledge of Git here. The *How notes are stored* section explained that notes are stored in the object database of the repository using special references in the refs/notes/ namespace. The contents of the note is stored as a blob, referenced through this special ref. Commit notes (notes in refs/notes/commits) store the history of notes, though Git allows you to store notes without history as well. So, what you need to do is get this special ref. The contents of the notes will follow. This is the usual mechanism of repository synchronization (object transfer).

This means that to publish your notes, you need to configure appropriate push lines in the appropriate remote repository configuration (see *Chapter 6, Collaborative Development with Git*). Assuming that you are using a separate public remote (if you are the maintainer, you will probably simply use origin), which is perhaps set as remote.pushDefault, and that you would like to publish notes in any category, you can run the following:

```
$ git config --add remote.public.push '+refs/notes/*:refs/notes/*'
```

If push.default is set to matching (or Git is old enough to have this as the default behavior), or the push lines use special refspecs such as : or +:, it is enough to push notes refs the first time, as they would be pushed automatically each time after:

```
$ git push origin 'refs/notes/*'
```

The process of **fetching notes** is only slightly more involved. If you don't produce specified types of notes yourself, you can fetch notes in the "mirror-like" mode to the ref with the same name:

```
$ git config --add remote.origin.fetch '+refs/notes/*:refs/notes/*'
```

However, if there is a possibility of conflict, you would need to fetch notes from the remote into the remote-tracking notes reference, and then use `git notes merge` to join them into your notes. Please see the documentation for details.

> **Tip**
>
> If you want to make it easy to merge Git notes, perhaps even automatically, then following the convention of the **Key: Value** entries on separate lines for the content of notes, with duplicates removed, will help.

There is no standard naming convention for remote-tracking notes references, but you can use either `refs/notes/origin/*` (so that the shortened `commits` notes category from the `origin` remote is `origin/commits`, and so on), or go whole works and fetch `refs/*` from the `origin` remote into `refs/remotes/origin/refs/*` (so that the `commits` category lands in `refs/remotes/origin/refs/notes/commits`).

Using git replace

The original idea for the replace- or replacement-like mechanism was to make it possible to join the history of two different repositories.

The original impulse was to be able to switch from the other version control system to Git by creating two repositories: the first one for the current work, starting with the most recent version in the empty repository, and the second one for the historical data, storing the conversion from the original system. That way, it would be possible to take time doing the faithful conversion of the historical data, and even fix it if the conversion were incorrect, without affecting the current work.

What was needed was some mechanism to connect the histories of those two repositories, to have a full history for inspection going back to the creation of a project (for example, for blame, that is, line-history annotation).

The replacements mechanism

The modern incarnation of such a tool is a replace (or replacements) mechanism. With it, you can replace any object with any object, or rather create a virtual history (virtual object database of a repository) by creating an overlay, so that most Git commands return a replacement in place of the original object.

However, the original object is still there, and Git's behavior with respect to the replacement mechanism was done in such a way as to eliminate the possibility of losing data. You can get the original view with the `--no-replace-objects` option passed to the `git` wrapper, added before the command. You can also use the `GIT_NO_REPLACE_OBJECTS` environment variable, instead. For example, to view the original history, you can use `git --no-replace-objects log`.

The information about replacements is saved in the repository by storing the ref named after the SHA-1 of the replaced object in the `refs/replace/` namespace, with the SHA-1 of replacement as its sole content. However, there is no need to edit it by hand or with low-level plumbing commands – you can use the `git replace` command.

Almost all the commands use replacements unless they are told not to, as explained previously. The exception is reachability analysis commands. This means that Git would not remove the replaced objects because they are no longer reachable if we take replacement into account. Of course, replacement objects are reachable from the replace refs.

> **Important note**
> Currently, some of the mechanisms that are used to make Git faster for very large repositories (see *Chapter 12, Managing Large Repositories*) don't work if `git replace` is used.

You can replace any object with any other object, though changing the type of an object requires telling Git that you know what you are doing with `git replace -f <object> <replacement>`. This is because such a change might lead to troubles with Git, since it was expecting one type of object, and getting another.

With `git replace --edit <object>`, you can edit its contents interactively. What really happens is that Git opens the editor with the object contents, and after editing, Git creates a new object and a replacement ref. The object format (in particular, the commit object format, as one would almost always edit commits) was described at the beginning of this chapter. You can change the commit message, commit parents and authorship, and so on.

Example – joining histories with git replace

Let's assume that you have split the repository into two, as described in an earlier section about `filter-repo`, perhaps for performance reasons. However, let's say that you want to be able to treat the joined history as if it were one.

Or perhaps there was a natural history split after changing the version control system to Git, with the fresh repository with the current work (started after switching from the current state of a project, with an empty history) and the converted historical repository kept separate. This could be done to make the switch faster. This technique has the advantage of allowing you to improve the conversion after the split.

This situation is shown in *Figure 10.12*, with the historical repository added as a remote to the current work repository (one with new commits).

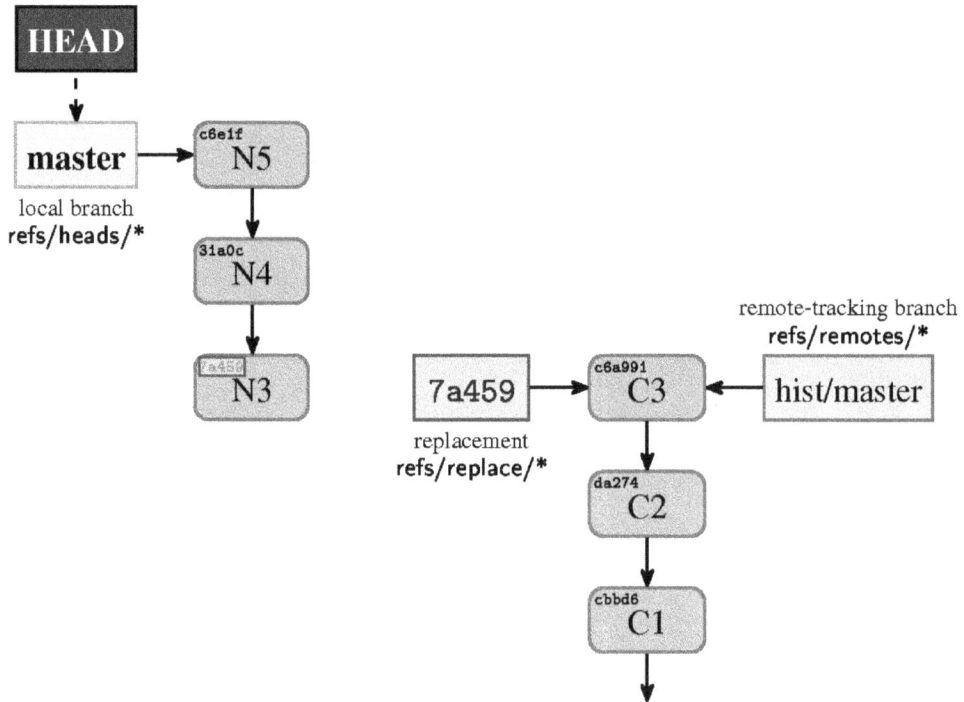

Figure 10.12 – The view of a split history, with the replacements turned off (git --no-replace-objects). The shortened SHA-1 in the left upper corner of a commit denotes its identifier.

In many cases, you might want to create a kind of informational commit on top of the "historical" repository (the one with the older part of the history), for example, adding the notification where one can find the current work repository to the README file. Such a commit is, for simplicity, not shown in *Figure 10.12*.

How to join history depends somewhat on whether the history was originally split or joined. If it was originally joined, then split, just tell Git to replace the post-split version with the pre-split version using `git replace <post-split> <pre-split>`. If the repository was split from beginning, use the `--edit` or `--graft` option of `git replace`.

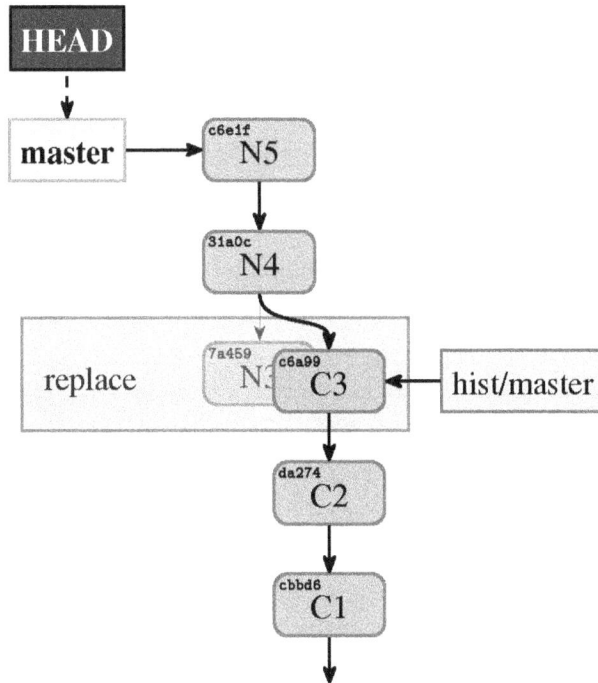

Figure 10.13 – The view of a split history, joined using replacements

The split history is there, it is just hidden from view. For all Git commands, the history looks like in *Figure 10.13*. You can, as described earlier, turn it off using replacements; in this case, you would see the history as in *Figure 10.12*.

Historical note – grafts

The first attempt to create a mechanism to make it possible to join lines of history came about in the form of **grafts**. It is a simple `.git/info/grafts` file with the SHA-1 identifier of the affected commit and its replacement parents in line, separated by spaces.

This mechanism was only for commits, and allowed only to change the parentage of the commit. There was no support for transport, that is, for propagating this information from inside of Git. You could not turn the grafts mechanism off temporarily, at least not easily. Moreover, it was inherently unsafe because there were no exceptions for reachability-checking commands, making it possible for Git to remove needed objects by accident during pruning (garbage collecting).

However, you can find its use in examples. Nowadays, it is obsolete, especially with the existence of the `git replace --graft` option. If you use grafts, consider replacing them with replacements objects; the `contrib/convert-grafts-to-replace-refs.sh` script can help with this in the Git sources.

> **Other graft-like files in Git**
>
> The **shallow clone** (the result of `git clone --depth=<N>`, a clone with the shortened history) is managed with a graft-like `.git/shallow` file. This file is managed by Git, however, not by the user.

Publishing and retrieving replacements

How can you publish replacements, and how do you get them from the remote repository? Since replacements use references, this is quite simple.

Each replacement is a separate reference in the `refs/replaces/` namespace. Therefore, you can get all the replacements with the globing `fetch` or `push` line:

```
+refs/replace/*:refs/replace/*
```

There can be only one replacement for an object, so there are no problems with merging replacements. You can only choose between one replacement or the other.

Theoretically, you could also request individual replacements by fetching (and pushing) individual replacement references instead of using the '*' wildcard.

Summary

This chapter, along with *Chapter 8, Advanced Branching Techniques*, provided all the tools required to manage a clean, readable, and easy-to-review history of a project.

You learned how to make history cleaner by rewriting it in this chapter. You also learned what rewriting history means in Git, when and why to avoid it, and how to recover from an untimely upstream rewrite. You have learned to use an interactive rebase to delete, reorder, squash and split commits, and how to test each commit during the rebase. You know how to do a large-scale scripted rewrite with `filter-repo`, as well as how to edit commits and commit metadata and how to permanently change history, such as by splitting it in two. You also got to know some third-party external tools, which can help with these tasks.

You learned what to do if you cannot rewrite history: how to fix mistakes by creating commits with appropriate changes (for example, with `git revert`), how to add extra information to the existing commits with notes, and how to change the virtual view of the history with replacements. You learned to handle reverting a faulty merge and how to re-merge after a reverted merge. You learned how to fetch and publish both notes and replacements.

To really understand advanced history rewriting and the mechanism behind notes and replacements, this chapter explained the basics of Git internals and low-level commands that are usable for scripting (including scripted rewrite).

The following chapter, *Chapter 11*, *Managing Subprojects*, will explain and show different ways to connect different subprojects in one repository, from submodules to subtrees.

In the subsequent chapter, *Chapter 12*, *Managing Large Repositories*, you will also learn techniques to manage (or mitigate managing) large-size assets inside a repository, or large numbers of files in a repository. Splitting a large project into submodules is one, but not the only, way to handle this issue.

Questions

Answer the following questions to test your knowledge of this chapter:

1. When working on a series of commits to implement a feature, how can you mark a bugfix commit for later squashing into the original commit before publishing the series?

2. Why should you not rewrite (rebase or amend) published history if you are using merging to integrate changes?

3. How can you recover from the upstream rebase?

4. What can you do when you notice that you accidentally included some large file that should not be put in version control in a commit?

5. How can you undo the effect of the commit if you cannot rewrite history?

6. What mechanisms exist to amend history, or a view of history, without rewriting it?

Answers

Here are the answers to the questions given above:

1. You can use `git commit --fixup` when creating a bugfix, and then later `git rebase --interactive --autosquash` before publishing the series.

2. You should not rewrite published history because other developers can do their work based on the version before the changes, and then merging would bring older versions (from before the rewrite) back into existence.

3. Rebase your own changes on top of the new, rebased version of the upstream.

4. If the problem is in the most recent commit, you can amend it with `git commit --amend`. If you need to rewrite the whole history of the project, you can use the `git filter-repo` tool. Note, however, the caveat that comes with rewriting published history, namely that it can cause problems for other developers when they will try to integrate their changes.

5. You can use `git revert` to create the commit that undoes changes brought by an unwanted commit.

6. You can use `git notes` to add extra information to commit objects after the fact, and you can use `git replace` to change the effective shape of the history.

Further reading

To learn more about the topics that were covered in this chapter, take a look at the following resources:

- Scott Chacon and Ben Straub: *Pro Git, 2nd Edition* (2014) `https://git-scm.com/book/en/v2`.

  - *Chapter 7.6, Git Tools - Rewriting History*

  - *Chapter 7.13, Git Tools - Replace*

- Aske Olsson and Rasmus Voss: *Git Version Control Cookbook (2014)*, Packt Publishing Ltd

  - *Chapter 5, Storing Additional Information in Your Repository*

  - *Chapter 8, Recovering From Mistakes*

- Git Documentation HOWTOs

  - *How to revert a faulty merge* `https://github.com/git/git/blob/master/Documentation/howto/revert-a-faulty-merge.txt`

  - *How to revert an existing commit* `https://github.com/git/git/blob/master/Documentation/howto/revert-branch-rebase.txt`

- Tyler Cipriani: *Git Notes: Git's Coolest, Most Unloved Feature* (2022) `https://tylercipriani.com/blog/2022/11/19/git-notes-gits-coolest-most-unloved-feature/`

- Elijah Newren: *git filter-repo* `https://github.com/newren/git-filter-repo`

- Stacked Git: *StGit Tutorial* `https://stacked-git.github.io/guides/tutorial/`

- Jackson Gabbard: *Stacked Diffs Versus Pull Requests* (2018) `https://jg.gg/2018/09/29/stacked-diffs-versus-pull-requests/`

Part 3 – Managing, Configuring, and Extending Git

This part describes how to manage Git, how to customize and extend it, and how to configure it. You will also learn how to automate things with Git hooks. You can also find information about setting up serving repositories if you don't need fully-featured repository hosting software. Finally, you will be presented with a collection of version-control, generic, and Git-specific recommendations and best practices. This will cover issues such as managing a working directory, creating commits and a series of commits (pull requests), submitting changes for inclusion, and a peer review of changes.

This part has the following chapters:

- *Chapter 11, Managing Subprojects*
- *Chapter 12, Managing Large Repositories*
- *Chapter 13, Customizing and Extending Git*
- *Chapter 14, Git Administration*
- *Chapter 15, Git Best Practices*

11

Managing Subprojects

In *Chapter 6*, *Collaborative Development with Git*, we learned how to manage multiple repositories, while *Chapter 8*, *Advanced Branching Techniques*, taught us various development techniques utilizing multiple branches, and multiple lines of development in these repositories. Up until now, these multiple repositories were all being developed independently of each other. Repositories of the different projects were autonomous.

This chapter will explain and show different ways to connect different subprojects in one single repository of the framework project, from the strong inclusion by embedding the code of one project in the other (subtrees) to the light connection between projects by nesting repositories (submodules). You will learn how to add a subproject to a master project, how to update the superproject state, and how to update a subproject. We will find out how to send our changes upstream, backport them to the appropriate project, and push them to the appropriate repository. Different techniques of managing subprojects have different advantages and drawbacks here.

In this chapter, we will cover the following topics:

- Managing library and framework dependencies
- Dependency management tools: managing dependencies outside of Git
- Importing code into a superproject as a subtree
- Using subtree merges; the `git-subtree` and `git-stree` tools
- Nested repositories (submodules): a subproject inside a superproject
- Internals of submodules: gitlinks, `.gitmodules`, and the `.git` file
- Use cases for subtrees and submodules; comparison of approaches
- Alternative third-party solutions and tools/helpers

Building a living framework

There are various reasons to join an external project to your own project. As there are different reasons to include a project (let's call it a **subproject**, or a **module**) inside another project (let's call it a **superproject**, or a **master project**), there are different types of inclusions geared toward different circumstances. They all have their advantages and disadvantages, and it is important to understand these to be able to choose the correct solution for your problem.

Let's assume that you work on a web application and that your web app uses JavaScript (perhaps working as a single-page app). To make it easier to develop, you probably use some JavaScript library or a web framework, such as React.

Such a library is a separate project. You would want to be able to pin it to a known working version (to avoid problems where future changes to the library would make it stop working for your project), while also being able to review changes and automatically update it to the new version. Perhaps you would want to make your own changes to the library and send the proposed changes to the upstream. Of course, you would want users of your project to be able to use the library with your out-of-tree fixes, even if they are not yet accepted by original developers. Conceivably, you might have customizations and changes that you don't want to publish (send to the upstream), but you might still make them available.

This is all possible in Git. There are two main solutions for including subprojects: importing code into your project with the **subtree** merge strategy and linking subprojects with **submodules**.

Both submodules and subtrees aim to reuse the code from another project, which usually has its own repository, putting it somewhere inside your own repository's working directory tree. The goal is usually to benefit from the central maintenance of the reused code across a number of container repositories, without having to resort to clumsy, unreliable manual maintenance (usually by copy-pasting).

Sometimes, it is more complicated. The typical situation in many companies is that they use many in-house produced applications, which depend on the common utility library, or a set of libraries. You would often want to develop each of such applications separately, use it together with others, branch and merge, and apply your own changes and customizations, all in their own separate Git repositories. But there are also advantages to having a single **monolithic repository** (**monorepo**), such as simplified organizations, dependencies, cross-project changes, and tooling if you can get away with it.

The mechanism used by submodules and subtrees solutions (of having separate Git repositories for each application, framework, or library) is not without problems. The development gets more complex because you now have multiple repositories to interact with. If the library gets improved, you would want to update your subproject and need to test whether this new version correctly works with your code, then decide whether to use it in your superproject. On the other hand, at some point in time, you would want to send your changes to the library itself to share their changes with other developers, if only to share the burden of maintaining these features (the out-of-tree patches bring maintenance costs to keep them current).

What to do in those cases? This chapter describes a few strategies used to manage subprojects. For each technique, we will detail how to add such subprojects to superprojects, how to keep them up to date, how to create your own changes, and how to publish selected changes upstream.

> **Subdirectory requirement**
>
> Note that all the solutions require that all the files of a subproject be contained in a single subdirectory of a superproject. No currently available solution allows you to mix the subproject files with other files, or have them occupy more than one directory.

However you manage subprojects, be it subtrees, submodules, third-party tools, or dependency management outside Git, you should strive for the module code to remain independent of the particularities of the superproject (or at least, handle such particularities using an external, possibly non-versioned configuration). Using superproject-specific modifications goes against modularization and encapsulation principles, unnecessarily coupling the two projects.

On the other hand, sharing common components, libraries, and tooling, and keeping them the same for all the distinct but related projects might be more important than the autonomy of those projects (for example, if they are all developed by the same company). It might be the case with the polyrepo setup that introducing a new feature always makes it necessary to create multiple commits in multiple repositories, instead of requiring only a single commit. In those cases, monorepo might be a better solution.

Managing dependencies outside of Git

In many cases, the development stack used allows you to simply use **packaging** and **formal dependency management**. If it is possible, it is usually preferable to go this route. Using dependency management solutions lets you split your code base better and avoid a number of side effects, complications, and pitfalls that litter the submodule and subtree solution space (with different complications for different techniques). It removes the version control systems from the managing modules. It also lets you benefit from versioning schemes, such as **semantic versioning**, for your dependencies.

As a reminder, here's a partial list (in alphabetical order) of the main languages and development stacks, and their dependency management/packaging systems and registries (see the full comparison at www.modulecounts.com):

- Go has GoDoc
- Java has Maven Central (Maven and Gradle)
- JavaScript has npm (for Node.js) and Bower
- .NET has NuGet
- Objective-C has CocoaPods

- Perl has **Comprehensive Perl Archive Network (CPAN)** and carton

- PHP has Composer, Packagist, and good old PEAR and PECL

- Python has **Python Package Index (PyPI)** and pip

- Ruby has Bundler and RubyGems

- Rust has Crates

Sometimes, just using the official package registry is not enough. You might need to apply some out-of-tree patches (changes) to customize the module (subproject) for your needs. Sometimes, however, for many reasons, you might be unable to publish these changes upstream to have them accepted. Perhaps the changes are relevant only to your specific project, or the upstream is slow to respond to the proposed changes, or perhaps there are license considerations. Maybe the subproject in question is an in-house module that cannot be made public, but which you are required to use for your company projects.

In all these cases, you need the **custom package registry** (the package repository) to be used in addition to the default one, or you need to let subprojects be managed as private packages, which these systems often allow. If there is no support for private packages, a tool to manage the private registry, such as Pinto or CPAN::Mini for Perl, would be also needed.

Manually importing the code into your project

Sometimes, the library or a tool that you want to include in your project is not available in the package registry (perhaps because of the software stack; for example, package registries for C++ such as Conan or vcpkg are quite a new thing).

Therefore, let's take a look at one of the other possibilities: why don't we simply import the library into some subdirectory in our project? If you need to bring it up to date, you can just copy the new version as a new set of files. In this approach, the subproject code is embedded inside the code of the superproject.

The simplest solution would be to just overwrite the contents of the subproject's directory each time we want to update the superproject to use the new version. If the project you want to import doesn't use Git, or if it doesn't use a **version control system** (**VCS**) at all, or if the repository it uses is not public, this will indeed be the only possible solution.

> **Using repositories from a foreign VCS as a remote**
>
> If the project you want to import (to embed) uses a VCS other than Git but there is a good conversion mechanism (for example, with a fast import stream), you can use **remote helpers** to set up a foreign VCS repository as a remote repository (via automatic conversion). You can check *Chapter 6, Collaborative Development with Git*, and *Chapter 13, Customizing and Extending Git*, for more information.
>
> This can be done, for example, with the Mercurial and Bazaar repositories, thanks to the `git-remote-hg` and `git-remote-bzr` helpers.

Moving to the new version of the imported library is quite simple (and the mechanism is easy to understand). Remove all the files from the directory, add files from the new version of the library (for example, by extracting them from the archive), then use the `git add` command to the directory:

```
$ rm -rf mylib/
$ git rm mylib
$ tar -xzf /tmp/mylib-0.5.tar.gz
$ mv mylib-0.5 mylib
$ git add mylib
$ git commit
```

This method works quite well in simple cases with the following caveats:

- In the Git history of your project, you have only the versions of the library at the time of import. On the one hand, this makes your project history clean and easy to understand; on the other hand, you don't have access to the fine-grained history of a subproject. For example, when using `git bisect`, you would only be able to find that it was introduced by upgrading the library, but not the exact commit in the history of the library that introduced the bug in question.

- If you want to customize the code of the library, fitting it to your project by adding the changes dependent on your application, you would need to reapply that customization in some way after you import a new version. You could extract your changes with `git diff` (comparing it to the unchanged version at the time of import) and then use `git apply` after upgrading the library. Or, you could use a rebase, an interactive rebase, or some patch management interface; see *Chapter 10, Keeping History Clean*. Git won't do this automatically.

- Each importing of the new version of the library requires running a specific sequence of commands to update the superproject: removing the old version of files, adding new ones, and committing the change. It is not as easy as running `git pull`, though you can use scripts or aliases to help.

A Git subtree solution for embedding the subproject code

In a slightly more advanced solution, you can use the **subtree merge** to join the history of a subproject to the history of a superproject. This is only somewhat more complicated than an ordinary pull (at least, after the subproject is imported), but provides a way to automatically merge changes together.

Depending on your requirements, this method might fit well with your needs. It has the following advantages:

- You would always have the correct version of the library, never using the wrong library version by accident.

- The method is simple to explain and understand, using only the standard (and well-known) Git features. As you will see, the most important and most commonly used operations are easy to do and easy to understand, and it is hard to go wrong.

- The repository of your application is always self-contained; therefore, cloning it (with plain old `git clone`) will always include everything that's needed. This means that this method is a good fit for the *required dependencies*.

- It is easy to apply patches (for example, customizations) to the library inside your repository, even if you don't have the commit rights to the upstream repository.

- Creating a new branch in your application also creates a new branch for the library; it is the same for switching branches. That's the behavior you expect. This is contrasted with the submodule's behavior (the other technique for managing subprojects).

- If you are using the `subtree` merge strategy (described shortly in *Chapter 9, Merging Changes Together*), for example, with `git pull -s subtree`, then getting a new library version will be as easy as updating all the other parts of your project.

Unfortunately, however, this technique is not without its disadvantages. For many people and for many projects, these disadvantages do not matter. The simplicity of the subtree-based method usually prevails over its faults.

Here are the problems with the subtree approach:

- Each application using the library doubles its files. There is no easy and safe way to share its objects among different projects and different repositories. (See the following callout about the possibility of sharing the Git object database.)

- Each application using the library has its files checked out in the working area, though you can change it with the help of the **sparse checkout** (which will be described in the next chapter: *Chapter 12, Handling Large Repositories*).

- If your application introduces changes to its copy of the library, it is not that easy to publish these changes and send them upstream. Third-party tools such as `git subtree` or `git stree` can help here. They have specialized subcommands to extract the subproject's changes.

- Because of the lack of separation between the subproject files and the superproject files, it is quite easy to mix the changes to the library and the changes to the application in one commit. In such cases, you might need to rewrite the history (or the copy of a history), as described in *Chapter 10, Keeping History Clean*.

The first two issues mean that subtrees are not a good fit to manage the subprojects that are *optional dependencies* (needed only for some extra features) or *optional components* (such as themes, extensions, or plugins), especially those that are installed by a mere presence in the appropriate place in the filesystem hierarchy.

> **Sharing objects between forks (copies) with alternates**
>
> You can mitigate the duplication of objects in the repository with **alternates** or, in other words, with `git clone --reference`. However, then you would need to take greater care about garbage collection. The problematic parts are those parts of the history that are referenced in the borrower repository (that is, one with alternates set up), but are not referenced in the lender reference's repository. The description and explanation of the alternate mechanisms will be presented in *Chapter 14, Git Administration*.

There are different technical ways to handle and manage the subtree-imported subprojects. You can use classic Git commands, just using the appropriate options while affecting the subproject, such as `--strategy=subtree` (or the `subtree` option to the default `recursive` merge strategy, `--strategy-option=subtree=<path>`) for `merge`, `cherry-pick`, and related operations. This manual approach works everywhere, is actually quite simple in most cases, and offers the best degree of control over operations. However, it requires a good understanding of the underlying concepts.

In modern Git (since version 1.7.11), there is the `git subtree` command available among installed binaries. It comes from the `contrib/` area and is not fully integrated (for example, with respect to its documentation). This script is well tested and robust, but some of its notions are rather peculiar or confusing, and this command does not support the whole range of possible subtree operations. Additionally, this tool supports only the *import with history* workflow (which will be defined later), which some say clutters the history graph.

There are also other third-party scripts that help with subtrees; among them is `git-subrepo`.

Creating a remote for a subproject

Usually, while importing a subproject, you would want to be able to update the embedded files easily. You would want to continue interacting with the subproject. For this, you would add that subproject (for example, the common library) as a **remote reference** in your own (super)project and fetch it:

```
$ git remote add mylib_repo https://git.example.com/mylib.git
$ git fetch mylib_repo
warning: no common commits
From https://git.example.com/mylib.git
 * [new branch]      master      -> mylib_repo/master
```

Note that, in this example, progress messages were removed for simplicity.

You can then examine the `mylib_repo/master` remote-tracking branch, which can be done either by checking it out into the detached `HEAD` with `git checkout mylib_repo/master`, or by creating a local branch out of it and checking this local branch out with `git checkout -b mylib_branch mylib_repo/master`. Alternatively, you can just list its files with `git` `ls-tree -r --abbrev mylib repo/master`. You will see then that the subproject has a different project root from your superproject. Additionally, as can be seen from the **warning: no common commits** message, this remote-tracking branch contains a completely different history coming from a separate project.

Adding a subproject as a subtree

If you are not using specialized tools such as `git subtree` but a manual approach, the next step will be a bit complicated and will require you to use some advanced Git concepts and techniques. Fortunately, it needs to be done only once.

First, if you want to import the *subproject history*, you would need to create a merge commit that will import the subproject in question. You need to have the files of the subproject in the given directory in a superproject. Unfortunately (at least, with the current version of Git as of writing this chapter), using the `-Xsubtree=mylib/` merge strategy option would not work as expected. We would have to do it in two steps: prepare the parents and then prepare the contents.

The first step would then be to prepare a merge commit using the `ours` merge strategy, but without creating it (writing it to the repository). This strategy joins histories, but takes the current version of the files from the current branch:

```
$ git merge --no-commit --strategy=ours --allow-unrelated-histories
mylib_repo/master
Automatic merge went well; stopped before committing as requested
```

If you want to have *simple history*, similar to the one we get from just copying files, you can skip this step.

We now need to update our index (the staging area for the commits) with the contents of the `master` branch from the library repository and update our working directory with it. All this needs to happen in the proper subfolder too. This can be done with the low-level (plumbing) `git read-tree` command:

```
$ git read-tree --prefix=mylib/ -u mylib_repo/master
$ git status
On branch master
All conflicts fixed but you are still merging.
  (use "git commit" to conclude merge)

Changes to be committed:

        new file:   mylib/README [...]
```

We have used the `-u` option, so the working directory is updated along with the index. We then need simply to finalize the merge with `git commit`, as Git tells us.

> **Important note!**
> It is important to not forget the trailing slash in the argument of the `--prefix` option.
> Checked-out files are literally prefixed with it.

This set of steps is described in the HOWTO section of the Git documentation, namely, in **How to use the subtree merge strategy**. This HOWTO is available at `https://kernel.org/pub/software/scm/git/docs/howto/using-merge-subtree.html`.

It is much easier to use tools such as `git subtree`:

```
$ git subtree add --prefix=mylib mylib_repo master
git fetch mylib_repo master
From https://git.example.com/mylib.git
* branch          master      -> FETCH_HEAD
Added dir 'mylib'
```

The `git subtree` command would fetch the subtree's remote when necessary; there's no need for the manual fetch that you had to perform in the manual solution.

If you examine the history, for example, with `git log --oneline --graph --decorate`, you will see that this command merged the library's history with the history of the application (of the superproject). If you don't want this, tough luck. The `--squash` option that `git subtree` offers on its `add`, `pull`, and `merge` subcommands won't help here. One of the peculiarities of this tool is that this option doesn't create a **squash merge**, but simply merges the squashed subproject's history (as if it were squashed with an interactive rebase). The commit message would look like this: **Squashed 'mylib/' content from commit 5e28a71**. See *Figure 11.2(b)* later in this chapter.

If you want a subtree without its history attached to the superproject history, as in *Figure 11.2(c)*, consider using the external tool, `git-subrepo`. It has the additional advantage that it remembers the subtree settings:

```
$ git subrepo clone \
   https://git.example.com/mylib.git mylib/
Subrepo 'https://git.example.com/mylib.git' (master) cloned into
'mylib'.
```

The information about the subproject repository URL, the main branch, the original commit, and so on, is stored in the `.gitrepo` file in the directory with the subproject. All subsequent `git subrepo` commands refer to the embedded subproject by the name of the directory it is in (which is `mylib/` in the preceding example).

You can achieve similar results with the external `git-stree` tool, which was deprecated in favor of `git-subrepo`.

Cloning and updating superprojects with subtrees

All right! Now that we have our project with a library embedded as a subtree, what do we need to do to get it? Because the concept behind subtrees is to have just one repository (the container), you can simply clone this repository.

To get an up-to-date repository, you just need a regular pull; this would bring both the superproject (the container) and subproject (the library) up to date. This works regardless of the approach taken, the tool used, and the manner in which the subtree was added. It is a great advantage of the subtrees approach.

Getting updates from subprojects with a subtree merge

Let's see what happens if there are some new changes in the subproject since we imported it. It is easy to bring the version embedded in the superproject up to date:

```
$ git pull --strategy subtree mylib_repo master
From https://git.example.com/mylib.git
 * branch            master      -> FETCH_HEAD
Merge made by the 'subtree' strategy.
```

You could have fetched and then merged instead, which allows for greater control. Or, you could have rebased instead of merging, if you prefer; that works too.

> **Importance of selecting subtree merge strategy**
>
> Don't forget to select the merge strategy with `-s subtree` while pulling a subproject. Merging could work even without it because Git does rename detection and would usually be able to discover that the files were moved from the root directory (in the subproject) to a subdirectory (in the superproject we are merging into). The problematic case is when there are conflicting files inside and outside of the subproject. Potential candidates are Makefiles and other standard filenames.
>
> If there are some problems with Git detecting the correct directory to merge into, or if you need advanced features of an ordinary `ort` merge strategy (which is the default), you can instead use `-Xsubtree=<path/to/subproject>`, the `subtree` option of the `ort` merge strategy.

You may need to adjust other parts of the application code to work properly with the updated code of the library.

Note that, with this solution, you have a subproject history attached to your application history, as you can see in *Figure 11.1*:

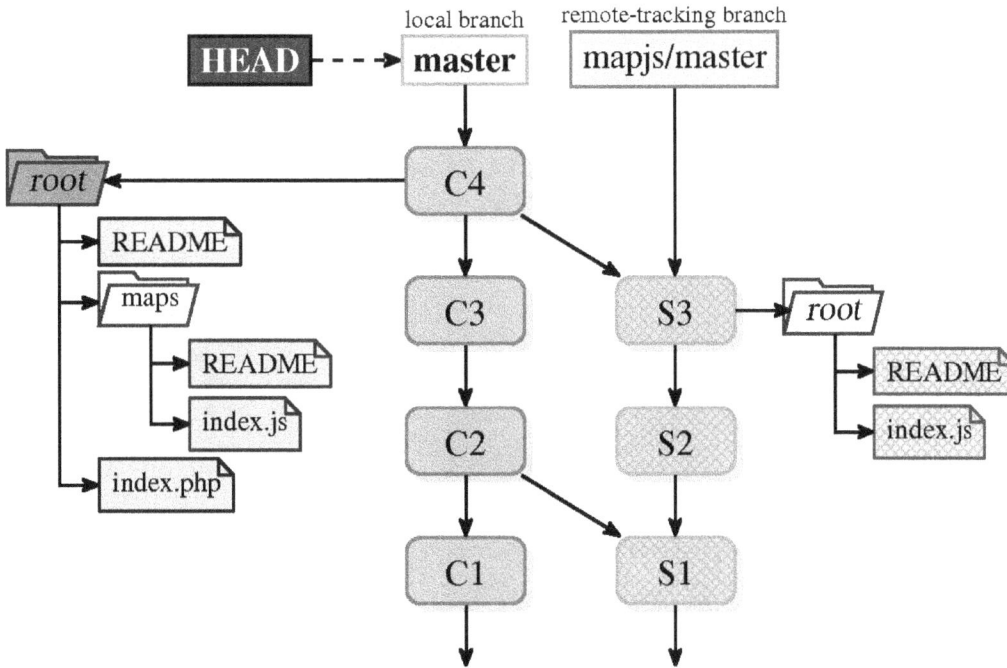

Figure 11.1 – History of a superproject with a subtree-merged subproject inside the 'maps/' directory. Subproject history is available in the superproject via relevant remote-tracking branch

If you don't want to have the history of a subproject entangled in the history of a master project and prefer a simpler-looking history (as shown in *Figure 11.1*), you can use the --squash option of the git merge (or git pull) command to squash it:

```
$ git merge -s subtree --squash mylib_repo/master
Squash commit -- not updating HEAD
Automatic merge went well; stopped before committing as requested
$ git commit -m "Updated the library"
```

Squash merge is described in *Chapter 9, Merging Changes Together*.

In this case, in the history, you would have only the fact that the version of the subproject had changed, which has its advantages and disadvantages. You get *simpler* history but also *simplified* history.

With the git subtree or git subrepo tools, it is enough to use their pull subcommand; they supply the subtree merge strategy themselves. However, currently, git subtree pull requires you to respecify --prefix and the entire subtree settings.

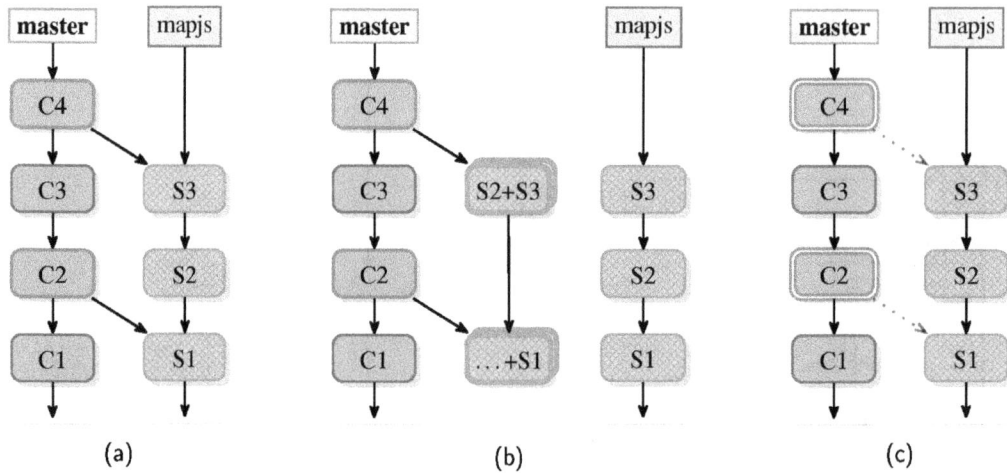

Figure 11.2 – Different types of subtree merges: (a) subtree merge, (b) subtree merge of squashed commits, (c) squashed subtree merge

Note that the `git subtree` command always merges, even with the `--squash` option; it simply squashes the subproject commits before merging (such as the `squash` instruction in the interactive rebase). In turn, `git subrepo pull` always squashes the merge (such as `git merge --squash`), which keeps the superproject history and subproject history separated without polluting the graph of the history. All this can be seen in *Figure 11.2*. Note that the dotted line in *(c)* denotes how commits **C2** and **C4** were made, and not that it is the parent commit.

Showing changes between a subtree and its upstream

To find out the differences between the subproject and the current version in the working directory, you need nontypical selector syntax for `git diff`. This is because all the files in the subproject (for example, in the `mylib_repo/master` remote-tracking branch) are in the root directory, while they are in the `mylib/` directory in the superproject (for example, in `master`). We need to select the subdirectory to be compared with `master`, putting it after the revision identifier and the colon (skipping it would mean that it would be compared with the root directory of the superproject).

The command looks as follows:

```
$ git diff master:mylib mylib_repo/master
```

Similarly, to check after the subtree merge whether the commit we just created (HEAD) has the same contents in the `mylib/` directory as the merged in the commit, that is, HEAD^2, we can use the following:

```
$ git diff HEAD:mylib HEAD^2
```

Sending changes to the upstream of a subtree

In some cases, the subtree code of a subproject can only be used or tested inside the container code; most themes and plugins have such constraints. In this situation, you'll be forced to evolve your subtree code straight inside the master project code base, before you finally backport it to the subproject upstream.

These changes often require adjustments in the rest of the superproject code; though it is recommended to make two separate commits (one for the subtree code change and one for the rest), it is not strictly necessary. You can tell Git to extract only the subproject changes. The problem is with the commit messages of the split changes, as Git is not able to automatically extract relevant parts of the changeset description.

Another common occurrence, which is best avoided but is sometimes necessary, is the need to customize the subproject's code in a container-specific way (configure it specifically for a master project), usually without pushing these changes back upstream. You should carefully distinguish between both situations, keeping each use case's changes (backportable and non-backportable) in their own commits.

There are different ways to deal with this issue. You can avoid the problem of extracting changes to be sent upstream by requiring that all the subtree changes have to be done in a separate module-only repository. If it is possible, we can even require that all the subproject changes have to be sent upstream first, and we can get the changes into the container only through upstream acceptance.

If you need to be able to extract the subtree changes, then one possible solution is to utilize `git filter-branch --directory-filter` (or `--index-filter` with the appropriate script). Another simple solution is to just use `git subtree push`. Both of the methods, however, backport *every* commit that touches the subtree in question.

If you want to send upstream only a selection of the changes to the subproject of those that made it into the master project repository, then the solution is a bit more complicated. One possibility is to create a local branch meant specifically for backporting out of the subproject remote-tracking branch. Forking it from said subtree-tracking branch means that it has the subtree as the root and it would include only the submodule files.

This branch, intended for backporting changes to the subproject, would need to have the appropriate branch in the remote of the subproject upstream repository as its upstream branch. With such a setup, we would then be able to use `git cherry-pick --strategy=subtree` the commits we're interested in sending to the subproject's upstream onto it. Then, we can simply `git push` this branch into the subproject's repository.

> **Cherry picking and submodules**
>
> It is prudent to specify `--strategy=subtree` even if `cherry-pick` would work without it, to make sure that the files outside the subproject's directory (outside `subtree`) will get quietly ignored. This can be used to extract the subtree changes from the mixed commit; without this option, Git will refuse to complete the cherry-pick.

This requires much more steps than ordinary `git push`. Fortunately, you need to face this problem only while sending the changes made in the superproject repository back to the subproject. As you have seen, fetching changes from the subproject into the superproject is much, much simpler.

Well, using `git-stree` would make this trivial; you just need to list the commits to be pushed to backport:

```
$ git stree push mylib_repo master~3 master~1
• 5e28a71 [To backport] Support for creating debug symbols
• 5b0aa4b [To backport] Timestamping (requires application tweaks)
✓  STree 'mylib_repo' successfully backported local changes to its
remote
```

In fact, this tool internally uses the same technique, creating and using a backport-specific local branch for the subproject.

The Git submodules solution – a repository inside a repository

The subtrees method of importing the code (and possibly also the history) of a subproject into the superproject has its disadvantages. In many cases, the subproject and the container are two different projects: your application depends on the library, but it is obvious that they are separate entities. Joining the histories of the two doesn't look like the best solution.

Additionally, the embedded code and imported history of a subproject are always here. Therefore, the subtrees technique is not a good fit for optional dependencies and components (such as plugins or themes). It also doesn't allow you to have different access controls for the subproject's history, with the possible exception of restricting write access to the subproject (actually to the subdirectory of a subproject), by using Git repository management solutions such as `gitolite` (you can find more in *Chapter 14, Git Administration*).

The submodule solution is to keep the subproject code and history in its own repository and to embed this repository inside the working area of a superproject, but not to add its files as superproject files.

Gitlinks, .git files, and the git submodule command

Git includes the command named `git submodule`, which is intended to work with submodules. However, to utilize it correctly, you need to understand at least some of the details of its operation. It is a combination of two distinct features: the so-called **gitlinks** and the `git submodule` tool itself.

Both in the subtree solution and the submodule solution, subprojects need to be contained in their own folder inside the working directory of the superproject. But while, with subtrees, the code of the subproject belongs to the superproject repository, this is not the case for submodules. With submodules, each subproject has instead its own repository somewhere inside its container repository. The code of the submodule belongs to its repository, and the superproject itself simply stores the meta-information required to get appropriate revisions of the subproject files.

In practice, in modern Git, submodules use a simple .git file with a single gitdir: line containing a relative path to the actual repository folder. The submodule repository is actually located inside the superproject's .git/modules folder (and has core.worktree set up appropriately). This is done mostly to handle the case when the superproject has branches that don't have a submodule at all. It allows us to avoid having to scrap the submodule's repository while switching to the superproject revision without it.

> **Tip**
>
> You can think of the .git file with the gitdir: line as a symbolic reference equivalent for the .git directories, an OS-independent symbolic link replacement. The path to the repository doesn't need to be a relative path:

```
$ ls -aloF plugins/demo/
total 10
drwxr-xr-x 1 user  0 Jul 13 01:26 ./
drwxr-xr-x 1 user  0 Jul 13 01:26 ../
-rw-r--r-- 1 user 32 Jul 13 01:26 .git
-rw-r--r-- 1 user  9 Jul 13 01:26 README
[…]
$ cat plugins/demo/.git
gitdir: ../../.git/modules/plugins/demo
```

Be that as it may, the contained superproject and the subproject module truly act as (and, in fact, are) independent repositories: they have their own history, their own staging area, and their own current branch. Therefore, you should take care while typing commands, minding whether you're inside the submodule or outside it, as the context and impact of your commands differ drastically!

The main idea behind the typical use of submodules is that the superproject commit remembers the *exact* revision of the subproject; this reference uses the SHA1 identifier of the subproject commit. Instead of using a manifest-like file as in some dependency management tools, the submodules solution stores this information in a tree object using so-called gitlinks. **Gitlink** is a reference from a **tree object** (in the superproject repository) to a **commit object** (usually, in the submodule repository); see *Figure 11.3*. The faint shade of submodule files on the left-hand side denotes that they are present as files in the working directory of the superproject, but are not in the superproject repository themselves.

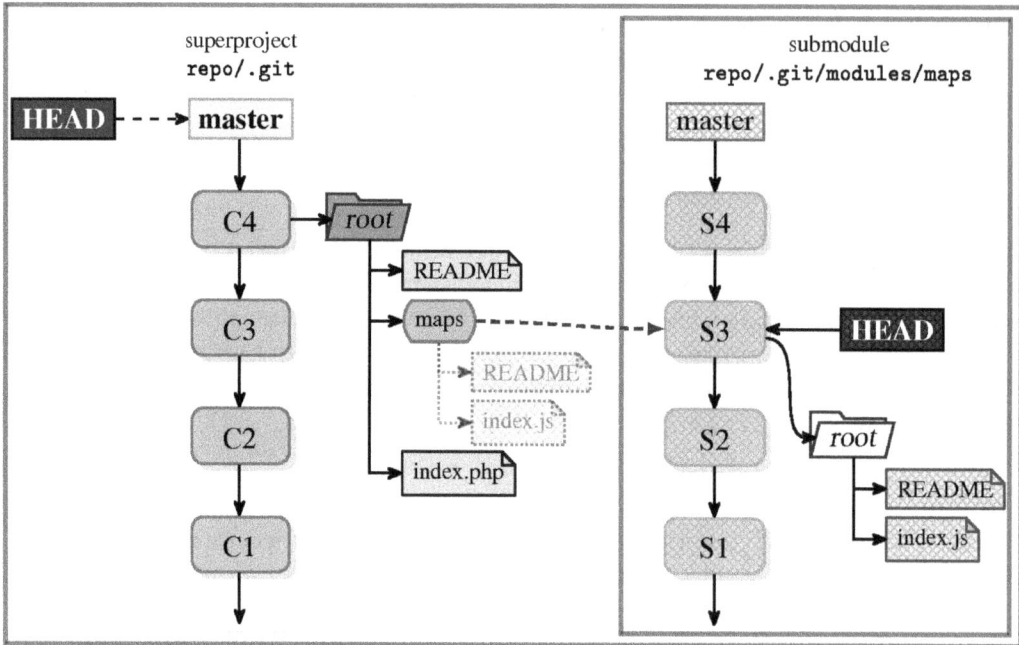

Figure 11.3 – The history of a superproject with a subproject linked as a submodule
inside the 'maps/' subdirectory. Subproject history is separate

Recall that, following the description of the types of objects in the repository database from *Chapter 10, Keeping History Clean*, each commit object (representing a revision of a project) points exactly to one tree object with the snapshot of the repository contents. Each tree object references blobs and trees, representing file contents and directory contents, respectively. The tree object referenced by the commit object uniquely identifies the set of file contents, filenames, and file permissions contained in a revision associated with the commit object.

Let's remember that the commit objects themselves are connected with each other, creating the **Directed Acyclic Graph** (**DAG**) of revisions. Each commit object references zero or more parent commits, which together describe the history of a project.

Each type of reference mentioned earlier took part in the reachability check. If the object pointed to was missing, it means that the repository is corrupt.

It is not so for gitlinks. Entries in the tree object pointing to the commits refer to the objects in the other separate repository, namely, in the subproject (submodule) repository. The fact that the submodule commit being unreachable is not an error is what allows us to optionally include submodules: no submodule repository, no commit referenced in gitlink.

The results of running `git ls-tree --abbrev HEAD` on a project with all the types of objects is as follows:

```
040000 tree 573f464     docs
100755 blob f27adc2     executable.sh
100644 blob 1083735     README.txt
040000 tree ef9bcb4     subdirectory
160000 commit 5b0aa4b    submodule
120000 blob 3295d66     symlink
```

Compare it with the contents of the working area (with `ls -l -o -F`):

```
drwxr-xr-x    5 user     12288 06-28 17:18 docs/
-rwxr-xr-x    1 user     36983 02-20 20:11 executable.sh*
-rw-r--r--    1 user      2628 2015-01-03  README.txt
drwxr-xr-x    3 user      4096 06-28 17:19 subdirectory/
drwxr-xr-x   48 user     36864 06-28 17:19 submodule/
lrwxrwxrwx    1 user        32 06-28 17:18 symlink -> docs/toc.html
```

Adding a subproject as a submodule

To manage submodules there is the `git submodule` command. It was created to help manage the filesystem contents, the metadata, and the configuration of your submodules, as well as inspect their status and update them.

With subtrees, the first step was usually to add a subproject repository as a remote, which meant that objects from the subproject repository were fetched into the *superproject* object database.

To add the given repository as a submodule at a specific directory in the superproject, use the `add` subcommand of the `git submodule`:

```
$ git submodule add https://git.example.com/demo-plugin.git plugins/
demo
Cloning into 'plugins/demo'...
done.
```

> **Note about adding subprojects via a path to their repository**
>
> While using paths instead of URLs for remotes, you need to remember that the relative paths for remotes are interpreted relative to our main remote, not to the root directory of our repository.

This command stores the information about the submodule, for example, the URL of the repository, in the `.gitmodules` file. It creates a `.gitmodules` file if it does not exist:

```
[submodule "plugins/demo"]
    url = https://git.example.com/demo-plugin.git
```

Note that a submodule gets a name equal to its path. You can set the name explicitly with the `--name` option (or by editing the configuration); `git mv` on a submodule directory will change the submodule path but keep the same name.

> **Reuse of authentication while fetching submodules**
>
> While storing the URL of a remote repository, it is often acceptable and useful to store the username with the subproject information (for example, storing the username in a URL, such as `user@git.company.com:mylib.git`).
>
> However, remembering the username as a part of the URL is undesirable in `.gitmodules`, as this file must be visible to other developers (which often use different usernames for authentication). Fortunately, the commands that descend into submodules can reuse the authentication from cloning (or fetching) a superproject.

The `add` subcommand also runs an equivalent of `git submodule init` for you, assuming that if you have added a submodule, you are interested in it. This adds some submodule-specific settings to the local configuration of the master project:

```
[submodule "plugins/demo"]
        url = https://git.example.com/demo-plugin.git
```

Why the duplication? Why store the same information in `.gitmodules` and in `.git/ config`? Well, because while the `.gitmodules` file is meant for all developers, we can fit our local configuration to specific local circumstances. The other reason for using two different files is that while the presence of the submodule information in `.gitmodules` means only that the subproject is available, having it also in `.git/ config` implies that we are interested in a given submodule (and that we want it to be present).

You can create and edit the `.gitmodules` file by hand or with `git config -f .gitmodules`.

This file is usually committed to the superproject repository (similar to `.gitignore` and `.gitattributes` files), where it serves as the list of possible subprojects.

All the other subcommands require the `.gitmodules` file to be present; for example, if we would run `git submodule update` before adding it, we would get the following:

```
$ git submodule update
No submodule mapping found in .gitmodules for path 'plugins/demo'
```

That's why `git submodule add` stages both the `.gitmodules` file and the submodule itself:

```
$ git status
On branch master
Changes to be committed:
  (use "git reset HEAD <file>..." to unstage)


        new file:   .gitmodules
        new file:   plugins/demo
```

Note that the whole submodule, which is a directory, looks to `git status` like the new file. By default, most Git commands are limited to the active container repository only and do not descend to the nested repositories of the submodules. As we will see, this is configurable.

Cloning superprojects with submodules

One important issue is that, by default, if you clone the superproject repository, you will not get any submodules. All the submodules will be missing from the working duplicated directory; only their base directories are here. This behavior is the basis of the optionality of submodules.

We need then to tell Git that we are interested in a given submodule. This is done by calling the `git submodule init` command. What this command does is copy the submodule settings from the `.gitmodules` file into the superproject's repository configuration, namely, `.git/config`, registering the submodule:

```
$ git submodule init plugins/demo
Submodule 'plugins/demo' (https://git.example.com/demo-plugin.git)
registered for path 'plugins/demo'
```

The `init` subcommand adds the following two lines to the `.git/config` file:

```
[submodule "plugins/demo"]
    url = https://git.example.com/demo-plugin.git
```

This separate local configuration for the submodules you are interested in allows you also to configure your local submodules to point to a different location URL (perhaps, a per-company reference clone of a subproject's repository) from the one that is present in the `.gitmodules` file.

This mechanism also makes it possible to provide a new URL if the repository of a subproject has moved. That's why the local configuration overrides the one that is recorded in `.gitmodules`; otherwise, you would not be able to fetch from the current URL when switched to the version before the URL change. On the other hand, if the repository moved and the `.gitmodules` file was updated accordingly, we can re-extract the new URL from `.gitmodules` into local configuration with `git submodule sync`.

We have told Git that we are interested in the given submodule. However, we have still not fetched the submodule commits from its remote and neither have we checked it out and have its files present in the working directory of the superproject. We can do this with `git submodule update`.

> **Shortcut command**
>
> In practice, while dealing with submodules using repositories, we usually group the two commands (`init` and `update`) into one with `git submodule update --init`; unless we need to customize the URL.

If you are interested in all the submodules, you can use `git clone --recursive` (or `git clone --recurse-submodules`) to automatically initialize and update each submodule right after cloning.

To temporarily remove a submodule, retaining the possibility of restoring it later, you can mark it as not interesting with `git remote deinit`. This just affects `.git/config`. To permanently remove a submodule, you need to first deinitialize it, and then remove it from `.gitmodules` and from the working area (with `git rm`).

Updating submodules after superproject changes

To update the submodule so that the working directory contents reflect the state of a submodule in the current version of the superproject, you need to perform `git submodule update`. This command updates the files of the subproject or, if necessary, clones the initial submodule repository:

```
$ rm -rf plugins/demo    # clean start for this example
$ git submodule update
Submodule path 'plugins/demo': checked out '5e28a713d8e87…'
```

The `git submodule update` command goes to the repository referenced by `.git/config`, fetches the ID of the commit found in the index (`git ls-tree HEAD -- plugins/demo`), and checks out this version into the directory given by `.git/config`. You can, of course, specify the submodule you want to update, giving the path to the submodule as a parameter.

Because we are here checking out the revision given by gitlink, and not by a branch, `git submodule update` detaches the subproject's HEAD (see *Figure 11.3*). This command rewinds the subproject straight to the version recorded in the supermodule.

There are a few more things that you need to know:

- If you are changing the current revision of a superproject in any way, either by changing a branch, importing a branch with `git pull`, or rewinding the history with `git reset`, you need to run `git submodule update` to get the matching content to submodules. This is not done automatically by default, because it could lead to potentially losing your work in a submodule.

- Conversely, if you switch to another branch, or otherwise change the current revision in a superproject, and do not run `git submodule update`, Git would consider that you changed your submodule directory deliberately to point to a new commit (while it is really an old commit that you used before but forgot to update). If, in this situation, you would run `git commit -a`, then by accident, you will change the gitlink, leading to having an incorrect version of a submodule stored in the superproject history.

- You can upgrade the gitlink reference simply by fetching (or switching to) the version of a submodule you want to have by using ordinary Git commands inside the subproject and then committing this version in the supermodule. You don't need to use the `git submodule` command here.

You can have Git automatically fetch the initialized submodules while pulling the updates from the master project's remote repository. This behavior can be configured using `fetch.recurseSubmodules` (or `submodule.<name>.fetchRecurseSubmodules`). The default value for this configuration is `on-demand` (to fetch if gitlink changes and the submodule commit that it points to is missing). You can set it to `yes` or `no` to turn recursively fetching submodules on or off unconditionally. The corresponding command-line option is `--recurse-submodules`.

You can pass the `--recurse-submodules` command-line option to many Git commands, including the `git pull` command, which would then fetch initialized modules and update working trees of active submodules.

> **Always recursing into active submodules**
>
> To make those Git commands that support it use the `--recurse-submodules` option by default, you can set the `submodule.recurse` configuration option to `true`. The `checkout`, `fetch`, `grep`, `pull`, `push`, `read-tree`, `reset`, `restore`, and `switch` commands are supported.

Note that instead of checking out the gitlinked revision on the detached HEAD, we can merge the commit recorded in the superproject into the current branch in the submodule with `--merge`, or rebase the current branch on top of the gitlink with `--rebase`, just like with `git pull`. The submodule repository branch used defaults to `master`, but the branch name may be overridden by setting the `submodule.<name>.branch` option in either `.gitmodules` or `.git/config`, with the latter taking precedence.

As you can see, using gitlinks and the `git submodule` command is quite complicated. Fundamentally, the concept of gitlink might fit well with the relationship between subprojects and your superproject, but using this information correctly is harder than you think. On the other hand, it gives great flexibility and power.

Examining changes in a submodule

By default, the status, logs, and `diff` output are based solely on the state of the active repository and do not descend into submodules. This is often problematic; you would need to remember to run `git submodule summary`. It is easy to miss a regression if you are limited to this view: you can see that the submodule has changed but you can't see how.

You can, however, set up Git to make it use a **submodule-aware status** with the `status.submoduleSummary` configuration variable. If it is set to a nonzero number, this number will provide the `--summary-limit` restriction; a value of `true` or `-1` will mean an unlimited number.

After setting this configuration, you would get something like the following:

```
$ git status
On branch master
Changes to be committed:
  (use "git reset HEAD <file>..." to unstage)

        new file:   .gitmodules
        new file:   plugins/demo

Submodule changes to be committed:

* plugins/demo 0000000...5e28a71 (3):
  > Fix repository name in a README file
```

The status extends the always present information that the submodule changed (**new file: plugins/demo**), adding the information that the submodule present at `plugins/demo` got three new commits, and showing the summary for the last one (**Fix repository name in a README file**). The right-pointing angle bracket (>) preceding the summary line means that the commit was added, that is, present in the working area but not (yet) in the superproject commit.

Trivia

Actually, this added part is just the `git submodule summary` output.

For the submodule in question, a series of commits in the submodule between the submodule version in the given superproject's commit and the submodule version in the index or the working tree (the former shown by using `--cached`) are listed. There is also `git submodule status` for short information about each module.

The `git diff` command's default output also doesn't tell much about the change in the submodule, just that it is different:

```
$ git diff HEAD -- plugins/demo
diff --git a/plugins/demo b/plugins/demo
new file mode 160000
index 0000000..5e28a71
--- /dev/null
+++ b/plugins/demo
@@ -0,0 +1 @@
+Subproject commit 5e28a713d8e875f2cf1060c2580886dec3e5b04c
```

Fortunately, there is the `--submodule=log` command-line option (which you can enable by default with the `diff.submodule` configuration setting) that lets us see something more useful:

```
$ git diff HEAD --submodule=log -- plugins/demoSubmodule subrepo
0000000...5e28a71 (new submodule)
```

Instead of using `log`, we can use the `short` format that shows just the names of the commits, which is the default if the format is not given (that is, with just `git diff --submodule`). Alternatively, we can use the `diff` format to show an inline `diff` of the changed contents of the submodule.

Getting updates from the upstream of the submodule

To remind you, the submodule commits are referenced in gitlinks using the SHA1 identifier, which always resolves to the same revision; it is not a volatile (inconstant) reference such as a branch name. Because of this, a submodule in a superproject does not automatically upgrade (which could possibly be breaking the application). However, sometimes, you may want to update the subproject to its upstream.

Let's assume that the subproject repository got new revisions published and, for our superproject, we want to update to the new version of a submodule.

To achieve this, we need to update the local repository of a submodule, move the version we want to the working directory of the superproject, and, finally, commit the submodule change in the superproject.

We can do this manually, starting by first changing the current directory to be inside the working directory of the submodule. Then, inside the submodule, we perform `git fetch` to get the data to the local clone of the repository (in `.git/modules/` in the superproject). After verifying what we have with `git log`, we can then update the working directory. If there are no local changes, you can simply check out the desired revision. Finally, you need to create a commit in a superproject.

In addition to the finer-grained control, this approach has the added benefit of working regardless of your current state (whether you are on an active branch or on a detached HEAD).

Another way to go about this would be, working from the container repository, to explicitly upgrade the submodule to its tracked remote branch with `git submodule update --remote`. Similarly to the ordinary update command, you can choose to merge or rebase instead of checking out a branch; you can configure the default way of updating with the `submodule.<name>.update` configuration variable, and the default upstream branch with `submodule.<name>.branch`.

> **Variants of git submodule update**
>
> In short, `submodule update --remote --merge` will merge the upstream's subproject changes into the submodule, while `submodule update --merge` will merge the superproject gitlink changes into the submodule.

The `git submodule update --remote` command would fetch new changes from the submodule remote site automatically unless told not to with `--no-fetch`.

Sending submodule changes upstream

One of the major dangers in making changes live directly in a submodule (and not via its standalone repository) is forgetting to push the submodule. A good practice for submodules is to commit changes to the submodule first, push the module changes, and only then get back to the container project, commit it, and push the container changes.

If you only push to the supermodule repository, forgetting about the submodule push, then other developers will get an error while trying to get the updates. Though Git does not complain while fetching the superproject, you would see the problem in the `git submodule summary` output (and in the `git status` output, if properly configured) and while trying to update the working area:

```
$ git submodule summary
* plugins/demo 12e3a52...0e90143:
  Warn: plugins/demo doesn't contain commit 12e3a529698c519b2fab790…
$ git submodule update
fatal: reference is not a tree: 12e3a529698c519b2fab790…
Unable to checkout '12e3a529698c519b2fab790…' in submodule path
'plugins/demo'
```

You can plainly see how important it is to remember to push the submodule. You can ask Git to automatically push the submodules while pushing the superproject, if it is necessary, with `git push --recurse-submodules=on-demand` (the other option is just to check). You can also use the `push.recurseSubmodules` configuration option.

Transforming a subfolder into a subtree or submodule

The first issue that comes to mind while thinking of the use cases of subprojects in Git is about having the source code of the base project ready for such division.

Submodules and subtrees are always expressed as subdirectories of the superproject (the master project). You can't mix files from different subsystems in one directory.

Experience shows that most systems use such a directory hierarchy, even in monolithic repositories, which is a good beginning for modularization efforts. Therefore, transforming a subfolder into a real submodule/subtree is fairly easy and can be done in the following sequence of steps:

1. Move the subdirectory in question outside the working area of a superproject to have it beside the top directory of the superproject. If it is important to keep the history of a subproject, consider using `git subtree split`, or `git filter-branch --subdirectory-filter` or its equivalent, perhaps together with tools such as `reposurgeon` to clean up the history. See *Chapter 10, Keeping History Clean*, for more details.

2. Rename the directory with the subproject repository to better express the essence of the extracted component. For example, a subdirectory originally named `refresh` could be renamed `refresh-client-app-plugin`.

3. Create the public repository (upstream) for the subproject as a first-class project (for example, create a new project on GitHub to keep extracted code, either under the same organization as a superproject, or under a specialized organization for application plugins).

4. Initialize a self-sufficient and standalone plugin as a Git repository with `git init`. If, in *step 1*, you have extracted the history of the subdirectory into some branch, then push this branch into the just-created repository. Set up the public repository created in *step 3* as a default remote repository and push the initial commit (or the whole history) to the just-created URL to store the subproject code.

5. In the superproject, read the subproject you have just extracted but, this time, as a proper submodule or subtree, whichever solution is a better fit and whichever method you prefer to use. Use the URL of the just-created public repository for the subproject.

6. Commit the changes in the superproject and push them to its public repository, in the case of submodules, including the newly created (or the just modified) `.gitmodules` file.

The recommended practice for the transformation of a subdirectory into a standalone submodule is to use a read-only URL for cloning (adding back) a submodule. This means that you can use either the `git://` protocol (warning: in this case, the server is unauthenticated) or `https://` without a username. The goal of this recommendation is to enforce separation by moving the work on a submodule code to a standalone separate subproject repository. In order to ensure that the submodule commits are available to all other developers, every change should go through the public repository for a subproject.

If this recommendation (best practice) is met with a categorical refusal, in practice, you could work on the subproject source code directly inside the superproject, though it is more error-prone. You would need to remember to commit and push in the submodule first, doing it from inside of the nested submodule subdirectory; otherwise, other developers would be not able to get the changes. This combined approach might be simpler to use, but it loses the true separation between implementing and consuming changes, which should be better assumed while using submodules.

Subtrees versus submodules

In general, subtrees are easier to use and less tricky. Many people go with submodules, because of the better built-in tooling (they have their own Git command, namely, `git submodule`), detailed documentation, and similarity to the Subversion externals, making them feel falsely familiar. Adding a submodule is very simple (just run `git submodule add`), especially compared to adding a subtree without the help of third-party tools such as `git subtree` or `git subrepo`.

The major difference between subtrees and submodules is that, with subtrees, there's only one repository, which means just one life cycle. Submodules and similar solutions use nested repositories, each with its own lifeline.

Though submodules are easy to set up and fairly flexible, they are also fraught with peril, and you need to practice vigilance while working with them. The fact that the submodules are opt-in also means that the changes touching the submodules demand a manual update by every collaborator. Subtrees are always there, so getting the superproject's changes means getting the subproject's too.

Commands such as `status`, `diff`, and `log` display precious little information about submodules, unless properly configured to cross the repository boundary; it is easy to miss a change. With subtrees, `status` works normally, while `diff` and `log` need some care because the subproject commits have a different root directory. The latter assumes that you did not decide to not include the subproject history (by squashing subtree merges). Then, the problem is only with the remote-tracking branches in the subproject's repository, if any.

Because the life cycles of different repositories are separate, updating a submodule inside its containing project requires two commits and two pushes. Updating a subtree-merged subproject is very simple: only one commit and one push. On the other hand, publishing the subproject changes upstream is much easier with submodules, while it requires changeset extraction with subtrees (here, tools such as `git subtree` help a lot).

The next major issue, and a source of problems, is that the submodule has two sources of the current revision: the gitlink in the superproject and the branches in the submodule's clone of the repository. This means that `git remote update` works a bit like a sideways push into a non-bare repository (see *Chapter 8, Advanced Branching Techniques*). Submodule heads are, therefore, generally detached, so any local update requires various preparatory actions to avoid creating a lost commit. There is no such issue with subtrees. All the revision changing commands work as usual with subtrees, bringing the subproject directory to the correct version without the requirement of any additional action. Getting changes from the subproject repository is just a subtree merge away. The only difference between ordinary pull is the `-s subtree` option.

Still, sometimes, submodules are the right choice. Compared to subtrees, they allow for a subproject (a module) to be not fetched, which is helpful when your code base is massive. Submodules are also useful when the heavy modularization is not natively handled, or not well natively handled, by the development stack's ecosystem.

Submodules might also themselves be superprojects for other submodules, creating a hierarchy of subprojects. Using nested submodules is made easier thanks to the `git submodule status`, `update`, `foreach`, and `sync` subcommands all supporting the `--recursive` switch.

Use cases for subtrees

With subtrees, there is only one repository (no nested repositories), just like a regular code base. This means that there is just one life cycle. One of the key benefits of subtrees is being able to mix container-specific customizations with general-purpose fixes and enhancements.

Projects can be organized and grouped together in whatever way you find to be most logically consistent. Using a single repository also reduces the overhead of managing dependencies.

The basic example of using subtrees is managing the customized version of a library, a *required dependency*. It is easy to get a development environment set up to run builds and tests. Monorepo makes it also viable to have one universal version number for all the projects. Atomic cross-submodule commits are possible; therefore, a repository can always be in a consistent state.

You can also use subtrees for *embedding related projects*, such as a GUI or a web interface, inside a superproject. In fact, many use cases for submodules can also apply to the subtrees solution, with the exception of the cases where there is a need for a subproject to be optional, or to have different access permissions than a master project. In those cases, you need to use submodules.

Use cases for monorepo

If all subprojects are managed by a single organization or a company, then it might be advantageous to have all those inter-related projects in a single repository, which we call monorepo.

One of the advantages is simplified organization. You can group and organize projects in whatever way you find to be most logically consistent. You don't need to consider how to split them into separate repositories, and how to join them into a superproject. It is also easier to navigate and search the history and the contents if all is in the single repository.

Because atomic cross-project commits are possible with monorepo, the repository can be always in a consistent state. It is easier to ensure that everything uses the same version of a specific component. Making cross-repository/cross-project changes in a polyrepo setting (multiple repositories, one per project, managed with a subtree or submodule strategy) is much more difficult than in a monorepo.

It is also easier to keep consistent tooling and a common **continuous integration** (**CI**) infrastructure.

Use cases for submodules

The strongest argument for the use of submodules is the issue of modularization. Here, the main area of use for submodules is handling plugins and extensions. Some programming ecosystems, such as ANSI C and C++ and also Objectve-C, lack good and standard support for managing version-locked multimodule projects. In this case, a plugin-like code can be included in the application (superproject) using submodules, without sacrificing the ability to easily update to the latest version of a plugin from its repository. The traditional solution of putting instructions about how to copy plugins in the README disconnects it from the historical metadata.

This schema can be extended also to the non-compiled code, such as the Emacs Lisp settings, configuration in dotfiles, (including frameworks such as `oh-my-zsh`), and themes (also for web applications). In these situations, what is usually needed to use a component is the physical presence of a module code at conventional locations inside the master project tree, which is mandated by the technology or framework being used. For instance, themes and plugins for WordPress, Magento, and so on are often de facto installed this way. In many cases, you need to be in a superproject to test these optional components.

Yet another particular use case for submodules is the division based on *access control and visibility restriction* of a complex application. For example, the project might use a cryptographic code with license restrictions, limiting access to it to a small subset of developers. With this code in a submodule with restricted access to its repository, other developers would simply be unable to clone this submodule. In this solution, the common build system needs to be able to skip cryptographic components if it is not available. On the other hand, the dedicated build server can be configured in such a way that the client gets the application built with crypto enabled.

A similar visibility restriction purpose, but in reverse, is making the source code of examples available long before it was to be published. This allows for better code thanks to the social input. The main repository for a book itself can be closed (private), but having an `examples/` directory contain a submodule intended for a sample source code allows you to make this subrepository public. While generating the book in the PDF and EPUB (and perhaps also MOBI) formats, the build process can then embed these examples (or fragments of them), as if they were ordinary subdirectories.

Third-party subproject management solutions

If you don't find a good fit in either `git subtree` or `git submodule`, you can try to use one of the many third-party projects to manage dependencies, subprojects, or collections of repositories.

One such tool is `repo` (`https://android.googlesource.com/tools/repo/`) used by the Android open source project to unify the many Git repositories for cross-network operations.

Another tool is `gil` (gitlinks) (`https://github.com/chronoxor/gil`) to manage complex recursive repositories dependencies, with cross references and cycles. Compared to submodules, `gil` avoids including the same dependency multiple times if the superproject and its subproject use the same library as a dependency. This tool also makes it easier to contribute changes upstream than with `git subtree`.

If you need to split a single monolithic repository into many standalone repositories, besides `git subtree split`, you can use a third-party `splitsh-lite` tool. If, on the other hand, you have multiple separate repositories that you want to merge into a single monorepo, you can use the `tomono` tool.

You can find many other such tools.

> **Important consideration**
>
> When choosing between native support and one of the many tools to manage many repositories together, you should check whether the tool in question uses a subtree-like or submodule-like approach to find whether it would be a good fit for your project.

Summary

This chapter provided all the tools you need to manage multicomponent projects with Git, from libraries and graphical interfaces, through plugins and themes, to frameworks.

You learned about the concept behind the subtrees technique and how to use it to manage subprojects. You know how to create, update, examine, and manage subprojects using subtrees.

You got to know the submodule approach of nested repositories for optional dependencies. You learned the ideas behind gitlinks, `.gitmodules`, and `.git` files. You encountered the pitfalls and traps for the unwary that you need to be vigilant about while using submodules. You know the reason for these problems and understand the notions behind them. You know how to create, update, examine, and manage subprojects using submodules.

You learned when to use subtrees and submodules, and their advantages and disadvantages. You know a few use cases for each technique.

Now that you know how to use Git effectively in a variety of circumstances and have learned the high-level ideas behind Git behavior that help you understand it, it's time to tackle how to make Git easier to use in *Chapter 13, Customizing and Extending Git*.

Questions

Answer the following questions to test your knowledge of this chapter:

1. What are subtrees, and what are their advantages and disadvantages?
2. What are submodules, and what are their advantages and disadvantages?
3. Why is the information about submodules duplicated between `.gitmodules` and project configuration?

Answers

Here are the answers to the questions given above:

1. With subtree merging, the history of the subproject (or its summary) is included in the superproject repository, and subproject files are put directly in a subirectory of the superproject and are superproject files. Subtrees can be used only for required dependencies, as they are embedded in a superproject. They are simpler to understand and use.

2. With submodules, repositories and the histories of superproject and subproject are kept separate. A superproject includes a link to the commit in a subproject. Submodules can be initialized and active, but can also be kept inactive, thus they can be used for optional dependencies. To include a change, you need to make a change in a subproject and to include it in a commit in a superproject.

3. The information in the project configuration file about submodules is local to the repository and, among others, defines which submodules are active and which are not.

Further reading

To learn more about the topics that were covered in this chapter, take a look at the following resources:

- *git-submodule* – Initialize, update, or inspect submodules: `https://git-scm.com/docs/git-submodule`

- *git-subtree* – Merge subtrees together and split the repository into subtrees: `https://github.com/git/git/blob/master/contrib/subtree/git-subtree.txt`

- Git documentation HOWTO – *How to use the subtree merge strategy*: `https://github.com/git/git/blob/master/Documentation/howto/using-merge-subtree.txt`

- Scott Chacon, Ben Straub: *Pro Git*, 2nd Edition (2014) `https://git-scm.com/book/en/v2`

  - Chapter *7.11 Git Tools - Submodules*

- Eric Pidoux, *Git Best Practices Guide* (2014), Packt Publishing Ltd

  - *Chapter 4, Going Deeper into Git, Managing Git Submodules*

- Johan Abildskov, *Practical Git: Confident Git Through Practice* (2020), Apress

  - *Chapter 8, Additional Git Features – Git Submodules*

- Everything you need to know about monorepos and the tools to build them: `https://monorepo.tools/`

12

Managing Large Repositories

Because of its distributed nature, Git includes the full change history in each copy of the repository. Every clone gets not only all the files but every revision of every file ever committed. This allows for efficient development (local operations not involving a network are usually fast enough so that they are not a bottleneck) and efficient collaboration with others (their distributed nature allows for many collaborative workflows).

But what happens when the repository you want to work on is huge? Can we avoid taking a large amount of disk space for version control storage? Is it possible to reduce the amount of data that end users need to retrieve while cloning the repository? Do we need to have all files present to be able to work on a project?

If you think about it, there are broadly three main reasons for repositories to become massive: they can accumulate a very long history (every revision direction), they can include huge binary assets that need to be managed together with code, the project can include a large number of files (every file direction), or any combination of those. For those scenarios, the techniques and workarounds are different and can be applied independently, though modern Git also includes a one-stop solution.

Submodules (presented in the previous chapter, *Managing Subprojects*) are sometimes used to manage large-size assets. This chapter will describe how this can be done while also presenting alternate solutions to the problem of handling large binary files and other large assets in Git.

In this chapter, we will cover the following topics:

- Git and large files
- Handling repositories with a very long history with a shallow clone
- Storing large binary files in a submodule or outside the repository
- Reducing the size of the working directory with sparse checkout
- How to make a local repository smaller with a sparse clone
- Which operations will require network access in different variants of sparse clone
- Faster operations with filesystem monitor

Scalar – Git at scale for everyone

The simplest way to configure Git so that it works better with large repositories, apart from enabling the relevant Git features, is to use the built-in `scalar` tool. This executable has been present in Git since version 2.38, which was released in 2022. Earlier, it was a separate project, then part of Microsoft's fork of Git.

Using it is very simple: instead of using `git clone`, you use `scalar clone`. If the repository has already been cloned, you can run `scalar register` to achieve the same result. One of the things that the command does is schedule background maintenance; you can stop this and remove the repository from the list of repositories that have been registered with `scalar` by using the `scalar unregister` command. The `scalar delete` command unregisters the repository and removes it from the filesystem.

After a `scalar` upgrade (which might be caused by moving to newer Git), you can run `scalar reconfigure --all` to upgrade all repositories registered with Scalar.

By registering the repository with Scalar (or the top-level directory of the project, which is called the **enlistment** in the Scalar documentation), you can turn on **partial clone** and **sparse-checkout**, configure Git to use **filesystem monitor**, and turn on **background maintenance** tasks such as **repository prefetching**.

All these features will be described in the following sections, as will some other features for handling large Git repositories that are more specific to users' needs.

Handling repositories with a very long history

Even though Git can effectively handle repositories with a long history, very old projects spanning a huge number of revisions can become a pain to clone. In many cases, you aren't interested in ancient history and don't want to pay the time to get all the revisions of a project and the disk space to store them. In this section, we will talk about techniques that you can use to clone truncated history, or how to make Git fast despite the long history.

For example, if you want to propose a new feature or a bug fix, you might not want to wait for the full clone to finish, which may take a while.

> **Editing project files online**
>
> Some Git repository hosting services, such as GitHub, offer a web-based interface to manage repositories, including in-browser file management and editing. They may even automatically create a fork of the repository so that you can write and propose changes.
>
> But a web-based interface doesn't cover everything, and you might be using self-hosted repositories or a service that doesn't provide this feature.

However, fixing the bug might require running git bisect on your machine, where the regression bug is easily reproducible (see *Chapter 4, Exploring Project History*, for how to use bisection). If you're tight on space and time, you might want to try to do either a shallow clone (described in the following subsection) or a sparse clone (described later in this chapter).

Using shallow clones to get truncated history

The simple solution to a fast clone and to save disk space is to perform a **shallow clone** using Git. This operation allows you to get a local copy of the repository with the history truncated to a particular specified **depth** – that is, the number of latest revisions.

How do you do it? Just use the --depth option:

```
$ git clone --depth=1 https://git.company.com/project
```

The preceding command only clones the most recent revision of the primary branch. This trick can save quite a bit of time and relieve a great deal of load from the servers. Often, a shallow clone finishes in seconds rather than minutes, which is a significant improvement. This can be useful if you're only interested in checking out project files, and not in the whole history, such as what's inside Git hooks or GitHub Actions – that is, the case of builds where you delete the clone immediately after the action.

Since version 1.9, Git supports pull and push operations even with shallow clones, though some care is still required. You can change the depth of a shallow clone by providing the --depth=<n> option to git fetch (however, note that tags for the deepened commits aren't fetched). To turn a shallow repository into a complete one, use --unshallow.

> **Important note**
>
> Since the commit history in a shallow clone is truncated, commands such as git merge-base and git log show different results than they would in a full clone. This will happen if you try to go outside the depth of the clone. Also, because of how the Git server is optimized, incremental fetch in a shallow repository might take longer than using fetch in a full repository. Fetch might also unexpectedly make the repository not so shallow.

Note that git clone --depth=1 may still get all the branches and all the tags. This can happen if the remote repository doesn't have HEAD, so it doesn't have a primary branch selected; otherwise, only the tip of the said single branch is fetched. Long-lived projects usually had many releases during their long history. To save time, you would need to combine shallow clone with the next solution: branch limiting.

With modern Git, it might make more sense to use the **partial clone** feature instead.

Cloning only a single branch

By default, Git clones all the branches and tags (if you want to fetch notes or replacements, you need to specify them explicitly). You can limit the amount of history you clone by specifying that you want to **clone only a single branch**:

```
$ git clone --branch master --single-branch \
  https://git.company.com/project
```

Because most of the project history (most of the DAG of revisions) is shared among branches, with very few exceptions, you probably won't see a huge difference using this technique.

This feature might be quite useful if you don't want detached orphan branches or the opposite: you only want an orphan branch (for example, with a web page for a project, or a branch used for GitHub Pages). Single-branch cloning works well in regard to saving disk space when they're used together with a very shallow clone (with so short a history that most branches don't have time to converge).

Making operations faster in repositories with a long history

One of the features that makes Git faster on repositories with a very long history is the **commit-graph** file. Using this feature, which is turned on by default as of Git 2.24, configures Git to periodically write or update a helper file with a serialized (and easy-to-access) graph of revisions. This makes Git operations that query project history much faster.

You can turn this feature off by setting the `core.commitGraph` configuration variable to `false`. If you need to refresh the helper file, you can do this with the `git commit-graph write` command.

> **Avoiding doing the work**
>
> One unexpected place that might get slower with long history is running `git status`. This is caused by the command in question computing detailed ahead/behind counts for the current branch (how many commits you have on the local branch ahead of the upstream branch in the remote repository, how many commits in the remote repository you are behind).
>
> You can turn off computing this information with the `--no-ahead-behind` option, or by setting the `status.aheadBehind` configuration variable to `false`. Nowadays, `git status` will print this advice when it is slowed by ahead/behind calculations.

Handling repositories with large binary files

In some specific circumstances, you might need to track **huge binary assets** in the code base. For example, gaming teams have to handle huge 3D models, and web development teams might need to track raw image assets or Photoshop documents. Both gaming development and web development might require video files to be under version control. Additionally, sometimes, you might want the convenience of including large binary deliverables that are difficult or expensive to generate – for example, storing a snapshot of a virtual machine image.

There are some tweaks you can make to improve how binary assets are handled by Git. For binary files that change significantly from version to version (and not just change some metadata headers), you might want to turn off the **delta compression** by adding `-delta` explicitly for specific types of files in a `.gitattributes` file (see *Chapter 3*, *Managing Your Worktrees*, and *Chapter 13*, *Customizing and Extending Git*). Git will automatically turn off delta compression for any file above the `core.bigFileThreshold` size, which is 512 MiB by default. You may also want to turn the compression off (for example, if a file is in the compressed format already). However, because `core.compression` and `core.looseCompression` are global for the whole repository, it makes more sense if binary assets are in a separate repository (submodule).

Splitting the binary asset folder into a separate submodule

One possible way of handling large binary asset *folders* is to split them into a separate repository and pull the assets into your main project as a **submodule**. The use of submodules gives you a way to control when assets are updated. Moreover, if a developer doesn't need those binary assets to work, they can simply exclude the submodule with assets from fetching.

The limitation is that you need to have a separate folder with these huge binary assets that you want to handle this way. Additionally, the service hosting the submodule repository with those large assets needs to be able to store those large files; many Git hosting sites impose hard limits on the maximum size of a file, or of the repository.

Storing large binary files outside the repository

Another solution is to use one of the many third-party tools that try to solve the problem of handling large binary files in Git repositories. Many of them use a similar paradigm, namely storing the contents of huge binary files outside the repository while providing some kind of pointers to the contents in the checkout.

There are three parts to each such implementation:

- How they store the information about the contents of the managed files inside the repository
- How they share large binary files between a team
- How they integrate with Git (and their performance penalty)

While choosing a solution, you need to take this data into account, along with the operating system support, ease of use, and the size of the community.

What's stored in the repository and what's checked in might be a *symlink* to the file or the key, or it might be a *pointer file* (often plain text), which acts as a reference to the actual file contents (by name or by the cryptographic hash of file contents). The tracked files need to be stored in some kind of backend for collaboration (cloud service, rsync, shared directory, and so on). Backends might be accessed directly by the client, or there might be a separate server with a defined API into which the blobs are written, which would, in turn, offload the storage elsewhere.

The tool might either require the use of separate commands for checking out and committing large files and for fetching from and pushing to the backend, or it might be integrated into Git. The integrated solution uses the `clean/smudge` filters to handle check-out and check-in transparently, and the `pre-push` hook to send large file contents transparently together. You only need to state which files to track and, of course, initialize the repository for the tool use.

The advantage of a filter-based approach is its ease of use; however, there is a performance penalty because of how this approach works. Using separate commands to handle large binary assets makes the learning curve a bit steeper but provides better performance. Some tools provide both interfaces.

Among different solutions, there's **git-annex**, which has a large community and support for various backends, and **Git-Large File Storage** (**Git-LFS**), created by GitHub, which provides good Microsoft Windows support, a client-server approach, and transparency (with support for a filter-based approach). The Git-LFS extension is supported not only by GitHub but also by other Git hosting sites and software forges, such as GitLab, Bitbucket, and Gitea. Specialized services and projects for implementing Git-LFS also exist.

There are many other such tools, but those two are the most popular, and both are still maintained.

Versioning data files for data analysis and machine learning

Machine learning projects often process large files or large numbers of files. Those include the raw dataset, but also the results of various pre-processing steps, as well as the trained model.

You want to store those large files or directories somewhere to avoid having to re-download or re-compute them. On the other hand, you also want to be able to recreate everything from scratch, to make the science reproducible. Those requirements are different enough from the ones that are encountered in typical software projects that need to handle large assets, where specialized solutions for integrating data handling and version control are necessary. Among such solutions, there's **Data Version Control** (**DVC**) and **Pachyderm**.

Handling repositories with a large number of files

The rise in the use of monorepos (this concept was explained in detail in *Chapter 11, Managing Subprojects*) has led to the need to handle repositories with large amounts of files. In a monorepo – that is, a repository composed of many interconnected subprojects – you would usually work on a single subproject and access and change files only within a specific subdirectory.

Limiting the number of working directory files with sparse checkout

Git includes the **sparse checkout** technique, which allows you to explicitly detail which files and folders you want to populate on checkout. This mode can be turned on by setting the `core.sparseCheckout` configuration variable to `true` and uses the `.git/info/sparse-checkout` file with the gitignore-like syntax to specify what is to appear in the working directory. The index (also known as the staging area) is populated in full, with the `skip-worktree` flag set for files missing from checkout.

While it can be helpful if you have a huge tree of folders, it doesn't affect the overall size of the local repository itself. To reduce the size of the repository, it needs to be used together with **sparse clone** (which will be described later).

However, sparse checkout definitions are extremely generic. This makes the feature very flexible but at the cost of bad performance for large definitions and large amounts of files. With a monorepo, you don't need that flexibility as each subproject is contained in its own subdirectory – you only directory matches in sparse checkout definitions.

To achieve this, you need to use **"cone mode" sparse-checkout**. Note that since Git 2.37, the non-cone mode of the `sparse-checkout` feature is deprecated (see, for example, the `git sparse-checkout` command's documentation). This mode has the additional advantage that it is much easier to use. Everything is managed with the help of the `git sparse-checkout` command.

To restrict your working directory to a given set of directories, run the following command:

```
$ git sparse-checkout set <directory_1> <directory_2>
```

Earlier versions of the feature required you to run `git sparse-checkout init --cone` first, but using this command is no longer needed, and the `init` subcommand is itself being deprecated.

> **Tip**
>
> If you're cloning a repository with a large number of files, you can avoid filling out the working directory with them by using the `--no-checkout` or `--sparse` option of `git clone` (the second option will only check out files in the top directory of the project). You can add the `--filter=blob:none` option for even more speed (turning on blobless sparse clone).

At any point, you can check which directories are included in your `sparse-checkout` definitions, and are present in your working directory, using the following command:

```
$ git sparse-checkout list
```

You can add a new directory to your existing sparse checkout with the `add` subcommand, as shown in the following example:

```
$ git sparse-checkout add <new_directory>
```

At the time of writing, there's no `remove` subcommand. To remove a directory from the list of checked-out files, you would need to edit the contents of the `.git/info/sparse-checkout` file and then run the following subcommand:

```
$ git sparse-checkout reapply
```

This subcommand reapplies the existing sparse directory specifications to make the working directory match. It can also be used when some operation updates the working directory without fully respecting `sparse-checkout` definitions. This might be caused by using tools external to Git, or by running Git commands that do not fully support sparse checkouts.

You can turn off this feature and restore the working directory so that it includes all files by running the `git sparse-checkout disable` command.

Reducing the local repository size with sparse clone

The initial section of *Chapter 11*, *Managing Subprojects*, described how Git stores the history of the project, which includes a description of changes, directory structure, and file contents at each revision. This data is stored using different types of objects: tag objects, commit objects, tree objects, and blob objects. Objects reference other objects: tags point to commits, commits point to a parent commit(s), trees represent the state of the project at a given revision, and trees point to other trees and blobs.

When running the ordinary `git clone` command, the client asks the server for the latest commits (representing the latest revisions). The server provides those objects, all objects they point to, all objects those objects point to, and so on. In short, the server provides those commit objects and every other reachable object (excluding possibly those objects that the client already has). The result is that you have the whole history of the whole project available locally.

Nowadays, however, many developers have network connections available as they work. Modern Git only allows you to download a subset of objects via **partial clone**. In this case, Git remembers where it can get the rest of the objects, and later asks the server for more data when it turns out to be necessary.

Git's partial clone feature can be enabled by specifying the `--filter` option when running the `git clone` command. There are several filters available, but the server that hosts the repository you're cloning can choose to deny your filter and revert to creating a full clone.

Running `git fetch` in sparse clone preserves sparse clone filters, and it doesn't download those types of objects that would not be downloaded by the initial clone.

The two most commonly used filters that should be supported by most Git hosting sites are as follows:

- **Blobless clone**: `git clone --filter=blob:none <url>`
- **Treeless clone**: `git clone --filter=tree:0 <url>`

When using the `--filter=blob:none` option, the initial `git clone` command will download everything but the blob objects (which ordinarily contain different versions of file contents for different files). The checkout part of the `clone` operation (if not suppressed) will download blobs for current versions of project files. The Git client knows how to batch those download requests to ask the server only for the missing blobs.

With **blobless clone**, you will trigger a blob download whenever you need the *contents* of the file. This means that `git log`, `git merge-base`, and other commands that do not examine file contents run without the need for additional download of blob objects.

Moreover, to examine if the file has been changed, Git can simply compare object IDs, and it doesn't need to access the actual contents. Therefore, examining file history with `git log -- <path>` doesn't need to download any objects either. This command runs with the same performance as in a full clone. This is the result of the fact that the object ID is based on the cryptographic hash of file contents (Git currently uses SHA-1 for this purpose).

Git commands such as `git checkout/git switch`, `git reset --hard <revision>`, and `git merge` need to download blobs to populate the working directory, the index (the staging area), or both. To compute diffs, Git also needs to have blobs to compare; therefore, commands such as `git diff` or `git blame <path>` might trigger blob downloads the first time they are run with specific arguments.

In some repositories, the tree data might be a significant portion of the repository's size. This might happen if the repository has a large amount of files and directories and deep and wide directory hierarchies. In such cases, using a **treeless clone** with the `--filter=tree:0` option might offer a better solution.

Note that any objects that are only referenced by those objects that were skipped due to the selected filter will also be missing. This means that the treeless clone is more sparse than the blobless clone (as only trees can point to blobs… well, a tag object can point to a blob object, but you won't typically encounter this).

The advantage of a treeless clone over a blobless clone is a much faster initial clone and faster subsequent fetches. The disadvantage is that working in a treeless clone is more difficult because downloading a missing tree when needed is more expensive. It is also more difficult for the server to notice that the client already has some tree objects locally, so the request might send more data than necessary. Additionally, more commands require additional data to be downloaded. An example of this is `git log -- <file>`, which in the blobless clone could be run without the need to download anything extra. In a treeless clone, the command will start downloading trees for almost every commit in the history.

> **Treeless clones and submodules**
>
> The repositories that contain submodules (see *Chapter 11, Managing Subprojects*) may behave poorly with treeless clones. If you get too many tree download requests, you can either *turn off the automatic fetching of submodules* by ensuring that the `fetch.recurseSubmodules` configuration variable is set to `false` (or by using the `--no-recurse-submodules` option) or *also filter submodules* by setting the `clone.filterSubmodules` config option (or using the `--recurse-submodules --filter=tree:0 --also-filter-submodules` combination of command-line options).

Treeless clones are helpful for automatic builds, when you want to quickly clone the project, check out a single revision, compile it and/or run a test, and then throw away the repository (instead of using shallow clone). They are also useful if all you're interested in is examining the history of the whole project.

The treeless clone is a special case of the **depth-limited clone**; the full syntax of the tree filter specification is `--filter=tree:<depth>`. In this case, the clone omits all blobs and trees whose depth from the root tree (from the top directory of the project) is greater than or equal to the specified limit. It can be easily seen that with `<depth>` being equal to 0 (that is, `--filter=tree:0`), the clone will not include any trees or blobs (except for those required for initial checkout).

Omitting large file contents with sparse clone

The partial clone can also work as a tool to help you work with large files. This requires that the Git server (the Git hosting site) supports the specific type of filter. It also doesn't remove the requirement that at least one remote repository must include those large files and their history so that you can download them on demand.

You can do this by providing the `--filter=blob:limit=<size>` option when you're cloning, where `<size>` can include the **k**, **m**, and **g** suffixes. This will make Git omit blobs of the size at least `<size>` bytes, be it KiB, MiB, or GiB (depending on the suffix). For example, `blob:limit=1k` is the same as `blob:limit=1024`.

Matching clone sparsity to checkout sparsity

Modern Git includes basic support for the sparse clone filter, which makes it omit all blobs that would be not required for sparse checkout. For security reasons, however, support for the easier-to-use form of `--filter=sparse:path=<path>` was dropped from Git. The supported form is `--filter=sparse:oid=<blob-ish>`. This form is safe against the time of check to time of use problem, as opposed to the path-based form, because `<blob-ish>` (that is, a reference to a blob object) ultimately resolves to the object ID that defines its contents.

At the time of writing, you would be hard to find a Git server that supports this filter and doesn't respond with **warning: filtering not recognized by server, ignoring**. But when it starts getting widely supported, one possible solution would be to create a tag for every sparse checkout pattern of interest, and then use the selected tag for blob-ish:

```
$ git sparse-checkout set <subdir>
$ git hash-object -t blob -w .git/info/sparse-checkout
bd177ff9527327c67f50c644c421d280bb8b55f5
$ git tag -a -m 'sparse-checkout pattern for <subdir>'
  sparse/<subdir> bd177ff9527327c67f50c644c421d280bb8b55f5
$ git push origin tag sparse-checkout/<subdirectory>
To <repository url>
* [new tag]              sparse/<subdir> -> sparse/<subdir>
```

Of course, in the third step, you need to use the SHA-1 that's output from the previous command.

In this case, cloning would use the `--filter=sparse:oid=sparse/<subdir>^{blob}` option (where you would need to use the name of the tag that was created by the sequence of commands shown previously).

Faster checking for file changes with filesystem monitor

When you run a Git command that operates on the worktree, such as `git status` or `git diff`, Git has to discover what changed relative to the index, or relative to the specified revision. It does that by searching the entire worktree, which for repositories with a large number of files can take a long time. It also has to rediscover the same information from scratch every time you run such a command.

Filesystem monitor is a long-running daemon or a service process that does the following:

- Registers with the operating system to watch specified directories and receive change notification events for directories and files of interest

- Keeps the pathnames of those changed watched files and directories in some (in-memory) data structure that can be queried quickly

- Responds to client requests for a list of files and directories that have been modified recently

Since version 2.37, Git includes the **built-in file system monitor** (FSMonitor) known as `git fsmonitor--daemon`. It is currently available on macOS and Windows. This daemon listens for IPC connections from client processes, such as `git status`, and sends a list of changed files over a Unix domain socket or a named pipe.

Turning it on is very simple; you just need to configure Git to use it. This can be done with the following command:

```
$ git config core.fsmonitor true
```

This monitor works well with `core.untrackedCache`, so it is recommended to set this configuration option to `true` as well.

You can query this daemon for the list of watched repositories:

```
$ git fsmonitor--daemon status
fsmonitor-daemon is watching 'C:/work/chromium'
```

If either the operating system or the filesystem the repository is on does not allow you to use this monitor, there is an option to use the **hook-based file system monitor**. This type of hook can be turned on by setting the `core.fsmonitor` config option to the path to the filesystem monitor hook. The hook must support the `fsmonitor-watchman` hook protocol, and when run return the list of changed files on standard output.

Git comes with the `fsmonitor-watchman.sample` file, which is installed inside the `.git/hooks/` directory. Before turning it on, as described in the previous paragraph, rename it by removing the `*.sample` suffix. If the file is missing, you can download it from `https://github.com/git/git/tree/master/templates`. This hook requires the **Watchman** file's watching service (`https://facebook.github.io/watchman/`) to be installed.

Summary

This chapter provided solutions to handling large Git repositories, from the use of the Scalar tool to specialized solutions.

First, you learned how to use shallow clone to download and operate on the selected shallow subset of the project history.

Then, you learned how to handle large files by storing them outside the repository or separating them into submodules. The problem of large data in data science projects was briefly mentioned, as were specialized solutions to this problem.

Finally, you learned how to manage large monorepos with sparse checkout, sparse clone, and filesystem monitor.

The next chapter will help you make Git easier to use and better fit it to your specific circumstances. This includes configuring repository maintenance, which is particularly important for making working with large repositories smooth.

Questions

Answer the following questions to test your knowledge of this chapter:

1. What is the simplest solution to handling large repositories?
2. How you can make cloning faster for repositories with a long history?

3. How can you handle large files that are needed only by some developers?

4. What techniques make working with repositories with large numbers of files faster?

5. What's the difference between shallow clone, sparse clone, and sparse checkout?

Answers

Here are the answers to this chapter's questions:

1. Use the built-in `scalar` tool, either using it to clone the repository or to register the given repository with the tool.

2. You can use shallow clone or blobless sparse clone. In the first case, you would get a shortened history, while in the second case, the repository's size will be smaller but some operations will require network access to download additional data.

3. You can store large files outside the repository with Git-LFS or git-annex (or a similar solution). You can clone the repository without downloading large file data with the sparse clone feature.

4. Use the sparse checkout feature if you're only working inside a specific subdirectory, use sparse clone to reduce repository size, and use filesystem monitor (if possible) to make operations faster.

5. Shallow clone only downloads selected part of the repository history, and all local operations are limited to this selection, though it is easy to change the depth of the history. Sparse clone reduces repository size by downloading only a selected subset of objects, fetching those objects on demand, as their presence becomes necessary to perform operations. Sparse checkout reduces the number of checked-out files, making the working directory smaller (and operations faster).

Further reading

To learn more about the topics that were covered in this chapter, take a look at the following resources:

- *Introducing Scalar: Git at scale for everyone*, by Derrick Stolee (2020): `https://devblogs.microsoft.com/devops/introducing-scalar/`

- *The Story of Scalar*, by Derrick Stolee and Victoria Dye (2022): `https://github.blog/2022-10-13-the-story-of-scalar/`

- *scalar(1) - A tool for managing large Git repositories*: `https://git-scm.com/docs/scalar`

- *Supercharging the Git Commit Graph*, by Derrick Stolee (2018): `https://devblogs.microsoft.com/devops/supercharging-the-git-commit-graph/`

- *git-commit-graph(1) - Write and verify Git commit-graph files*: `https://git-scm.com/docs/git-commit-graph`

- *Git LFS - Git Large File Storage*: `https://git-lfs.com/`

- *git-annex*: `https://git-annex.branchable.com/`

- *Get up to speed with partial clone and shallow clone*, by Derrick Stolee (2020): `https://github.blog/2020-12-21-get-up-to-speed-with-partial-clone-and-shallow-clone/`

- *Bring your monorepo down to size with sparse-checkout*, by Derrick Stolee (2020): `https://github.blog/2020-01-17-bring-your-monorepo-down-to-size-with-sparse-checkout/`

- *git-sparse-checkout(1) - Reduce your working tree to a subset of tracked files*: `https://git-scm.com/docs/git-sparse-checkout`

- *git-clone(1) - Clone a repository into a new directory*: `https://git-scm.com/docs/git-clone`

- *Improve Git monorepo performance with a file system monitor*, Jeff Hostetler (2022): `https://github.blog/2022-06-29-improve-git-monorepo-performance-with-a-file-system-monitor/`

- *git-fsmonitor--daemon - A Built-in Filesystem Monitor*: `https://git-scm.com/docs/git-fsmonitor--daemon`

- *githooks - Hooks used by Git: fsmonitor-watchman*: `https://git-scm.com/docs/githooks#_fsmonitor_watchman`

13

Customizing and Extending Git

Earlier chapters were designed to help you understand how Git works and master Git as a version control system. The following two chapters will help you set up and configure Git, so that you can use it more effectively for yourself (this chapter) and help other developers use it (the next chapter).

This chapter will cover configuring and extending Git to fit one's needs. First, it will show how to set up a Git command line to make it easier to use. For some tasks, though it is easier to use visual tools; the short introduction to graphical interfaces in this chapter should help you in choosing one. Next, there will be an explanation of how to change and configure Git behavior, from configuration files (with the selected configuration options described) to a per-file configuration with the `.gitattributes` file.

Then this chapter will cover how to automate Git with hooks, describing for example how to make Git check whether the commit being created passes coding guidelines for a project. This part will focus on the client-side hook, and will only touch upon the server-side hooks— those are left for *Chapter 14, Git Administration*. The last part of the chapter will describe how to extend Git, from the Git command aliases, through integrating new user-visible commands, to helpers and drivers (new backend abilities).

Many issues, such as gitattributes, remote and credential helpers, and the basics of the Git configuration should be known from the previous chapters. This chapter will gather this information in a single place, and expand it a bit.

In this chapter, we will cover the following topics:

- Setting up the shell prompt and Tab completion for a command line
- Types and examples of graphical user interfaces
- Configuration files and basic configuration options
- Installing and using various types of hooks
- Simple and complex aliases
- Extending Git with new commands and helpers

Git on the command line

There are a lot of different ways to use the Git version control system. There are many **graphical user interfaces** (**GUIs**) of varying use cases and capabilities, and there exist tools and plugins that allow integration with an **integrated development environment** (**IDE**) or a file manager.

However, the command line is the only place you can run all of the Git commands and which provides support for all their options. New features, which you might want to use, are developed for the command line first. Also, most of the GUIs implement only some subsets of the Git functionality. Mastering the command line always guarantees a deep understanding of tools, mechanisms, and their abilities. Just knowing how to use a GUI is probably not enough to get a founded knowledge.

Whether you use Git on a command line from choice, as a preferred environment, or you need it because it is the only way to access the required functionality, there are a few shell features that Git can tap into to make your experience a lot friendlier.

Git-aware command prompt

It's useful to customize your **shell prompt** to show information about the state of the Git repository we are in.

> Definition
>
> The **shell prompt** is a short text message that is written to the terminal or the console output to notify the user of the interactive shell that some typed input is expected (usually a shell command).

This information can be as simple or as complex as you want. Git's prompt might be similar to the ordinary command-line prompt (to reduce dissonance), or visibly different (to be able to easily distinguish that we are inside the Git repository).

There is an example implementation for bash and zsh shells in the contrib/ area. If you install Git from the sources, just copy the contrib/completion/git-prompt.sh file to your home directory; if you have installed Git on Linux via a package manager, you will probably have it at /etc/bash_completion.d/git-prompt.sh. This file provides the __git_ps1 shell function to generate a Git-aware prompt in the Git repositories, but first, you need to source this file in your .bashrc or .zshrc shell configuration file:

```
if [ -f /etc/bash_completion.d/git-prompt.sh ]; then
    source /etc/bash_completion.d/git-prompt.sh
fi
```

The shell prompt is configured using environment variables. To set up a prompt, you must change, directly or indirectly, the PS1 (prompt string one, the default interaction prompt) environment variable. Thus, one solution to create a Git-aware command prompt is to include a call to the __git_ps1 shell function in the PS1 environment variable, by using command substitution:

```
export PS1='\u@\h:\w$( git_ps1 " (%s)")\$ '
```

Note that, for zsh, you would also need to turn on the command substitution in the shell prompt with the setopt PROMPT_SUBST command.

Alternatively, for a slightly faster prompt and the possibility of color, you can use __git_ps1 to set PS1. This is done with the PROMPT_COMMAND environment variable in bash and with the precmd() function in zsh. You can find more information about this option in the comments in the git-prompt.sh file; for bash, it could be the following:

```
PROMPT_COMMAND='__git_ps1 "\\u@\\h:\\w" "\\\$ "'
```

With this configuration (either solution), the prompt will look as follows:

```
bob@host.company.org:~/random/src (master)$
```

The Git Bash command from Git for Windows comes out of the box with a similar prompt configured (though the Git Bash default prompt takes two lines, not one).

The bash and zsh shell prompts can be customized with the use of special characters that get expanded by a shell. In the example used here (you can find more for example in the *Bash Reference Manual*), we have the following:

- \u means the current user (**bob**)
- \h is the current hostname (**host.company.org**)
- \w means the current working directory (~/**random/src**)
- \$ prints the $ part of the prompt (# if you are logged in as the root user)

$(...) in the PS1 setup is used to call external commands and shell functions. __git_ps1 " (%s)" here calls the git_ps1 shell function provided by git-prompt.sh with a formatting argument: the %s token is the placeholder for the presented Git status. Note that you need to either use single quotes while setting the PS1 variable from the command line, as in the example shown here, or escape shell substitution, so it is expanded while showing the prompt and not while defining the variable.

If you are using the __git_ps1 function, Git will also display information about the current ongoing multistep operation: merging, rebasing, bisecting, and so on. For example, during an interactive rebase (-i) on the branch master, the relevant part of the prompt would be **master|REBASE-i**. It is very useful to have this information right here in the command prompt, especially if you get interrupted in the middle of the operation.

It is also possible to indicate in the command prompt the state of the working tree, the index, and so on. We can enable these features by exporting the selected subset of these environment variables (for some features you can additionally turn it off on a per-repository basis with provided boolean-valued configuration variables):

- `GIT_PS1_SHOWDIRTYSTATE` (with `bash.showDirtyState` for per-repository settings) shows "*" for unstaged changes and "+" for staged changes, if set to a non-empty value.

- `GIT_PS1_SHOWSTASHSTATE` shows "**$**" if something is stashed.

- `GIT_PS1_SHOWUNTRACKEDFILES` and `bash.showUntrackedFiles` show "**%**" if there are untracked files in the working directory.

- `GIT_PS1_SHOWUPSTREAM` and `bash.showUpstream` can be used to configure the ahead-behind state of the upstream repository, with a value of `auto` in a space-separated list of values makes the prompt show whether you are behind "<", up to date "=", or ahead ">" of the upstream, `name` shows the upstream name, and `verbose` details the number of commits you are ahead/behind (with sign; for example "+1" for being 1 commit ahead). `git` compares HEAD to `@{upstream}` and `svn` to the SVN upstream.

- `GIT_PS1_DESCRIBE_STYLE` can be set to configure how to show information about a detached HEAD situation; it can be set to one of the following values: `contains` uses newer annotated tags (v1.6.3.2~35), `branch` uses newer tag or branch (main~4), `describe` uses the older annotated tag (v1.6.3.1-13-gdd42c2f), `tag` uses any tag, `default` shows tag only if it is exactly matching the current commit.

- `GIT_PS1_SHOWCONFLICTSTATE` set to "`yes`" will notify the user if there are unresolved conflicts with |**CONFLICT**.

- `GIT_PS1_SHOWCOLORHINTS` can be used to configure colored hints about the current dirty state, that is about whether there are uncommitted changes (like `git status -sb` does).

- `GIT_PS1_HIDE_IF_PWD_IGNORED` or `bash.hideIfPwdIgnored` are used to not show a Git-aware prompt if the current directory is set to be ignored by Git, even if we are inside a repository.

If you are using the `zsh` shell, you can take a look at the `zsh-git` set of scripts, the `zshkit` configuration scripts, or the `oh-my-zsh` framework available for `zsh`, instead of using `bash`—first complete the prompt setup from the Git contrib/. Alternatively, you can use the `vcs_info` subsystem built in to `zsh`.

There are also alternative prompt solutions for `bash` (usually for multiple different shells), for example, `git-radar` or `powerline-shell`.

> **Tip**
> You can, of course, generate your own Git-aware prompt. For example, you might want to split the current directory into the repository path part and the project subdirectory path part with the help of the `git rev-parse` command.

Command-line completion for Git

Another shell feature that makes it easier to work with the Git command line is the programmable **command-line completion**. This feature can dramatically speed up typing Git commands. Command-line completion allows you to type the first few characters of a command, or a filename, and press the completion key (usually *Tab*) to fill the rest of the item. With the Git-aware completion, you can also fill in subcommands, command-line parameters, remotes, branches, and tags (ref names), each only where appropriate (for example, remote names are completed only if the command expects the remote name at a given position).

Git comes with built-in (but not always installed) support for the auto-completion of Git commands for the `bash` and `zsh` shells.

For `bash`, if the completion functionality is not installed with Git (at `/etc/bash_completion.d/git.sh` in Linux by default), you need to get a copy of the `contrib/completion/git-completion.bash` file out of the Git source code. Copy it somewhere accessible, such as your home directory, and source it from your `.bashrc` or `.bash_profile`:

```
. ~/git-completion.bash
```

Once the completion for Git is enabled, to test it, you can start to type a Git command, then press the *Tab* key. For example, you can type `git check` and then press *Tab*:

```
$ git check<TAB>
```

With Git completion enabled, the `bash` (or `zsh`) shell would autocomplete what you entered so far to `git checkout`.

Similarly, in an ambiguous case, a double *Tab* press shows all the possible completions (though this is not true for all shells; some instead cycle through different completions):

```
$ git che<TAB><TAB>
checkout        cherry          cherry-pick
```

The completion feature also works with options; this is quite useful if you don't remember the exact option but only the prefix:

```
$ git config --<TAB><TAB>
--add                   --get-regexp            --remove-section     --unset
[...]
```

> **Important note**
>
> Instead of the list of possible completions, some shells use (or can be configured to use) rotating completion, where with multiple possible completions, each *Tab* shows a different completion for the same prefix (cycling through them).

Note that command-line completion (also called **tab completion**) generally works only in the interactive mode, and is based on the unambiguous prefix, not on the unambiguous abbreviation.

Autocorrection for Git commands

An unrelated built-in Git tool, but similar to tab completion, is **autocorrection**. By default, if you type something that looks like a mistyped command, Git helpfully tries to figure out what you meant. It still refuses to perform the guessed operation, even if there is only one candidate:

```
$ git chekout
git: 'chekout' is not a git command. See 'git --help'.

The most similar command is
        checkout
```

However, with the `help.autoCorrect` configuration variable set to a positive number, Git will automatically correct and execute the mistyped commands after waiting for the given number of deciseconds (0.1 of second). You can use a negative value with this option for immediate execution, or zero to go back to the default:

```
$ git chekout
WARNING: You called a Git command named 'chekout', which does not
exist.
Continuing in 0.1 seconds, assuming that you meant 'checkout'.
Your branch is up-to-date with 'origin/master'.
```

If there is more than one command that can be deduced from the entered text, nothing will be executed. This mechanism works only for Git commands; you cannot autocorrect subcommands, parameters, and options (as opposed to tab completion).

Making the command line prettier

Git fully supports a colored terminal output, which greatly aids in visually parsing the command output. A number of options can help you set the coloring to your preference.

First, you can specify when to use colors, such as for the output of certain commands. There is a `color.ui` master switch to control output coloring to turn off all of Git's colored terminal outputs and set them to `false`. The default setting for this configuration variable is `auto`, which makes Git color the output when it's going straight to a terminal, but omit the color-control codes when the output is redirected to a file or a pipe.

You can also set `color.ui` to `always`, though you'd rarely want this: if you want color codes in your redirected output, simply pass a `--color` flag to the Git command; conversely, the `--no-color` option would turn off the colored output.

If you want to be more specific about which commands and which parts of the output are colored, Git provides appropriate coloring settings: `color.branch`, `color.diff`, `color.interactive`, `color.status`, and so on. Just as with the `color.ui` master switch, each of these can be set to `true`, `false`, `auto`, and `always`.

In addition, each of these settings has subsettings that you can use to set specific colors for specific parts of the output. The color value of such configuration variables – for example, `color.diff.meta` (to configure the coloring of meta information in your diff output) – consists of space-separated names of the foreground color, the background color (if set), and the text attribute.

You can set the color to any of the following values: `normal`, `black`, `red`, `green`, `yellow`, `blue`, `magenta`, `cyan`, or `white`. As for the attributes, you can choose from `bold`, `dim`, `ul` (underline), `blink`, and `reverse` (swap the foreground color with the background one).

The pretty formats for `git log` also include an option to set colors; see the `git log` documentation for more information.

> **External tools**
>
> There are diff syntax highlighters that can be used with Git. They can be set up to work as a pager with the `core.pager` config variable, or configured via an alias. Examples include **delta** (`https://dandavison.github.io/delta`) and **diff-highlight** from the contrib area of Git source code.

Alternative command line

To understand some of the rough edges of the Git user's interface, you need to remember that Git was developed to a large extent in a bottom-up fashion. Historically, Git began as a tool to write version-control systems (you can see how early Git was used in the *A Git core tutorial for developers* documentation that you can view with the `git help core-tutorial` command).

The first alternative "porcelain" (i.e., alternative user interface) for Git was *Cogito*. Nowadays, Cogito is no more; all of its features have long been incorporated into Git (or replaced by better solutions). There were some attempts to write wrapper scripts (alternative UIs) designed to make it easy to learn and use, for example, **Easy Git** (`eg`) and the newer **Gitless**. **Jujutsu** (`jj`) is a version control system in the early stages of development that can use Git repositories to store project history, and thus can be thought as a layer on top of Git, too.

There are also external Git porcelains that do not intend to replace the whole user interface, but either provide access to some extra features, or wrap Git to provide some restricted feature set. **Patch management interfaces**, such as **StGit**, **TopGit**, or **Guilt** (formerly Git Queues (`gq`)), are created to make it easy to rewrite, manipulate, and clean up selected parts of the unpublished history; these were mentioned as an alternative to an interactive rebase in *Chapter 10, Keeping History Clean*. Then, there are single-file version control systems, such as **Zit**, which use Git as a backend.

> **Alternative implementations**
>
> Beside alternative user interfaces, there are also different implementations of Git (defined as reading and writing Git repositories). They are at different stages of completeness. Besides the core C implementation, there is **JGit** in Java, and also the **libgit2** project—the modern basis of Git bindings for various programming languages.

Graphical interfaces

You have learned how to use Git on the command line. The previous section told you how to customize and configure it to make it even more effective. But the terminal is not the end. There are other kinds of environments you can use to manage Git repositories. Sometimes, a visual representation is what you need.

Now, we'll take a short look at the various kinds of user-centered graphical tools for Git; the tour of Git administrative tools is left for the next chapter, *Chapter 14, Git Administration*.

Types of graphical tools

Different tools and interfaces are tailored for different workflows. Some tools expose only a selected subset of the Git functionality or encourage a specific way of working with version control.

To be able to make an informed choice when selecting a graphical tool for Git, you need to know what types of operations the different types of tools support. Note that one tool can support more than one type of use.

First, there is the **graphical history viewer**. You can think of it as a powerful GUI over `git log`. This is the tool to be used when you are trying to find something that happened in the past, or you are visualizing and browsing your project's history and the layout of branches. Such tools usually accept revision selection command-line options, such as `--all`. Command-line Git has `git log --graph` and the less-used `git show-branch` that uses ASCII art to show the history.

A similar tool is **graphical blame**, showing the line-wise history of a file. For each line, it can show when that line was created and when it was moved or copied to the current place. You can examine the details of each of the commits shown and usually browse through the history of the lines in a file. Other tools with similar applications, namely examining the evolution of the line range (`git log -L`) and the so-called pickaxe search (`git log -S`), do not have many GUIs.

Next, there are **commit tools** meant primarily to craft (and amend) commits, though usually they also include some kind of worktree management (for example, ignoring files and switching branches) and functionality for **management of remotes**. Such tools usually show both unstaged and staged changes, allowing you to move files between these states. Some of those tools, such as the interactive versions of `git add`, `git reset`, and so on, even allow you to stage and unstage individual chunks of changes. A graphical version of an interactive add is described in *Chapter 3, Managing Your Worktrees*, and mentioned in *Chapter 2, Developing with Git*. There are also tools to craft commit messages following specified criteria.

Then, we have **file manager integration** (or **graphical shell integration**). These plugins usually show the status of the file in Git (tracked/untracked/ignored) using icon overlays. They can offer a context menu for a repository, directory, and file, often with accompanying keyboard shortcuts. They may also bring drag and drop support.

Programmer editors and IDE) often offer support for **IDE integration** with Git (or version control in general). These offer repository management (as a part of team project management), make it possible to perform Git operations directly from the IDE, show the status of the current file and the repository, and perhaps even annotate the view of the file with version control information. They often include the commit tool, remote management, the history viewer, and the diff viewer.

Git repositories' hosting sites often offer workflow-oriented **desktop clients**. These mostly focus on a curated set of commonly used features that work well together in the flow. They automate common Git tasks. They are often designed to highlight their service, offering extra features and integration, but they will work with any repository hosted anywhere.

There are even **specialized editors and pagers**, such as a graphical editor for the interactive rebase instruction sheet (see *Chapter 9, Merging Changes Together*), that can be set up with the `sequence.editor` config variable, or a syntax highlighting tool for diffs that can be set up as the default Git pager with `core.pager`.

Graphical diff and merge tools

Graphical diff tools and graphical merge tools are somewhat special cases. In these categories, Git includes the commands for integration with third-party graphical tools, namely, `git difftool` and `git mergetool`. These tools are then called from the Git repository. Note that this is different from the external diff or diff merge drivers, which replace ordinary `git diff` or augment it.

Although Git has an internal implementation of diff and a mechanism for merge conflict resolutions (see *Chapter 9, Merging Changes Together*), you can use an external graphical diff tool instead. These are often used to show the differences better (usually, as a side-by-side diff, possibly with refinements), and help resolve a merge (often with a three-pane interface).

Configuring the graphical diff or graphical merge tool requires configuring a number of custom settings. To tell which tool to use for diff and merge, respectively, you can set up `diff.tool` and `merge.tool`, respectively. Without setting, for example, the `merge.tool` configuration variable, the `git mergetool` command would print the information on how to configure it, and will attempt to run one of predefined tools:

```
$ git mergetool

This message is displayed because 'merge.tool' is not configured.
See 'git mergetool --tool-help' or 'git help config' for more details.
```

```
'git mergetool' will now attempt to use one of the following tools:
tortoisemerge emerge vimdiff
No files need merging
```

Running `git mergetool --tool-help` will show all the available tools, including those that are not installed. If the tool you use is not in $PATH, or it has the wrong version of the tool, you can use `mergetool.<tool>.path` to set or override the path for the given tool:

```
$ git mergetool --tool-help
'git mergetool --tool=<tool>' may be set to one of the following:
        vimdiff           Use Vim with a custom layout
[...]
The following tools are valid, but not currently available:
        araxis           Use Araxis Merge
[...]
Some of the tools listed above only work in a windowed
environment. If run in a terminal-only session, they will fail.
```

If there is no built-in support for your tool, you can still use it; you just need to configure it. The `mergetool.<tool>.cmd` configuration variable specifies how to run the command, while `mergetool.<tool>.trustExitCode` tells Git whether the exit code of that program indicates a successful merge resolution or not. The relevant fragment of the configuration file (for a graphical mergetool named `extMerge`) could look as follows:

```
[merge]
    tool = extMerge
[mergetool "extMerge"]
    cmd = extMerge "$BASE" "$LOCAL" "$REMOTE" "$MERGED"
```

There are a few config options that control `git mergetool` behavior, either globally or on a per-tool basis. One of those is `mergetool.hideResolved` (and its per-tool `mergetool.<tool>.hideResolved` variant), which makes Git resolve as many conflicts as possible by itself, and presents only unresolved conflicts to the merge tool. Note that some merge tools do this themselves.

Some merge tools, such as vimdiff, are text interface tools that can work without the need for a graphical session. If you want to run one tool in text mode (for example, when using plain SSH access to a remote host), and another one in graphical mode, you can do this by configuring `mergetool.tool` in one tool, and `mergetool.guitool` in another – and use `git mergetool --gui` to invoke the GUI one.

Graphical interface examples

In this section, you will be presented with a selection of tools around Git that you could use, or that might prompt you to research further. A nice way to help you start this research is to list some selected GUI clients.

There are two visual tools that are a part of Git and are usually installed with it, namely **gitk** and **git-gui**. They are written in Tcl/Tk. `gitk` is a **graphical history viewer**, while `git gui` is a **graphical commit tool**; there is also `git gui blame`, a visually **interactive line-history browser**. These tools are interconnected: for example, browsing history from `git gui` opens `gitk`.

Visual tools do not need to use the graphical environment. There is **tig** (short for *Text Interface for Git*) that uses a nurses-based text-mode interface (TUI) and functions as a repository browser and commit tool and can act as a Git pager.

Another TUI example is **git interactive-rebase-tool**, which can be set up as an interactive sequence editor for the interactive rebase instruction sheet.

There is **git cola**, developed in Python and available for all the operating systems, which includes commit tools and remotes management, and also a diff viewer. Then, there is the simple and colorful **Gitg** tool for GNOME; you will get a graphical history viewer, diff viewer, and file browser.

One of the more popular open source GUI tools for macOS is **GitX**. There are a lot of forks of this tool; one of the more interesting ones is **Gitbox**. It features both the history viewer and commit tools.

For MS Windows, there is **TortoiseGit** and **git-cheetah**, both of which offer integration into the Windows context menu, so you can perform Git commands inside Windows Explorer (the file manager integration and shell interface).

GitHub Inc. and Atlassian both released a desktop GUI tool that you can easily use with your GitHub or Bitbucket repository, respectively, but neither tool is not limited to only interacting with a single service (GitHub or Bitbucket, respectively). Both **GitHub Client** and **SourceTree** feature repository management and offer a range of other common facilities to enhance your development workflow.

Many programming editors and IDEs have support for managing Git repositories, and sometimes also for interacting with Git hosting sites. This can be either built in, or available as IDE plugins or extensions. Examples include **GitLens** for Visual Studio Code, **Magit** for GNU Emacs, and **Fugitive** for ViM. Those tools often show information such as which lines are added or changed, or who authored them, inside the editor pane.

Configuring Git

So far, while describing how Git works and how to use it, we have introduced a number of ways to change its behavior. In this section, it will be explained in a systematic fashion how to configure Git operations on a temporary and permanent basis. We will also see how you can make Git behave in a customized fashion by introducing and reintroducing several important configuration settings. With these tools, it's easy to get Git to work the way you want it to.

Command-line options and environment variables

Git processes the switches that change its behavior in a hierarchical fashion, from the least specific to the most specific one, with the most specific one (and shortest term) taking precedence.

The most specific one, overriding all the others, is the command-line options. They affect, obviously, only the current Git command.

Important note

One issue to note is that some command-line options, for example, `--no-pager` or `--no-replace-objects`, go to the `git` wrapper, not to the Git command itself. Examine, for example, the following line to see the distinction:

```
$ git --no-replace-objects log -5 --oneline --graph --decorate
```

You can find the conventions used through the Git command-line interface on the manpage.

The second way to change how the Git command works is to use environment variables. They are specific to the current shell, and you need to use the `export` built-in command (or its equivalent) to propagate the variables to the subprocesses if a replacement is used. There are some environment variables that apply to all core Git commands, and some that are specific to a given (sub)command.

Git also makes use of some nonspecific environment variables. These are meant as a last resort; they are overridden by their Git-specific equivalents. Examples include variables such as `PAGER` and `EDITOR`.

Git configuration files

The final way to customize how Git works is with the configuration files. In many cases, there is a command-line option to configure an action, an environment variable for it, and finally a configuration variable, in descending order of preference.

Git uses a series of configuration files to determine non-default behavior that you might want to have. There are four layers of these files that Git looks through for configuration values. Git reads all these files in order from the least specific to the most specific one. The settings in the later ones override those set in the earlier ones. You can access the Git configuration with the `git config` command: by default, it operates on the union of all the files, but you can specify which one you want to access with the command-line options. You can also access any given file following the configuration file syntax (such as the `.gitmodules` file mentioned in *Chapter 11, Managing Subprojects*) by using the `--file=<pathname>` option (or the `GIT_CONFIG` environment variable).

Tip

You can also read the values from any blob with configuration-like contents; for example, you may use `git config --blob=master:.gitmodules` to read from the `.gitmodules` file in the `master` branch.

The first place Git looks for configuration is the **system-wide configuration file**. If Git is installed with the default settings, it can be found in /etc/gitconfig. Well, at least, on Linux it is there, as the **Filesystem Hierarchy Standard** (**FHS**) states that /etc is the directory for storing the host-specific system-wide configuration files; Git for Windows puts this file in the subdirectory of its Program Files folder. This file contains the values for every user on the system and all their repositories. To make git config read from and write and to this file specifically (and to open it with --edit), pass the --system option to the git config command.

You can skip the reading settings from this file with the GIT_CONFIG_NOSYSTEM environment variable. This can be used to set up a predictable environment or to avoid using a buggy configuration you can't fix.

The next place Git looks is the **user-specific configuration file**: ~/.gitconfig, falling back to ~/.config/git/config if it exists (with the default configuration). This file is specific to each user and it affects all of the user's repositories. If you pass the --global option to git config, it will read and write from this file specifically. Reminder: here, as in the other places, ~ (the tilde character) denotes the home directory of the current user ($HOME).

Finally, Git looks for the configuration values in the **per-repository configuration file** in the Git repository you are currently using, which is (by default and for non-bare repositories) .git/config. Values set there are specific to that local single repository. You can make Git read and write to this file by passing the --local option.

With modern Git, if the extensions.worktreeConfig is set to true (the default value is false), there can also be a **per-worktree configuration file** in the .git/config.worktree file (see the git worktree command).

Each of these levels (system, global, and local) overrides the values from the previous level, so for example, values in .git/config trump those in ~/.gitconfig; well, unless the configuration variable is multivalued.

> **Tip**
>
> You can use the fact that the local (per-repository) configuration overrides the global (per-user) configuration to have your default identity in the per-user file and to override it if necessary on a per-repository basis with a per-repository configuration file.

Finally, you can set the config variable for an individual command with the -c option to the git wrapper:

```
$ git -c foo.int=1k config --get --type=int foo.int
1024
```

See the following sections for a full explanation of this result.

The syntax of Git configuration files

Git's configuration files are plain text, so you can also customize Git's behavior by manually editing the chosen file. The syntax is fairly flexible and permissive; whitespaces are mostly ignored (contrary to .gitattributes). The hash # and the semicolon ; characters begin comments, which last until the end of the line. Blank lines are ignored.

The file consists of sections and variables, and its syntax is similar to the syntax of INI files. Both the section names and variable names are case-insensitive. A section begins with the name of the section in square brackets [**section**] and continues until the next section. Each variable must begin at some section, which means that there must be a section header before the first setting of a variable. Sections can repeat and can be empty.

Sections can be further divided into subsections. Subsection names are case-sensitive and can contain any character except newline (double quotes " and backslash \ must be escaped as \" and \\, respectively). The beginning of the subsection will look as follows:

```
[section "subsection"]
```

All the other lines (and the remainder of the line after the section header) are recognized as a setting variable in the name = value form. As a special case, just name is a shorthand for name = true (boolean variables). Such lines can be continued to the next line by ending it with \ (the backslash character), that is by escaping the end-of-line character. Leading and trailing whitespaces are discarded; internal whitespaces within the value are retained verbatim. You can use double quotes to preserve leading or trailing whitespaces in values.

Includes and conditional includes

You can include one config file from another by setting the special variable include.path to the path of the file to be included. The included file will be expanded immediately, similar to the mechanism of #include in C and C++. The path is relative to the configuration file with the include directive. You can turn this feature off with the --no-includes option.

You can also conditionally include a config file from another similarly by setting an includeIf.<condition>.path variable. The condition starts with a keyword, followed by a colon :, and data relevant to the type of conditional included.

The supported keywords are as follows:

- **gitdir**, where the data that follows the keyword is used as a glob pattern to match the location of the .git directory (of the repo itself). For convenience, ~and ~/ at the beginning of the pattern are substituted with the location of the home directory, and ./ at the beginning of the pattern is replaced with the directory containing the current config file. There is also **gitdir/i** variant that does the matching in a case-insensitive way.

- **onbranch**, which can be used to match the currently checked-out branch against the glob pattern.

- **hasconfig:remote.*.url**, which checks whether, in any of the configuration, there exists at least one remote URL that matches the glob pattern.

For example, to use a different configuration for repositories inside the `work-repos/` subdirectory in your home directory, you could use the following:

```
[includeIf "gitdir:~/work-repos/"]
    path = ~/work.inc
```

Accessing the Git configuration

You can use the `git config` command to access the Git configuration, starting from listing the configuration entries in a canonical form, through examining individual variables, to editing and adding entries.

You can query the existing configuration with `git config --list`, adding an appropriate parameter if you want to limit to a single configuration layer. On a Linux box with the default installation, in the fresh empty Git repository just after `git init`, the local (per-repository) setting would look approximately like the following:

```
$ git config --list --local
core.repositoryformatversion=0
core.filemode=false
core.bare=false
core.logallrefupdates=true
```

You can also use `git var -l` to list all configuration and environment variables affecting Git.

You can also query a single key with `git config`, limiting (or not) the scope to the specified file, by giving the name of the configuration variable as a parameter (optionally preceded by `--get`), with the section, optional subsection, and variable name (key) separated by a dot:

```
$ git config user.email
```

This would return the last value, that is, the one with the greatest precedence. You can get all the values with `--get-all`, or specific keys with `--get-regexp=<match>`. This is quite useful while accessing a multivalued option such as `refspecs` for a remote.

Types of configuration variables and type specifiers

While requesting (or writing) a config variable, you can give a *type specifier* with the `--type=<type>` option. The type can be `bool`, which ensures that the returned value is **true** or **false**; `int`, which expands the optional value suffix of **k** (1,024 elements), **m** (1024k), or **g** (1024m); `path`, which expands ~ for the value of $HOME; and ~**user** for the home directory of the given user, and `expiry-date` to convert a fixed or relative date string to a timestamp.

There is also `bool-or-int`, and a few options related to storing colors and retrieving color escape codes; see the `git config` documentation.

With `--get`, `--get-all`, and `--get-regexp`, you can also limit the listing (and the settings for multiple-valued variables) to only those variables matching the value `regexp` (which is passed as an optional last parameter). For example, to find all configurations that affect proxying for a given host, you can use the following command:

```
$ git config --get core.gitproxy 'for kernel\.org$'
```

You can also use the `git config` command to set the configuration variable value. The local layer (per-repository file) is the default for writing if nothing else is specified. For example, to set the email address of the user, which is to be common to most of their repositories, you can run the following:

```
$ git config --global user.name "Alice Developer"
```

For multivalue configuration options (multivar), you can add multiple lines to it by using the `--add` option. To change a single entry of a multivar config variable, you can use something like the following command, where the first value denotes which value to change, and the second denotes the new value:

```
$ git config core.gitproxy '"ssh" for kernel.org' '"ssh" for kernel\.
org$'
```

It is also very easy to delete configuration entries with `git config --unset`.

Instead of setting all the configuration values on the command line, as shown in the preceding example, it is possible to set or change them just by editing the relevant configuration file directly. Simply open the configuration file in your favorite editor, or run the `git config --edit` command.

The local repository configuration file just after a fresh `git init` on Linux looks as follows:

```
[core]
        repositoryformatversion = 0
        filemode = true
        bare = false
        logallrefupdates = true
```

If you want to change a configuration by editing the configuration file, it might be prudent to first find out where the configuration variable you want to change came from.

Finding where configuration value came from

With three (or four) layers of configuration files, it might be difficult to find out where the given configuration variable was set, and whether it was overridden or added to in a more specific configuration file. Then there is the additional complication of taking into account the `include` and `includeIf` sections.

That is where the `--show-origin` and `--show-source` options passed to the `git config` command together with the `--list` or `--get`/`--get-all` options can help. The `git config` `--list` command will list all variables set in config files, along with their values. The `--show-scope` option augments the output of all queried config options with the scope of that value (worktree, local, global, system, command), while `--show-source` shows the origin type (file, standard input, command line, blob) and the actual origin (config file path, or blob ID, if applicable).

> **Debugging per-file configuration**
>
> You can use `git check-ignore` to examine why a file is ignored, and `git check-attr` to find out the attributes assigned to the file and where they came from.

For example, let's assume that user identity is defined in a per-user configuration file: `~/.gitconfig`

```
[user]
    name  = Joe Random
    email = joe@company.com
```

Let's also assume that the `git config` command was used to create the `work.inc` file in the top directory of the project, and to include it from the per-repository config file:

```
$ git config --file=conf.inc foo.bar val
$ git config --local include.path ./conf.inc
```

In that context, we would get the following query results, shown here in shortened form:

```
$ git config --show-scope --show-origin --list
global   file:/home/joe/.gitconfig   user.name=Joe Random
global   file:/home/joe/.gitconfig   user.email=joe@company.com
local    file:.git/config            core.repositoryformatversion=0
local    file:.git/config            core.filemode=false
[...]
local    file:.git/config            include.path=./../conf.inc
local    file:.git/./../conf.inc foo.bar=val
```

The first column shows the scope, the second column the origin, and the third the fully qualified config variable and its value.

Basic client-side configuration

You can divide the configuration options recognized by Git into two categories: client side and server side. The majority of the options are about configuring your personal working preferences; they are client side. The server-side configuration will be touched upon in more detail in *Chapter 14, Git Administration*; in this section, you will find only the basics.

There are many supported configuration options, but only a small fraction of them *needs* to be set; a large fraction of them has sensible defaults, and explicitly setting them is only useful in certain edge cases. There are a lot of options available; you can see a list of all the options with `git config --help`. Here we'll be covering only the most common and most useful options.

Two variables that really need to be set up are **user.email** and **user.name**. Those configuration variables define the user's identity (though in modern Git, you can set up separate identities for authoring changes and for committing them with **author.name** and **committer.name**). Also, if you are signing annotated tags or commits (as discussed in *Chapter 6, Collaborative Development with Git*), you might want to set up your GPG signing key ID. This is done with the **user.signingKey** configuration setting.

By default, Git uses whatever you've set on the system as your default text editor (defined with the `VISUAL` or `EDITOR` environment variables; the first only for the graphical desktop environment) to create and edit your commit and tag messages. It also uses whatever you have set as the pager (`PAGER`) for paginating and browsing the output of the Git commands. To change this default to something else, you can use the **core.editor** setting. The same goes for **core.pager**. Git would ultimately fall back on the `vi` editor and on the `less` pager.

With Git, the pager is invoked automatically. The default `less` pager supports not only pagination, but also incremental search and other features.

> **Important note**
>
> With the default configuration (the `LESS` environment variable is not set), `less` when invoked by Git works as if it was invoked with `LESS=FRX`. This means that it would skip pagination if there were less than one page of output, it would pass through ANSI color codes, and it would not clear the screen on exit.

Creating commit messages is also affected by **commit.template**. If you set this configuration variable, Git will use that file as the default message when you commit. The template is not distributed with the repository in general. Note that Git would add the status information to the commit message template unless it is forbidden to do it by setting **commit.status** to false.

Such a template is quite convenient if you have a commit-message policy, as it greatly increases the chances of this policy being followed. It can, for example, include the commented-out instructions for filling the commit message. You can augment this solution with an appropriate hook that checks whether the commit message matches the policy (see the *Commit process hooks* section later in this chapter).

The status of the files in the working area is affected by the ignore patterns and the file attributes (see *Chapter 3, Managing Your Worktrees*). You can put ignore patterns in your project's in-tree `.gitignore` file (usually, `.gitignore` is about which files are not to be tracked, and it is tracked itself by Git), or in the `.git/info/excludes` file for local and private patterns, to define which files are not interesting. These are project-specific; sometimes, you would want to write a kind of global (per-user) `.gitignore` file. You can use **core.excludesFile** to customize the path to the said file; in modern Git, there is a default value for this path, namely, `~/.config/git/ignore`. There is also a corresponding **core.attributesFile** for this kind of global `.gitattributes` files, which defaults to `~/.config/git/attributes`.

> **Trivia**
>
> Actually, it is `$XDG_CONFIG_HOME/git/ignore`; if the `$XDG_CONFIG_HOME` environment variable is not set or is empty, `$HOME/.config/git/ignore` is used.

Although Git has an internal implementation of diff, you can set up an external tool to be used instead with the help of **diff.external**. You would usually want to create a wrapper script that massages the parameters that Git passes to it and passes the ones needed in the order external diff requires. By default, Git passes the following arguments to the diff program:

```
path old-file old-hex old-mode new-file new-hex new-mode
```

See also the *Graphical diff and merge tools* section for the configuration of `git difftool` and `git mergetool`.

The rebase and merge setup, configuring pull

When performing `git pull` operation, Git needs to know whether you prefer to use the *merge* operation to join the local history and the history fetched from the remote, or the *rebase* operation to join histories. That is why it requires you to provide a value for the `pull.rebase` configuration variable. You can find more information on the topic of merge and rebase in *Chapter 9, Merging Changes Together*.

There are several configuration settings that can be used to configure the behavior of `git pull`. There is the `pull.rebase` configuration option and a branch-specific `branch.<name>.rebase` option that, when set to `true`, tells Git to perform a rebase during the pull operation (for the `<name>` branch only in the latter case). If set to `false`, then `git pull` performs a merge. Both can also be set to `merges` to run rebase with the `--rebase-merges` option, to have local merge commits not be flattened in the process of rebasing.

You can make Git automatically set up the per-branch "pull to rebase" configuration while creating specific kinds of new branches with `branch.autoSetupRebase`. You can set it to `never`, `local` (for locally tracked branches only), `remote` (for remote tracked branches only), or `always` (for local plus remote). There is also `branch.autoSetupMerge` to set up a branch to track another branch.

Preserving undo information – the expiry of objects

By default, Git will automatically remove unreferenced objects, clean **reflogs** of stale entries, and pack loose objects, all to keep the size of the repository down. You can also run the garbage collection manually with the `git gc` command. You should know about a repository's object-oriented structure from *Chapter 10, Keeping History Clean*.

Git will, for safety reasons, use a grace period of two weeks while removing unreferenced objects; this can be changed with the `gc.pruneExpire` configuration: the setting is usually a relative date (for example, `1.month.ago`; you can use dots as word separators). To disable the grace period (which is usually done from the command line), the `now` value can be used.

The branch tip history is kept for 90 days by default (or `gc.reflogExpire`, if set) for reachable revisions, and for 30 days (or `gc.reflogExpireUnreachable`) for reflog entries that are not a part of the current history. Both settings can be configured on a per-reframe basis, by supplying a pattern of the ref name to be matched as a subsection name, that is, `gc.<pattern>.reflogExpire`, and similar for the other setting. This can be used to change the expire settings for HEAD or for `refs/stash` (see *Chapter 3, Managing Your Worktrees*), or for remote-tracking branches `refs/remotes/*` separately. The setting is a length of time (for example, 6 months); to completely turn off reflog expiry, use the value of `never`. You can use the latter, for example, to switch off the expiring of `stash` entries.

Formatting and whitespace

Code formatting and whitespace issues are some of the more frustrating and subtle problems you may encounter while collaborating, especially with cross-platform development. It's very easy for patches and merges to introduce subtle and unnecessary whitespace changes, because editing the code can silently introduce such changes (which are often not visible), and because there are different notions of line endings on different operating systems: MS Windows, Linux, and macOS. Git has a few configuration options to help with these issues.

One important issue for cross-platform work is the notion of **line-ending**. This is because MS Windows uses a combination of a **carriage return** (**CR**) character and a **linefeed** (**LF**) character for new lines in text files, whereas macOS and Linux use only a linefeed character. Many editors on MS Windows will silently replace existing LF-style line endings with CRLF or use CRLF for new lines, which leads to subtle but annoying issues.

Git can handle this issue by auto-converting line endings into LF when you add a file to the index. If your editor uses CRLF line endings, Git can also convert line endings to the native form when it checks out code in your filesystem. There are two configuration settings that affect this matter: `core.eol` and `core.autocrlf`. The first setting, `core.eol`, sets the line ending to be used while checking out files into the working directory for files that have the `text` property set (see the following *Per-file configuration with gitattributes* section, which summarizes and recalls information about the file attributes from *Chapter 3, Managing Your Worktrees*).

The second and older setting, `core.autocrlf`, can be used to turn on the automatic conversion of line endings to CRLF. Setting it to `true` converts the LF line endings in the repository into CRLF when you check out files, and vice versa when you stage them; this is the setting you would probably want on a Windows machine. (This is almost the same as setting the `text` attribute to **auto** on all the files and `core.eol` to `crlf`.) You can tell Git to convert CRLF to LF on a commit but not the other way around by setting `core.autocrlf` to `input` instead; this is the setting to use if you are on a Linux or Mac system. To turn off this functionality, recording the line-endings in the repository as they are set this configuration value to `false`.

This handles one part of the whitespace issues – line-ending variance, and one vector of introducing them – editing files. Git also comes with a way to detect and fix some of the other whitespace issues. It can be configured to look for a set of common whitespace problems. The `core.whitespace` configuration setting can be used to activate them (for those disabled by default) or turn them off (for those enabled by default). The three that are turned on by default are the following:

- `blank-at-eol`: This looks for trailing spaces at the end of a line
- `blank-at-eof`: This notices blank lines at the end of a file
- `space-before-tab`: This looks for spaces immediately before the tabs at the initial (beginning) indent part of the line

The `trailing-space` value in `core.whitespace` is a shorthand to cover both `blank-at-eol` and `blank-at-eof`.

The three that are disabled by default but can be turned on are the following:

- `indent-with-non-tab`: This treats the line that is indented with space characters instead of the equivalent tabs as an error (where equivalence is controlled by the `tabwidth` option). This option enforces *indenting with Tab characters*.
- `tab-in-indent`: This watches for tabs in the initial indentation portion of the line (here, `tabwidth` is used to fix such whitespace errors). This option enforces *indenting with space characters*.
- `cr-at-eol`: This tells Git that carriage returns at the end of the lines are OK (allowing CRLF endings in the repository).

You can tell Git which of these you want enabled or disabled by setting `core.whitespace` to the comma-separated list of values. To disable an option, prepend it with the - prefix in front of the value. For example, if you want all but `cr-at-eol` and `tab-in-indent` to be set, and also while setting the *Tab* space value to 4, you can use:

```
$ git config --local core.whitespace \
    trailing-space,space-before-tab,indent-with-non-tab,tabwidth=4
```

You can also set these options on a per-file basis with the `whitespace` attribute. For example, you can use it to turn off checking for whitespace problems in test cases to handle whitespace issues or ensure that the Python 2 code indents with spaces:

```
*.py whitespace=tab-in-indent
```

> **EditorConfig**
>
> There exists the EditorConfig project (`https://editorconfig.org/`) that consists of a file format for defining coding styles, including the type of line endings, and a collection of text editor plugins that make editors adhere to the chosen style. The `.editorconfig` file should be tracked by Git.

Git will detect these issues when you run a `git diff` command and inform you about them using the `color.diff.whitespace` color, so you can notice them and possibly fix them before you create a new commit. While applying patches with `git apply`, you can ask Git to either warn about the whitespace issues with `git apply --whitespace=warn`, error out with `--whitespace=error`, or you can have Git try to automatically fix the issue with `--whitespace=fix`. The same applies to the `git rebase` command as well.

Server-side configuration

There are a few configuration options available for the server side of Git. They will be described in more detail in *Chapter 14*, *Git Administration*; here you will find a short summary of some of the more interesting parameters.

You can make the Git server check for object consistency, namely, that every object received during a push matches its SHA-1 identifier, and that it is a valid object, with a `receive.fsckObjects` Boolean-valued configuration variable. It is turned off by default because `git fsck` is a fairly expensive operation, and it might slow down operations, especially on large pushes (which are common in large repositories). This is a check against faulty or malicious clients.

If you rewrite commits that you have already pushed to a server (which is bad practice, as explained in *Chapter 10*, *Keeping History Clean*) and try to push again, you'll be denied. The client might, however, force-update the remote branch with the `--force` flag to the `git push` command. However, the server can be told to refuse force-pushes by setting `receive.denyNonFastForward` to `true`.

The `receive.denyDeletes` setting blocks one of the workarounds to the `denyNonFastForward` policy, namely, deleting and recreating a branch. This forbids the deletion of branches and tags; you must remove refs from the server manually.

All of these features could also be implemented via the server-side receive-like hooks; this will be covered in the *Installing a Git hook* section, and also to some extent in *Chapter 14*, *Git Administration*.

Per-file configuration with gitattributes

Some of the customizations can also be specified for a path (perhaps via glob) so that Git applies these settings only for a subset of files or for a subdirectory. These path-specific settings are called gitattributes.

The order of precedence of applying this type of settings starts with the per-repository local (per-user) per-path settings in the $GIT_DIR/info/attributes file. Then, the .gitattributes files are consulted, starting with the one in the same directory as the path in question, going up through the .gitattributes files in the parent directories, up to the top level of the worktree (the root directory of a project).

Finally, the global per-user attributes file (specified by core.attributesFile, or at ~/.config/git/attributes if this is not set) and the system-wide file (in /etc/gitattributes in the default installation) are considered.

Available Git attributes are described in detail in *Chapter 3, Managing Your Worktrees*. Using attributes, you can, among others, do things such as specify the separate merge strategies via merge drivers for the specific kind of files (for example, ChangeLog), tell Git how to diff non-text files, or have Git filter content during checkout (on writing to the working area, that is, to the filesystem) and commit (on staging contents and committing changes to the repository, that is, creating objects in the repository database).

Syntax of the Git attributes file

A gitattributes file is a simple text file that sets up the local configuration on a per-path basis. Blank lines and lines starting with the hash character (#) are ignored; thus, a line starting with # serves as a comment, while blank lines can serve as separators for readability. To specify a set of attributes for a path, put a pattern followed by an attributes list, separated by a horizontal whitespace:

```
pattern     attribute1 attribute2
```

When more than one pattern matches the path, a later line overrides an earlier line, just like for the .gitignore files (you can also think that the Git attributes files are read from the least specific system-wide file to the most specific local repository file).

Git uses a backslash (\) as an escape character for patterns. Thus, for patterns that begin with a hash, you need to put a backslash in front of the first hash (that is written as \#). Because the attributes information is separated by whitespaces, trailing spaces in the pattern are ignored and inner spaces are treated as the end of the pattern unless they are quoted with a backslash (that is, written as "\ ").

If the pattern does not contain a slash (/), which is a directory separator, Git will treat the pattern as a shell glob pattern and will check for a match against the pathname relative to the location of the .gitattributes file (or the top level for other attribute files). Thus, for example, the *.c patterns match the C files anywhere down from the place the .gitattributes file resides. A leading slash matches the beginning of the pathname. For example, /*.c matches bisect.c but not builtin/bisect--helper.c, while the *.c pattern would match both.

If the pattern includes at least one slash, Git will treat it as a shell glob suitable for consumption by the fnmatch(3) function call with the FNM_PATHNAME flag. This means that the wildcards in the pattern will not match the directory separator, that is, the slash (/) in the pathname; the match is anchored to the beginning of the path. For example, the `include/*.h` pattern matches `include/version.h` but not `include/linux/asm.h` or `libxdiff/includes/xdiff.h`. The shell glob wildcards are the following:

- `*` matching any string (including empty)

- `?` matching any single character

- `[...]` expression matching the character class (inside brackets, asterisks and question marks lose their special meaning); note that unlike in regular expressions, the complementation/negation of the character class is done with `!` and not `^`. For example, to match anything but a number, one can use the `[!0-9]` shell pattern, which is equivalent to `[^0-9]` in a regexp.

Two consecutive asterisks (`**`) in patterns may have a special meaning, but only between two slashes (`/**/`), or between a slash and at the beginning or the end of the pattern. Such a wildcard matches zero or more path components. Thus, a leading `**` followed by a slash (`**/`) means a match in all directories, while a trailing `/**` matches every file or directory inside the specified directory.

Each attribute can be in one of four states for a given path:

- First, it can be **set** (the attribute has a special value of true). This is specified by simply listing the name of the attribute in the attribute list, for example, `text`.

- Second, it can be **unset** (the attribute has a special value of false). This is specified by listing the name of the attribute prefixed with minus, for example, `-text`.

- Third, it can be **set to a specific value**; this is specified by listing the name of the attribute followed by an equal sign and its value, for example, `text=auto` (note that there cannot be any whitespace around the equal sign as opposed to the configuration file syntax).

- If no pattern matches the path, and nothing dictates whether the path has or does not have attributes, the attribute is said to be **unspecified** (you can override the setting for an attribute, forcing it to be explicitly unspecified with `!text`).

If you find yourself using the same set of attributes over and over for many different patterns, you should consider defining a macro attribute. This can be defined in the local, global, or system-wide attributes file, but (from all possible places for a repository-specific attributes file), macros can be defined only in the top level `.gitignore` file. The macro is defined using `[attr]<macro>` in place of the file pattern; the attributes list defines the expansion of the macro. For example, the built-in `binary` macro attribute is defined as follows:

```
[attr]binary -diff -merge -text
```

But command-line options, environment variables, configuration files, gitattributes, and gitignore files are not the only ways to change what Git is doing. There is also the hooks mechanism, which can be used to make Git trigger user-defined actions automatically at specific points in Git's execution.

Automating Git with hooks

There are usually certain prerequisites to the code that is produced, either self-induced or enforced externally. The code should always be able to compile and pass at least a fast subset of the tests. With some development workflows, each commit message may need to reference an issue ID (or match the message template), or include a digital certificate of origin in the form of the **Signed-off-by** line. In many cases, these parts of the development process can be automated by Git.

Like many programming tools, Git includes a way to fire custom functionality contained in the user-provided code (custom scripts), when certain important pre-defined actions occur, that is, when certain events trigger. Such a functionality invoked as an event handler is called a **hook**. It allows us to take additional action and, at least for some hooks, also to stop the triggered functionality.

Hooks in Git can be divided into client-side and server-side hooks. **Client-side hooks** are triggered by local operations (on the client) such as committing, applying a patch series, rebasing, and merging. **Server-side hooks** on the other hand run on the server when network operations occur, such as receiving pushed commits.

You can also divide hooks into prehooks and post hooks. **Pre hooks** are called before an operation is finished, usually before the next step while performing an operation. If they exit with a nonzero value, they will cancel the current Git operation. **Post hooks** are invoked after an operation finishes and can be used for notification and logs; they cannot cancel an operation.

Installing a Git hook

The hooks in Git are executable programs (usually scripts), which are stored in the `hooks/` subdirectory of the Git repository administrative area, that is, `.git/hooks/` for non-bare repositories. You can change the location of the directory that Git searches for hooks via `core.hooksPath` configuration variable.

Hook programs are each named after the event that triggers them. This means that if you want one event to trigger more than one script, you will need to implement multiplexing yourself.

When you initialize a new repository with `git init` (this is done also while using `git clone` to create a copy of the other repository; clone calls `init` internally), Git populates the `.git/hooks/` directory with a bunch of inactive example scripts. Many of these are useful by themselves, but they also document the hook's API. All the examples are written as shell or Perl scripts, but any properly named executable would work just fine. If you want to use bundled example hook scripts, you'll need to rename them, stripping the `.sample` extension and ensuring that they have the executable permission bit.

A template for repositories

Sometimes you would want to have the same set of hooks for all your repositories. You can have a global (per-user and system-wide) configuration file, a global attributes file, and a global ignore list. It turns out that it is possible to select hooks to be populated during the creation of the repository. The default sample hooks that get copied to the `.git/hooks` repository are populated from `/usr/share/git-core/templates`.

Also, the alternative directory with the repository creation templates can be given as a parameter to the `--template` command-line option (to `git clone` and `git init`), as the `GIT_TEMPLATE_DIR` environment variable, or as the `init.templateDir` configuration option (which can be set in a per-user configuration file). This directory must follow the directory structure of `.git` (of `$GIT_DIR`), which means that the hooks need to be in the `hooks/` subdirectory there.

Note, however, that this mechanism has some limitations. As the files from the template directory are only copied to the Git repositories on their initialization, updates to the template directory do not affect the existing repositories. Though you can re-run `git init` in the existing repository to reinitialize it, just remember to save any modifications made to the hooks.

> **Hook management tools**
>
> Maintaining hooks for a team of developers can be tricky. There are many tools and frameworks for Git hook management; examples include **Husky** and **pre-commit**. You can find more examples of such tools listed on the `https://githooks.com` site. Those tools often allow for easier skipping hooks, running common code for all the hooks, or running multiple scripts for a specific hook.

Client-side hooks

There are quite a few client-side hooks. They can be divided into the commit-workflow hooks (a set of hooks invoked by the different stages of creating a new commit), apply-email workflow hooks, and everything else (not organized into a multihook workflow).

> **Important note**
>
> It is important to note that hooks are *not* copied when you clone a repository. This is done partially for security reasons, as hooks run unattended and mostly invisibly. You need to copy (and rename) files themselves, though you can control which hooks get installed when creating or reinitializing a repository (see the previous subsection). This means that you cannot rely on the client-side hooks to enforce a policy; if you need to introduce some hard requirements, you'll need to do it on the server side.

Commit process hooks

There are four client-side hooks invoked (by default) while committing changes. They are as follows.

The pre-commit hook

The **pre-commit hook** is run first, even before you invoke the editor to type in the commit message. It is used to inspect the snapshot to be committed to check whether you haven't forgotten anything. A nonzero exit from this hook aborts the commit. You can bypass invoking this hook altogether with `git commit --no-verifies`. This hook takes no parameters.

This hook can, among others, be used to check for the correct code style, run the static code analyzer (linter) to check for problematic constructs, make sure that the code compiles and that it passes all the tests (and that the new code is covered by the tests), or check for the appropriate documentation on some new functionality. The default hook checks for whitespace errors (trailing whitespace by default) with `git diff --check` (or rather its plumbing equivalent), and optionally for non-ASCII filenames in the changed files. You can, for example, make a hook that asks for a confirmation while committing with a dirty work-arena (for the changes in the worktree that would not be a part of the commit being created); though it is an advanced technique. Or, you can have it check whether there is documentation and unit tests on the new methods.

There is also the **pre-merge-commit hook** that is invoked by `git merge`. By default the hook, when enabled, runs the `pre-commit` hook.

The prepare-commit-msg hook

The **prepare-commit-msg hook** is run after the default commit message is created (including the static text of the file given by `commit.template`, if any), and before the commit message is opened in the editor. It lets you edit the default commit message or create a template programmatically, before the commit author sees it. If the hook fails with a nonzero status, the commit will be aborted. This hook takes as parameters the path to the file that holds the commit message (later passed to the editor) and the information about source of the commit message (the latter is not present for ordinary `git commit`): **message** if the `-m` or `-F` option was given, **template** if the `-t` option was given or `commit.template` was set, **merge** if the commit is merged or the `.git/MERGE_MSG` file exists, **squash** if the `.git/SQUASH_MSG` file exists, or **commit** if the message comes from the other commit: the `-c`, `-C`, or `--amend` option was given. In the last case, the hook gets additional parameters, namely, a SHA-1 hash of the commit that is the source of the message.

The purpose of this hook is to edit or create the commit message, and this hook is not suppressed by the `--no-verify` option. This hook is most useful when it is used to affect commits where the default message is autogenerated, such as the templated commit message, merged commits, squashed commits, and amended commits. The sample hook that Git provides comments out the `Conflict:` part of the merge commit message.

Another example of what this hook can do is to use the description of the current branch given by `branch.<branch-name>.description`, if it exists, as a base for a branch-dependent dynamic commit template. Or perhaps, it can check whether we are on the topic branch, and then list all the issues assigned to you on a project issue tracker, to make it easy to add the proper artifact ID to the commit message.

The commit-msg hook

The **commit-msg hook** is run after the developer writes the commit message, but before the commit is actually written to the repository. It takes one parameter, a path to the temporary file with the commit message provided by the user (by default, `.git/COMMIT_EDITMSG`).

If this script exits with a nonzero status, Git aborts the commit process, so you can use it to validate that, for example, the commit message matches the project state, or that the commit message conforms to the required pattern. The sample hook provided by Git can check, sort, and remove duplicated Signed-off-by: lines (which might be not what you want to use, if signoffs are to be a chain of provenance). You could conceivably check in this hook whether the references to the issue numbers are correct (and perhaps expand them, adding the current summary of each mentioned issue).

Gerrit Code Review provides a commit-msg hook (which needs to be installed in the local Git repository) to automatically create, insert, and maintain a unique Change-Id: line above the signoffs during git commit. This line is used to track the iterations of coming up with a commit; if the commit message in the revision pushed to Gerrit lacks such information, the server will provide instructions on how to get and install that hook script.

The post-commit hook

The **post-commit hook** runs after the entire process is completed. It doesn't take any parameters, but at this point of the commit operation, the revision that got created during commit is available as HEAD. The exit status of this hook is ignored. There is also the **post-merge hook**.

Generally, this script (like most of the post-* scripts) is most often used for notifications and logging, and it obviously cannot affect the outcome of git commit. You can use it, for example, to trigger a local build in a continuous integration tool such as Jenkins. In most cases, however, you would want to do this with the post-receive hook on the dedicated continuous integration server.

Another use case is to list information about all the **TODO** and **FIXME** comments in the code and documentation (for example, the author, version, file path, line number, and message), printing them to the standard output of the hook, so that that they are not forgotten and remain up to date and useful.

Hooks for applying patches from emails

You can set up three client-side hooks for the email-based workflow (where commits are sent by email). They are all invoked by the git am command (the name of which comes from **apply mailbox**), which can be used to take saved emails with patches (created, for example, with git format-patch and sent with git sent-email) and turn them into a series of commits. We will cover these hooks next.

The applypatch-msg hook

The first hook to run is the **applypatch-msg hook**. It is run after extracting the commit message from the patch and before applying the patch itself. As usual, for a hook which is not a post-* hook, Git aborts applying the patch if this hook exists with a nonzero status. It takes a single argument: the name of the temporary file with the extracted commit message.

You can use this hook to make sure that the commit message is properly formatted, or to normalize the commit message by having the script alter the file. The example applypatch-msg hook provided by Git simply runs the commit-msg hook if it exists as a hook (the file exists and is executable).

The pre-applypatch hook

The next hook to run is the **pre-applypatch hook**. It is run after the patch is applied to the working area, but before the commit is created. You can use it to inspect the state of the project before making a commit; for example, by running tests. Exiting with a nonzero status aborts the `git am` script without committing the patch.

The sample hook provided by Git simply runs the `pre-commit` hook, if present.

The post-applypatch hook

The last hook to run is the **post-applypatch hook**, which runs after the commit is made. It can be used for notifying or logging, for example, notifying all the developers or just the author of the patch that you have applied it.

Other client-side hooks

There are a few other client-side hooks that do not fit into a series of steps in a single process.

The pre-rebase hook

The **pre-rebase hook** runs before you rebase anything. Like all the `pre-*` hooks, it can abort the rebase process with a nonzero exit code. You can use this hook to disallow rebasing (and thus rewriting) any commits that were already published. The hook is called with the name of the base branch (the upstream the series was forked from), and the name of the branch being rebased. The name of the branch being rebased is passed to the hook only if it is not the current branch. The sample `pre-rebase` hook provided by Git tries to do this, though it makes some assumptions specific to Git's project development that may not match your workflow (take note that amending commits also rewrites them, and that rebasing may create a copy of a branch instead of rewriting it).

The pre-push hook

The **pre-push hook** runs during the `git push` operation, after checking the remote status and identifying which revisions are missing on the server, but before any changes are pushed. The hook is called with the reference to the remote (the URL or the remote name) and the actual push URL (the location of remote) as the script parameters. Information about the commits to be pushed is provided on the standard input, one line per ref to be updated. You can use this hook to validate a set of ref updates before a push occurs; a nonzero exit code aborts the push. The example installed simply checks whether there are commits beginning with **WIP** in a set of revisions to be pushed or marked with the **nopush** keyword in the commit message, and when either of those is true, it aborts the push. You can even make a hook prompt the user to confirm they are sure. This hook compliments the server-side checks, avoiding data transfer that would fail validation anyway.

The post-rewrite hook

The **post-rewrite hook** is run by commands that rewrite history (i.e., that replace commits), such as `git commit --amend` and `git rebase`. Note, however, that this hook is not run by large-scale history rewriting, such as `git filter-repo`. The type of command that triggered the rewrite (**amend** or **rebase**) is passed as a single argument, while the list of rewrites is sent to the standard input. This hook has many of the same uses as the `post-checkout` and `post-merge` hooks, and it runs after the automatic copying of notes, which is controlled by the `notes.rewriteRef` configuration variable (you can find more about the notes mechanism in *Chapter 10, Keeping History Clean*).

The post-checkout and post-merge hooks

The **post-checkout hook** is run after a successful `git checkout` (or `git checkout <file>`) after having updated the worktree. The hook is given three parameters: the SHA-1 hashes of the previous and current HEAD (which may or may not be different) and a flag indicating whether it was a whole project checkout (you were changing branches; the flag parameter is 1) or a file checkout (retrieving files from the index or named commit; the flag parameter is 0). As a special case, during the initial checkout after `git clone`, this hook passes the all-zero SHA-1 as the first parameter (as a source revision). You can use this hook to set up your working directory properly for your use case. This may mean handling large binary files outside the repository (as an alternative to applying the `filter` Git attribute on a per-file basis) that you don't want to have in the repository, or setting the working directory metadata properties such as full permissions, owner, group, times, extended attributes, or ACLs. It can also be used to perform repository validity checks or enhance the `git checkout` output by auto-displaying the differences (or just the diff statistics) from the previous checked-out revision (if they were different).

The **post-merge hook** runs after a successful merge operation. You can use it in a way similar to `post-checkout` to restore data and metadata in the working tree that Git doesn't track, such as full permissions data (or just make it invoke `post-checkout` directly). This hook can likewise validate the presence of files external to Git control that you might want copied in when the working tree changes.

For Git, objects in the repository (for example, commit objects representing revisions) are immutable; rewriting history (even amending a commit) is in fact creating a modified copy and switching to it, leaving the pre-rewrite history abandoned.

The pre-auto-gc hook

Deleting a branch also leaves abandoned history. To prevent the repository from growing too much, Git occasionally performs garbage collection by removing old unreferenced objects. In all but the most ancient instances of Git, this is done as a part of normal Git operations by invoking `git gc --auto`. The **pre-auto-gc hook** is invoked just before garbage collection takes place and can be used to abort the operation, for example, if you are on battery power. It can also be used to notify you that garbage collection is happening.

Server-side hooks

In addition to the client-side hooks, which are run in your own repository, there are a couple of important **server-side hooks** that a system administrator can use to enforce nearly any kind of policy for your project.

These hooks are run before and after you do a push to the server. The pre hooks (as mentioned earlier) can exit nonzero to reject a push or part of it; messages printed by the pre hooks will be sent back to the client (sender). You can use these hooks to set up complex push policies. Git repository management tools, such as `gitolite` and Git hosting solutions, use these to implement more involved access control for repositories. The post hooks can be used for notification, starting a build process (or just to rebuild and redeploy the documentation), or running a full test suite, for example as a part of a CI solution.

When writing server-side hooks, you need to take into account where in the sequence of operations the hook takes place and what information is available there, in the form of parameters, on the standard input, and in the repository.

Let's review what happens on the server when it receives a push:

1. Simplifying it a bit, the first step is that all the objects that were present in the client and missing on the server are sent to the server and stored (but are not yet referenced). If the receiving end fails to do this correctly (for example, because of the lack of disk space), the whole push operation will fail.

2. The **pre-receive hook** is run. It takes a list describing the references that are being pushed on its standard input. If it exits with a nonzero status, it aborts the whole operation and none of the references that were pushed are accepted.

3. For each ref being updated, the built-in sanity checks may reject the push to the ref, including the check for an update of a checked-out branch, a non-fast-forward push (unless forced), and so on.

4. The **update** hook is run for each ref, passing ref to be pushed in arguments; if this script exits with a nonzero status, only this ref will be rejected.

5. For each pushed ref, the ref in question is updated (unless it was rejected in an earlier stage).

6. The **post-receive hook** is run, taking the same data as the `pre-receive` one. This one can be used to update other services (for example, to notify CI servers) or notify users (via an email or a mailing list, IRC, or a ticket-tracking system).

If the push is atomic, either all the refs are updated (if none were rejected), or none are updated.

For each ref that was updated, the **post-update hook** is run. This can also be used for logging. The sample hook runs `git update-server-info` to prepare a repository, saving extra information to be used over *dumb* transports, though it would work better if run once as `post-receive`.

If push tries to update the currently checked-out branch and the `receive.denyCurrentBranch` configuration variable is set to `updateInstead`, then the **push-to-checkout** hook is run.

> **Important note**
>
> You need to remember that in pre hooks, you don't have refs updated yet, and that post hooks cannot affect the result of an operation. You can use pre hooks for access control (permission checking), and post hooks for notification and updating side data and logs.

You will see example hooks (server-side and client-side) for the Git-enforced policy in *Chapter 14, Git Administration*. You will also learn how other tools use those hooks, for example, for use in access control and triggering actions on push.

Extending Git

Git provides a few mechanisms to extend it. You can add shortcuts and create new commands, and add support for new transports; all without requiring you to modify Git sources.

Command aliases for Git

There is one little tip that can make your Git command-line experience simpler, easier, and more familiar, namely, **Git aliases**. It is very easy in theory to create an alias. You simply need to create an `alias.<command-name>` configuration variable; its value is the expansion of the alias.

One of the uses for aliases is defining short abbreviations for commonly used commands and their arguments. Another is creating new commands. Here are a couple of examples you might want to set up:

```
$ git config --global alias.co checkout
$ git config --global alias.ps = '--paginate status'
$ git config --global alias.lg "log --graph --oneline --decorate"
$ git config --global alias.aliases 'config --get-regexp ^alias\.'
```

The preceding setup means that typing, for example, `git co` would be the same as typing `git checkout`, and `git aliases` would print all defined aliases. Aliases take arguments just as the regular Git commands do. Git does not provide any default aliases to define shortcuts for common operations, unless you use `git-fc` project, a friendly fork of Git by Felipe Contreras.

Arguments here are split by spaces and the usual shell quoting and escaping is supported. Notably, you can use a quote pair (`"a b"`) or a backslash (`a\ b`) to include a space in a single argument.

> **Important note**
>
> Note, however, that you cannot have an alias with the same name as a Git command. In other words, you cannot use aliases to change the behavior of commands. The reasoning behind this restriction is that it could make existing scripts and hooks fail unexpectedly. Aliases that hide existing Git commands (with the same name as Git commands) are simply ignored.

You might, however, want to run an external command rather than a Git command in an alias. Or, you might want to join together the result of a few separate commands. In this case, you can start the alias definition with the ! character:

```
$ git config --global alias.unmerged \
  '!git ls-files --unmerged | cut -f2 | sort -u'
```

Because here the first command of the expansion of an alias can be an external tool, you need to specify the `git` wrapper explicitly, as shown in the preceding example.

> **Note**
>
> Note that in many shells, for example, in `bash`, the exclamation character ! is the history expansion character and it needs to be escaped as \ ! or be within single quotes (').

Note that such shell commands will be executed from the top-level directory of a repository (after doing `cd` to a top-level), which may not necessarily be the current directory. Git sets the `GIT_PREFIX` environment variable to the current directory path relative to the top directory of a repository, that is, `git rev-parse --show-prefix`. As usual, `git rev-parse` (and some `git` wrapper options) may be of use here.

The fact mentioned earlier can be used while creating aliases. The `git serve` alias, running `git daemon` to serve (read-only) the current repository at `git://127.0.0.1/`, makes use of the fact that the shell commands in aliases are executed from the top-level directory of a repo:

```
[alias]
    serve = !git daemon --reuseaddr --verbose --base-path=. --export-
all ./.git
```

Sometimes, you need to reorder arguments, use an argument twice, or pass an argument to the command early in the pipeline. You would want to refer to subsequent arguments as $1, $2, and so on, or to all arguments as $@, just like in shell scripts. One trick that you can find in older examples is to run a shell with a -c argument, like in the first of the examples mentioned next; the final dash is so that the arguments start with $1, not with $0. A more modern idiom is to define and immediately execute a shell function, like in the second example (it is a preferred solution because it uses one level of quoting less, and lets you use standard shell argument processing):

```
[alias]
    record-1 = !sh -c 'git add -p -- $@ && git commit' -
    record-2 = !f() {  git add -p -- $@ && git commit }; f
```

Aliases are integrated with command-line completion. While determining which completion to use for an alias, Git searches for a `git` command, skipping an opening brace or a single quote (thus, supporting both of the idioms mentioned earlier). With modern Git you can use the null command "`:`" to declare the desired completion style. For example, alias expanding to the following:

```
!f() { : git commit ; ... } f
```

would use a command completion for `git commit`, regardless of the rest of the alias.

Git aliases are also integrated with the help system. If you use the `--help` option on an alias, Git tells you its expansion (so you can check the relevant man page):

```
$ git co --help
'git co' is aliased to 'checkout'
```

Adding new Git commands

Aliases are best at taking small one-liners and converting them into small useful Git commands. You can write complex aliases, but when it comes to larger scripts, you would probably like to incorporate them into Git directly.

Git subcommands can be standalone executables that live in the Git execution path (which you can find by running `git --exec-path`); on Linux, this is normally `/usr/libexec/git-core`. The `git` executable itself is a thin wrapper that knows where the subcommands live. If `git foo` is not a built-in command, the wrapper searches for the `git-foo` command first in the Git exec path, then in the rest of your `$PATH`. The latter makes it possible to write local Git extensions (local Git commands) without requiring access to the system's space.

This feature is what it makes possible to have a user interface more or less integrated with the rest of Git in projects such as `git imerge` (see *Chapter 9, Merging Changes Together*), or `git lfs` or `git annex` (see *Chapter 12, Managing Large Repositories*). It is also how projects such as **Git Extras**, providing extra Git commands, were made.

Note, however, that if you don't install the documentation for your command in typical places, or configure the documentation system to find the help page for a command, then `git foo --help` won't work correctly.

You can list all external commands installed this way with `git --list-cmds=others`, or you can use `git help --all`, and the following list will appear at the end of its command output:

```
$ git --list-cmds=others
   credential-helper-selector
   credential-manager
   lfs
```

Credential helpers and remote helpers

There is another place where simply putting an appropriately named executable enhances and extends Git. **Remote helper** programs are invoked by Git when it needs to interact with remote repositories and remote transport protocols not supported by Git natively. You can find more about them in *Chapter 6, Collaborative Development with Git*.

When Git encounters a URL of the form `<transport>://<address>`, where `<transport>` is a (pseudo)protocol that is not natively supported, it automatically invokes the `git remote-<transport>` command with a remote and full remote URL as arguments. A URL of the form `<transport>::<address>` also invokes this remote helper, but with just `<address>` as a second argument in the place of a URL. Additionally, with `remote.<remote-name>.vcs` set to `<transport>`, Git would explicitly invoke `git remote-<transport>` to access that remote.

The helpers mechanism in Git is about interacting with external scripts using a well-defined format.

Each remote helper is expected to support a subset of commands. You can find more information about the issue of creating new helpers on the `gitremote-helpers(1)` man page.

There is another type of helpers in Git, namely, **credentials helpers**. They can be used by Git to get the credentials from the user required, for example, to access a remote repository over HTTP. They are specified by the configuration, though, just like the merge and diff drivers and the clean and smudge filters.

Summary

This chapter provided all the tools you need to use Git effectively. You got to know how to make the command-line interface easier to use and more effective with the Git-aware dynamic command prompt, command-line completion, autocorrection for Git commands, and using colors. You learned of the existence of alternative interfaces, from alternative porcelains to the various types of graphical clients.

You were reminded of the various ways to change the behavior of Git commands. You discovered how Git accesses its configuration and learned about a selected subset of configuration variables. You have learned how to automate Git with hooks and how to make use of them. Finally, you have learned how to extend Git with new commands and support new URL schemes.

This chapter was mainly about making Git more effective for you; the next chapter, *Chapter 14, Git Administration*, explains how to make Git more effective for other developers. You will cover more about server-side hooks and see their usage. You will also learn about repository maintenance.

Questions

Answer the following questions to test your knowledge of this chapter:

1. How do you save and reuse your favorite combination of options for the Git command?
2. How can you find all created aliases?
3. How do you run a graphical tool to display a `git diff`, or to help with resolving a merge?

4. How can you find where a given configuration came from?

5. How can you help ensure that a commit matches the recommended best practices?

Answers

Here are the answers to the questions given above:

1. Use a Git alias, a shell alias, or a shell function.

2. You can use the `git config --get-regexp ^alias\.` command.

3. Use `git difftool` for displaying differences, or `git mergetool` to help with resolving merge conflicts. There is built-in support for many of the existing graphical tools.

4. If it is about configuration values, you can use `git config --show-origin` (or `--show-scope`). If it is about per-file attributes, use `git check-attr`. If it is about ignoring files, use `git check-ignore`.

5. Use the `pre-commit` hook (and other similar hooks) to warn if best practices are not being followed. There are many third-party tools that help with hook management and often support various helper tools such as linters and code formatting tools.

Further reading

To learn more about the topics that were covered in this chapter, take a look at the following resources:

- Scott Chacon, Ben Straub: *Pro Git, 2nd Edition* (2014), Apress `https://git-scm.com/book/en/v2`

  - *Chapter 2 - Git Basics, Section 2.1 - Git Aliases*

  - *Chapter 8 - Customizing Git*

  - *Appendix A: Git in Other Environments*

- Matthew Hudson: *Git Hooks - A Guide for Programmers* `https://githooks.com/`

- *bash/zsh git prompt support* `https://github.com/git/git/blob/master/contrib/completion/git-prompt.sh`

- *bash/zsh completion support for core Git* `https://github.com/git/git/blob/master/contrib/completion/git-completion.bash`

- Seth House: *Conflict resolution in various mergetools* (2020) `https://www.eseth.org/2020/mergetools.html`

- Julia Evans: *Popular git config options* (2024) `https://jvns.ca/blog/2024/02/16/popular-git-config-options/`

- Ricardo Gerardi: *8 Git aliases that make me more efficient* (2020) `https://opensource.com/article/20/11/git-aliases`

- *Git SCM Wiki* (archived): *Aliases* `https://archive.kernel.org/oldwiki/git.wiki.kernel.org/index.php/Aliases.html`

- *Git Homepage - GUI Clients* `https://git-scm.com/downloads/guis`

- *Git Rev News* `https://git.github.io/rev_news/`

14

Git Administration

The previous chapter, *Customizing and Extending Git*, among other things, explained how to use Git hooks for automation. The client-side hooks were described in detail, while the server-side hooks were only covered briefly. In this chapter, we will cover server-side hooks comprehensively and discuss client-side hooks' usage as helpers.

The earlier chapters helped master your work with Git as a developer, as a team member collaborating with others, and as a maintainer. When the book discussed setting up repositories and branch structure, it was from the point of view of a Git user.

This chapter is intended to help those of you who are in a situation of dealing with the administrative side of Git. This includes setting up remote Git repositories and configuring their access. This chapter covers the work required to make Git go smoothly (that is, Git maintenance) and finding and recovering from repository errors. It also describes how to use server-side hooks to implement and enforce a development policy. Additionally, you will find here a short description of the various types of tools that can be used to manage remote repositories, helping you to choose from them.

In this chapter, we will cover the following topics:

- Server-side hooks – implementing a policy and notifications
- How to set up Git on a server
- Third-party tools to manage remote repositories
- Signed pushes to assert updating refs and enable audits
- Reducing the size of hosted repositories with alternates and namespaces
- Improving server performance and helping the initial clone
- Checking for repository corruption and fixing a repository
- Recovering from errors with the help of reflogs and `git fsck`
- Git repository maintenance and repacking
- Augmenting development workflows with Git

Repository maintenance

Occasionally, you may need to do some cleanup of a repository, usually to make it more compact. Such cleanups are also a very important step after migrating a repository from another version control system.

Automatic housekeeping with git-gc

Modern Git (or, rather, all but ancient Git) from time to time runs the `git gc --auto` command in each repository. This command checks whether there are too many loose objects (objects stored as separate files, with one file per object, rather than those stored together in a packfile; objects are almost always created loosely), and if so, then it launches the garbage collection operation. Garbage collection means gathering up all the loose objects and placing them in packfiles, as well as consolidating many small packfiles into one large packfile. Additionally, it packs references into the `packed-refs` file. Objects that are unreachable even via reflog and are safely old are, by default, packed separately into a cruft pack. Git then deletes loose objects, cruft packs, and packfiles that got repacked (with some safety margin relating to the age of the loose objects files), thus pruning old unreachable objects. There are various configuration knobs in the `gc.*` namespace to control garbage collection operations.

You can run `auto gc` manually with `git gc --auto` or force garbage collection with `git gc`. The `git count-objects` command (sometimes with the help of the `-v` parameter) can be used to check whether there are signs that a repack is needed. You can even run individual steps of the garbage collection individually with `git repack`, `git pack-refs`, `git prune`, and `git prune-packed`.

By default, Git will try to reuse the results of an earlier packing to reduce CPU time spent on repacking, while still providing good disk space utilization. In some cases, you will want to more aggressively optimize the size of the repository at the cost of it taking more time; this is possible with `git gc --aggressive` (or by repacking the repository by hand with `git repack`, run with appropriate parameters). It is recommended to do this after importing from other version control systems, as the mechanism that Git uses for importing (namely, the `fast-import` stream) is optimized for the speed of the operation, not for the final repository size.

There are issues of maintenance not covered by `git gc` because of their nature. One of them is pruning (deleting) remote-tracking branches that were deleted in the remote repository. This can be done with `git fetch --prune` or `git remote prune`, or on a per-branch basis with `git branch --delete --remotes <remote-tracking branch>`. This action is left to the user and not run by `git gc`, as Git simply cannot know whether you have based your own work on the remote-tracking branch that will be pruned.

Periodic maintenance with git-maintenance

Git commands that add data to the repository, such as `git add` or `git fetch`, can trigger automatic garbage collection and perform some repository optimization. However, because they need to provide a responsive user interface, this does not trigger more costly repository optimizations. Those tasks include updating the commit graph data, prefetching from remote repositories (so that `git fetch` will have fewer objects to download), cleaning up loose objects, and doing an incremental repack. Such optimization tasks often scale with the full size of the repository.

A better solution is to run the maintenance tasks that are expensive in the background, periodically – hourly, daily, or weekly. With modern Git, you can schedule those tasks with the help of the `git maintenance` command. It will schedule those jobs differently depending on the operating system.

You can configure how often a given task is run. Note that `git maintenance run`, a process that performs scheduled tasks, puts a lock on the repository's object database, preventing competing processes from leaving the repository in an unpredicted state. This is not the case for `git gc`; therefore, if you do periodic maintenance, use `git maintenance run --task=gc` instead of the `git gc` command.

Data recovery and troubleshooting

It is almost impossible to never make any mistakes. This applies also to using Git. The knowledge presented in this book, and your experience with using Git, should help to reduce the number of mistakes. Note that Git tries quite hard not to help you avoid losing your work; many mistakes are recoverable. The next subsection will explain how you can try to recover from an error.

Recovering a lost commit

It may happen that you accidentally lost a commit. Perhaps you force-deleted an incorrect branch that you were going to work on, you rewound the branch to an incorrect place, or you were on an incorrect branch while starting an operation. Assuming something like this happened, is there any way to get your commits back and undo the mistake?

Because Git does not delete objects immediately and keeps them for a while, only deleting them if they are unreachable during the garbage collection phase, the commit you lost will be there; you just need to find it. The garbage collection operation has, as mentioned, its own safety margins; however, if you find that you need to troubleshoot, it is better to turn off automatic garbage collection temporarily with `git config gc.auto never` (and turning off the gc task if it is scheduled to run periodically with `git maintenance`, by setting `maintenance.gc.enabled` to false or by turning maintenance off with `git maintenance unregister`).

Often, the simplest way to find and recover lost commits is to use the `git reflog` tool. For each branch, and separately for the HEAD, Git silently records (logs) where the tip of the branch was in your local repository, what time it was there, and how it got there. This record is called the **reflog**. Each time you commit or rewind a branch, the reflog for the branch and the HEAD is updated. Each time you change the branches, the HEAD reflog is updated, and so on.

You can see where the tip of a branch has been at any time by running `git reflog` or `git reflog <branch>`. You can also run `git log -g`, where `-g` is a short way of saying `--walk-reflog`; this gives you a normal configurable log output. There is also `--grep-reflog=<pattern>` to search the reflog:

```
$ git reflog
6c89dee HEAD@{0}: commit: Ping asynchronously
d996b71 HEAD@{1}: rebase -i (finish): returning to refs/heads/ajax
d996b71 HEAD@{2}: rebase -i (continue): Ping asynchronously WIP
89579c9 HEAD@{3}: rebase -i (pick): Use Ajax mode
7c6d322 HEAD@{4}: commit (amend): Simplify index()
e1e6f65 HEAD@{5}: cherry-pick: fast-forward
eea7a7c HEAD@{6}: checkout: moving from ssh-check to ajax
c3e77bf HEAD@{7}: reset: moving to ajax@{1}
```

You should remember the `<ref>@{<n>}` syntax from *Chapter 4*, *Exploring Project History*. With the information from reflogs, you can rewind the branch in question to the version from before the set of operations, or you can start a new branch, starting with any commit in the list.

Let's assume that your loss was caused by deleting the wrong branch. Because of the way reflogs are implemented (e.g., logs for a branch named `foo` – that is, for the `refs/heads/foo` ref – are kept in the `.git/logs/refs/heads/foo` file), a reflog for a given branch is deleted, together with the branch. You might still have the necessary information in the HEAD reflog, unless you have manipulated the branch tip without involving the working area, but it might not be easy to find it.

In a case where the information is not present in reflogs, one way to find the necessary information to recover lost objects is to use the `git fsck` utility, which checks your repository for integrity. With the `--full` option, you can use this command to show all unreferenced objects:

```
$ git fsck --full
Checking object directories: 100% (256/256), done.
Checking objects: 100% (58/58), done.

dangling commit 50b836cb93af955ca99f2ccd4a1cc4014dc01a58
dangling blob 59fc7435baf79180a3835dddc52752f6044bab99
dangling blob fd64375c1f2b17b735f3145446d267822ae3ddd5
[...]
```

You can see the SHA1 identifiers of the unreferenced (lost) commits in the lines with the **dangling commit** string prefix. To examine all these dangling commits, you can filter the `git fsck` output for the commits with `grep "commit"`, extract their SHA1 identifiers with `cut -d' ' -f3`, and then feed these revisions into `git log --stdin --no-walk`, as shown here:

```
$ git fsck --full | grep "commit" | cut -d' ' -f3 | git log --stdin
--no- walk
```

> **Tip**
>
> The same technique, but with using `blob` command, can be used to recover accidentally deleted files – assuming that you have used `git add` with the version of the file you want to recover.

Troubleshooting Git

The main purpose of `git fsck` is to check for repository corruption. Besides having the option to find dangling objects, this command runs sanity checks for each object and tracks the reachability fully. It can find corrupted and missing objects; then, if the corruption was limited to your clone and the correct version can be found in other repositories (in backups and other archives), you can try to recover those objects from an uncorrupted source.

Sometimes, however, the error might be more difficult to recover from. You can try to find a Git expert outside your team, but often, the data in the repository is proprietary. Creating a minimal reproduction of the problem is not always possible. With modern Git, if the problem is structural, you can try to use `git fast-export --anonymize` to strip the repository from the data, while ensuring that the anonymized repository reproduces the issue. Reproducing some bugs may require referencing particular commits or paths; with modern Git, you can ask for a particular token to be left as-is, or mapped to a new value with a set of `--anonymize-map` options.

If the repository is fine but the problem is with the Git operations, you can try to use various tracking and debugging mechanisms built into Git, or you can try to increase the verbosity of the commands. You can turn on tracing with the appropriate environment variables (which we will show later). The trace output can be written to a standard error stream by setting the value of the appropriate environment variable to **1**, **2**, or **true**. The **0** or **false** value disables it. Other integer values between 2 and 10 will be interpreted as open file descriptors to be used for trace output. You can also set such environment variables to the absolute path of the file to write trace messages to.

These tracking-related variables include the following (see the manpage of the `git` wrapper for the complete list):

- `GIT_TRACE`: This enables general trace messages that do not fit into any specific category. This includes the expansion of Git aliases (see *Chapter 13*, *Customizing and Extending Git*), built-in command execution, and external command execution (such as pager, editor, or helper).

- `GIT_TRACE_PACKET`: This enables packet-level tracking of the network operations for the "smart" transport protocols. This can help to debug protocol issues or any troubles with the remote server that you set up. To debug and fetch from shallow repositories, there is `GIT_TRACE_SHALLOW`.

- `GIT_TRACE_CURL` (possibly with `GIT_TRACE_CURL_NO_DATA`): This enables a `curl` full trace dump of the HTTP(S) transport protocol, similar to running the `curl --trace-ascii` option.

- GIT_TRACE_SETUP: This enables trace messages, printing information about the location of the administrative area of the repository, the working area, the current working directory, and the prefix (the last one is the subdirectory inside the repository directory structure).

- GIT_TRACE_PERFORMANCE: This shows the total execution time of each Git command.

With modern Git, you can enable more detailed trace messages from the trace2 library, either in a simple text-based format meant for human consumption with GIT_TRACE2, or in the JSON-based format meant for machine interpretation with GIT_TRACE2_EVENT. In addition to redirecting the output from a standard error, to a given file descriptor, or to a given file, you can also ask to write output files to a given directory (one file per process) and even ask to open the path as a Unix domain socket. The Trace2 API replacement for GIT_TRACE_PERFORMANCE is GIT_TRACE2_PERF. Instead of environment variables, you can use the trace2.normalTarget, trace2.eventTarget, and trace2.perfTarget configuration variables, respectively.

There is also GIT_CURL_VERBOSE to emit all the messages generated by the curl library for the network operations over HTTP, and GIT_MERGE_VERBOSITY to control the amount of output shown by the recursive merge strategy.

Git on the server

The previous chapters should have given you enough knowledge to master most of the day-to-day version control tasks in Git. *Chapter 6, Collaborative Development with Git*, explained how you can lay out repositories for collaboration. Here, we will explain how to set up Git repositories to enable remote access on a server, allowing you to fetch from and push to them.

The topic of administration of the Git repositories is a large one. There are books written about specific repository management solutions, such as Gitolite, Gerrit, GitHub, or GitLab. Here, you will hopefully find enough information to help you choose a solution or your own.

Let's start with the tools and mechanisms to manage remote repositories themselves, and then move on to the ways of serving Git repositories (i.e., putting Git on the server).

Server-side hooks

Hooks that are invoked on the server can be used for server administration; among others, these hooks can control access to the remote repository by performing the authorization step, and they can ensure that the commits entering the repository meet certain minimal criteria. The latter is best done with the additional help of client-side hooks, which were discussed in *Chapter 13, Customizing and Extending Git*. That way, users are not notified that their commits do not pass muster only when they want to publish them. Conversely, client-side hooks implementing validation are easy to skip with the --no-verify option (so server-side validation is necessary), and you need to remember to install them.

> **Important note**
>
> Note, however, that server-side hooks are invoked only during the push operation; you need other solutions for access control to the fetch (and clone) operation.
>
> Hooks are also obviously not run while using "dumb" protocols – there is no Git on the server invoked then.

While writing hooks to implement some Git-enforced policy, you need to remember at what stage the hook in question is run and what information is available then. It is also important to know how the relevant information is passed to the hook; however, you can find the last quite easily in the Git documentation on the `githooks` man page. The previous chapter included a simple summary of server-side hooks. Here, we will expand a bit on this topic.

All the server-side hooks are invoked by `git receive-pack`, which is responsible for receiving published commits (which are received in the form of a packfile, hence the name of the command). If a hook, except for a `post-*` one, exits with the non-zero status, then the operation is interrupted and no further stages are run. The post hooks are run after the operation finishes, so there is nothing to interrupt.

Both the standard output and the standard error output are forwarded to `git send-pack` at the client end, so the hooks can simply pass messages for the user by printing them (for example, with `echo`, if the hook was written as a shell script). Note that the client doesn't disconnect until all the hooks complete their operation, so be careful if you try to do anything that may take a long time, such as automated tests. It is better to have a hook simply start such long operations asynchronously and exit, allowing the client to finish.

You need to remember that, with pre-hooks, you don't have refs updated yet, and that post-hooks cannot affect the result of an operation. You can use pre-hooks for access control (permission checking),and post-hooks for notification, updating the side data, and logging. Hooks are listed in the order of operation.

The pre-receive hook

The first hook to run is the **pre-receive hook**. It is invoked just before you start updating refs (branches, tags, notes, and so on) in the remote repository, but after all the objects are received. It is invoked once for the receive operation. If the server fails to receive published objects (for example, because of a lack of disk space or incorrect permissions), the whole `git push` operation will fail before Git invokes this hook.

This hook receives no arguments; all the information is received on the standard input of the script. For each ref to be updated, it receives a line in the following format:

```
<old-SHA1-value> <new-SHA1-value> <full-ref-name>
```

Refs that need to be created will have the old SHA1 value of 40 zeros, while refs that need to be deleted will have a new SHA1 value equal to the same. The same convention is used in all the other places, where the hooks receive the old and new state of the updated ref.

> **Push options**
>
> You can pass additional data to the server with `git push --push-option=<option>` or the `push.pushOption` configuration variable. Both can be given multiple times. This data is then passed to pre-receive and post-receive hooks via environment variables – `GIT_PUSH_OPTION_COUNT` and `GIT_PUSH_OPTION_0`, `GIT_PUSH_OPTION_1`, and so on.

This hook can be used to quickly abort the operation if the update cannot to be accepted – for example, if the received commits do not follow the specified policy or if the signed push (more on this later) is invalid. Note that to use it for access control (i.e., authorization) you need to get the authentication token somehow, be it with the `getpwuid` command or with an environment variable such as `USER`. However, this depends on the server setup and the server configuration.

The push-to-checkout hook to push to non-bare repositories

When pushing to the non-bare repositories, if a push operation tries to update the currently checked-out branch, then the **push-to-checkout hook** will be run. This is done if the `receive.denyCurrentBranch` configuration variable is set to the `updateInstead` value (instead of one of the `true` or `refuse`, `warn` or `false`, or `ignore` values). This hook receives the SHA1 identifier of the commit that will be the tip of the current branch that is going to be updated.

This mechanism is intended to synchronize working directories when one side is not easily accessible interactively (for example, accessible via interactive `ssh`), or as a simple deployment scheme. It can be used to deploy to a live website or to run code tests on different operating systems.

If this hook is not present, Git will refuse the update of the ref if either the working tree or the index (the staging area) differs from `HEAD` – that is, if the status is "not clean." This hook should be used to override this default behavior.

You can craft this hook to have it make changes to the working tree and the index that are necessary to bring them to the desired state. For example, the hook can simply run `git read-tree -u -m HEAD "$1"` to switch to the new branch tip (the `-u` option updates the files in the working tree), while keeping the local changes (the `-m` option makes it perform a fast-forward merge with two commits/trees). If this hook exits with a nonzero status, then Git will refuse to push to the currently checked-out branch.

The update hook

The next to run is the **update hook**, which is invoked *separately* for each ref that is updated. This hook is invoked after the non-fast-forward check (unless the push is forced) and the per-ref built-in sanity checks that can be configured with `receive. denyDeletes`, `receive.denyDeleteCurrent`, `receive.denyCurrentBranch`, and `receive.denyNonFastForwards`.

Note that exiting with nonzero refuses the ref to be updated; if the push is *atomic* (`git push --atomic`), then refusing any ref to be updated will abandon the whole push operation. With an ordinary push, only the update of a single ref will be refused; the push of other refs will proceed normally.

This hook receives the information about the ref to be updated as its parameters, in order:

- The full name of the ref that is updated,

- The old SHA1 object name stored in the ref before the push operation

- The new SHA1 object name to be stored in the ref after the push operation

The `update.sample` hook example can be used to block unannotated tags from entering the repository, and also to allow or deny deleting and modifying tags and deleting and creating branches. All the configurable of this sample hook is done with the appropriate `hooks.*` configuration variables, rather than being hardcoded. There is also the `update-paranoid` Perl script in `contrib/hooks/`, which can be used as an example of how to use this hook for access control. This hook is configured with an external configuration file, where, among other options, you can set up access so that only commits and tags from specified authors are allowed, and authors are required to have correct access permissions.

Many repository management tools, such as Gitolite, set up and use this hook for their work. You need to read the tool documentation if you want, for some reason, to run your own `update` hook together with the one provided by such a tool, perhaps with the help of some hook management tool (see, for example, a list of such tools on `https://githooks.com/`).

The post-receive hook

Then, after all the refs are updated, the **post-receive hook** is run. It takes the same data as the `pre-receive` one. Only now do all the refs point to the new SHA1s. It can happen that another user has modified the ref after it was updated but before the hook was able to evaluate it. This hook can be used to update other services (for example, notify the continuous integration server), notify users (via an email or a mailing list, a chat channel, or a ticket-tracking system), or log the information about the push for audit (for example, about signed pushes). It supersedes the `post-update` hook, and should be used instead.

There is no default `post-receive` hook, but you can find the simple `post-receive-email` script, and its replacement, `git-multimail`, in the `contrib/hooks/` area.

These two example hooks are actually developed separately from Git itself, but for convenience, they are provided with the Git source. `git-multimail` sends one email summarizing each changed ref, one email for each new commit with the changes – threaded (as a reply) to the corresponding ref change email, and one announcement email for each new annotated tag. Each of these is separately configurable with respect to the email address used and, to some extent, also with respect to the information included in the emails.

To provide an example of third-party tools, `irker` includes the script to be used as Git's `post-receive` hook to send notifications about the new changes to the appropriate IRC channel, using the irker daemon (set up separately).

The post-update hook (a legacy mechanism)

Then, the **post-update hook** is run. Each ref that was actually successfully updated passes its name as one of parameters; this hook takes a variable number of parameters. This is only partial information; you don't know what the original (old) and updated (new) values of the updated refs were, and the current position of the ref is prone to race conditions (as explained before). Therefore, if you actually need the position of the refs, the `post-receive` hook is a better solution.

The sample hook runs `git update-server-info` to prepare a repository for use over the dumb transports(described in the *Legacy (dumb) transports* section of *Chapter 7, Publishing Your Changes*, and in the *Dumb protocols* section later in this chapter), by creating and saving some extra information. If the directory with the repository is to be accessible via plain HTTP or other walker-based transport like FTP, you may consider enabling it. However, in modern Git, it is enough to simply set `receive.updateServerInfo` to `true` so that a hook is no longer necessary.

Using hooks to implement Git-enforced policy

The only way to truly enforce a policy is to implement it using server-side hooks, either `pre-receive` or `update`; if you want a per-ref decision, you need to use the latter. Client-side hooks can be used to help developers pay attention to the policy, but these can be disabled, skipped, or not enabled.

Enforcing the policy with server-side hooks

One part of the development policy could be requiring that each commit message adheres to a specified template. For example, you could require each non-merge commit message to include the *digital certificate of origin* in the form of the **Signed-off-by:** line, or that each commit refers to the issue tracker ticket by including a string that looks like **ref: 2387**. The possibilities are endless.

To implement such a hook, you first need to turn the old and new values for a ref (that you got by either reading them line by line from the standard input in `pre-receive`, or as the `update` hook parameters) into a list of all the commits that are being pushed. You need to take care of the corner cases – deleting a ref (no commits pushed), creating a new ref, and a possibility of non-fast-forward pushes (where you need to use the merge base as the lower limit of the revision range – for example, with the `git merge-base` command), pushes to tags, pushes to notes, and other non-branch pushes. The operation of turning a revision range into a list of commits can be done with the `git rev-list` command, which is a low-level equivalent (plumbing) of the user-facing `git log` command (`porcelain`); by default, this command prints out only the SHA1 values of the commits in the specified revision range, one per line, and no other information.

Then, for each revision, you need to grab the commit message and check whether it matches the template specified in the policy. You can use another plumbing command, called `git cat-file`, and then extract the commit message from this command output by skipping everything before the first blank line. This blank line separates commit metadata in the raw form from the commit body:

```
$ git cat-file commit a7b1a955
tree 171626fc3b628182703c3b3c5da6a8c65b187b52
parent 5d2584867fe4e94ab7d211a206bc0bc3804d37a9
author Alice Developer  1440011825 +0200
committer Alice Developer  1440011825 +0200

Added COPYRIGHT file
```

Alternatively, you can use `git show -s` or `git log -1`, which are both porcelain commands, instead of `git cat-file`. However, you would then need to specify the exact output format – for example, `git show -s --format=%B <SHA1>`.

When you have these commit messages, you can then use the regular expression match or another tool on each of the commit messages caught to check whether they match the policy.

Another part of the policy may be the restrictions on how branches are managed. For example, you may want to prevent the deletion of long-lived development stage branches (see *Chapter 8, Advanced Branching Techniques*), while allowing the deletion of topic branches. To distinguish between them – that is, to find out whether the branch being deleted is a topic branch or not – you can either include a configurable list of branches to manage strictly, or you can assume that topic branches always use the `<user>/<topic>` naming convention. The latter solution can be enforced by requiring the newly created branches, which should be topic branches only, to match this naming convention.

Conceivably, you could make a policy that topic branches can be fast-forwarded only if they are not merged in, although implementing checks for this policy would be nontrivial.

Usually, only specific people have permission to push to the official repository of a project (holding a so-called commit bit). With server-side hooks, you can configure the repository so that it allows anyone to push, but only to the special mob branch; all the other push access is restricted.

You can also use server-side hooks to require that only annotated tags are allowed in the repository, that tags are signed with a public key that is present in the specified key server (and, thus, can be verified by other developers), and that tags cannot be deleted or updated. If needed, you can restrict signed tags to those coming from the selected (and configured) set of users – for example, enforcing a policy that only one of the maintainers can mark a project for a release (by creating an appropriately named tag – e.g., `v0.9`).

Early notices about policy violations with client-side hooks

It would be not a good solution to have strict enforcement of development policies and not provide users with a way to help watch and fulfill those policies. Having your work rejected during a push can be frustrating; to fix the issue preventing one from publishing the commit, you would have to edit your local history of the project (that is, rewrite your changes). See *Chapter 10*, *Keeping History Clean*, for details on how to do it.

The answer to that problem is to provide some client-side hooks that users can install and have Git notify them immediately when they violate the policy, which would make their changes get rejected by the server. The intent is to help correct any problem as fast as possible, usually before committing the changes. These client-side hooks must be distributed somehow, as hooks are not copied when cloning a repository. Various ways to distribute these hooks are described in *Chapter 13*, *Customizing and Extending Git*.

If there are any limitations on the contents of the changes (for example, some files might be changed only by specified developers), a warning message can be created with `pre-commit` hook. The `prepare-commit-msg` hook (and the `commit.template` configuration variable) can provide the developer with a customized template to be filled in while working on a commit message. You can also make Git check the commit message, just before the commit is recorded, with the `commit-msg` hook. This hook would find out and inform you whether you have correctly formatted the commit message and whether it includes all the information required by the policy. This hook can also be used instead of or in addition to `pre-commit`, checking whether you are modifying the files you are not allowed to.

The `pre-rebase` hook can be used to verify that you don't try to rewrite history in a manner that would lead to a non-fast-forward push (with `receive.denyNonFastForwards` on the server, forcing a push won't work anyway).

As a last resort, there is a `pre-push` hook, which can check for correctness before trying to connect to the remote repository.

Signed pushes

Chapter 6, *Collaborative Development with Git*, includes a description of various mechanisms that a developer can use to ensure the integrity and authenticity of their work – signed tags, signed commits, and signed merges (merging signed tags). All these mechanisms assert that the objects (and the changes they contain) came from the signer.

However, signed tags and commits do not assert that the developer wanted to have a particular revision at the tip of a particular branch. Authentication done by the hosting site cannot be easily audited later, and it requires you to trust the hosting site and its authentication mechanism. Modern Git (version 2.2 or newer) allows you to **sign pushes** for this purpose.

Signed pushes require the server to set up `receive.certNonceSeed` and the client to use `git push --signed`. Handling of signed pushes is done with the server-side hooks. Currently, none of the Git forges such as GitHub, GitLab, Bitbucket, or Gitea support signed pushes; there are tools such as **gittuf** or **Kernel.org Transparency Log Monitor** that provide transparency logs for push operations.

The signed push certificate sent by the client is stored in the repository as a blob object and is verified using the **GPG** (**GNU Privacy Guard**). The `pre-receive` hook can then examine various `GIT_PUSH_CERT_*` environment variables (see the `git-receive-pack` man page for the details) to decide whether to accept or deny a given signed push.

Logging signed pushes for audit can be done with the `post-receive` hook. You can have this hook send an email notification about the signed push or have it append information about the push to a log file. The **push certificate** that is signed includes an identifier for the client's GPG key, the URL of the repository, and the information about the operations performed on the branches or tags, in the same format as the `pre-receive` and `post-receive` input.

Serving Git repositories

In *Chapter 6, Collaborative Development with Git*, we examined four major protocols used by Git to connect with remote repositories – local, HTTP, **SSH** (**Secure Shell**), and Git (the native protocol). This was done from the point of view of a client connecting to the repository, discussing what these protocols are and which one to use if the remote repository offers more than one.

This chapter will offer the administrator's side of view, explaining how to set up and later move rephrased Git repositories to be served with different transport protocols. Here, we will also examine, for each protocol, what authentication and authorization look like.

Local protocol

This is the most basic protocol, where a client uses a path to the repository or the `file://` URL to access remotes. You just need to have a shared filesystem, such as an NFS or SMB/CIFS mount, which contains Git repositories to serve. This is a nice option if you already have access to a networked filesystem, as you don't need to set up any server.

Access to repositories using a file-based transport protocol is controlled by the existing file permissions and network access permissions. You need read permissions to fetch and clone and write permissions to push.

In the latter case, if you want to enable a push, you'd better set up a repository in such a way that pushing does not screw up the permissions. This can be helped by creating a repository with the `--shared` option to use `git init` (or `git clone`). This option allows users belonging to the same group to push into the repository by using the sticky group ID, ensuring that the repositories stay available to all the group members.

The disadvantage of this method is that shared access to a networked filesystem is, generally, more difficult to set up and reach safely from multiple remote locations than basic network access and setting up an appropriate server. Mounting the remote disk over the internet can be difficult and slow.

This protocol does not protect the repository against accidental damage. Every user has full access to the repository's internal files, and there is nothing preventing them from accidentally corrupting the repository.

The SSH protocol

SSH is a common transport protocol (commonly used by Linux users) to self-host Git repositories. SSH access to servers is often already set up in many cases as a way to safely log in to the remote machine; if not, it is generally quite easy to set up and use. SSH is an authenticated and encrypted network protocol.

Conversely, you can't serve anonymous access to Git repositories over SSH. People must have at least limited access to your machine over SSH; this protocol does not allow anonymous read-only access to published repositories.

Generally, there are two ways to give access to Git repositories over SSH. The first is to have a separate account on the server for each client trying to access the repository (although such an account can be limited and does not need full shell access, you can, in this case, use `git-shell` as a login shell for Git-specific accounts). This can be used both with ordinary SSH access, where you provide the password, and with a public-key login. In a one-account-per-user case, the access control situation is similar to the local protocol – namely, access is controlled with filesystem permissions.

A second method is to create a single shell account, which is often the `git` user, specifically to access Git repositories and use public-key login to authenticate users. Each user who will have access to the repositories would then need to send their SSH public key to the administrator, who would then add this key to the list of authorized keys. The actual user is identified by the key they use to connect to the server.

Another alternative is to have the SSH server authenticated from an LDAP server or some other centralized authentication scheme (often to implement single sign-on). As long as the client can get (limited) shell access, any SSH authentication mechanism can be used.

Anonymous Git protocol

Next is the Git protocol. This is served by a special and really simple TCP daemon, which listens on a dedicated port (by default, port `9418`). This is (or was) a common choice for fast, anonymous, and unauthenticated read-only access to Git repositories.

The Git protocol server, `git daemon`, is relatively easy to set up. Basically, you need to run this command, usually in a daemonized manner. How to run the daemon (the server) depends on the operating system you use. It can be a `systemd` unit file, an Upstart script, or a `sysvinit` script. A common solution is to use `inetd` or `xinetd`.

You can remap all the repository requests relative to the given path (a project root for the Git repositories) with `--base-path=<directory>`. There is also support for virtual hosting; see the `git-daemon` documentation for more details. By default, `git daemon` will export only the repositories that have the `git-daemon-export-ok` file inside `gitdir`, unless the `--export-all` option is used. Usually, you would also want to turn on `--reuseaddr`, allowing the server to restart without waiting for the connection to time out.

The downside of the Git protocol is the lack of authentication and the obscure port that it runs on (which may require you to punch a hole in the firewall). The lack of authentication is because, by default, it is used only for read access – that is, for fetching and cloning repositories. Generally, it is paired with either SSH (always authenticated and never anonymous) or HTTPS for pushing.

You can configure it to allow for a push (by starting the `receive-pack` service with the `--enable=<service>` command-line option or, on a per-repository basis, by setting the `daemon.receive-Pack` configuration to `true`), but it is generally not recommended. The only information available to hooks to implement access control is the client address, unless you require all the pushes to be signed. You can run external commands in an access hook, but this would not provide much more information about the client.

> **Tip**
> One service you might consider enabling is `upload-archive`, which serves `git archive --remote`.

This lack of authentication means that not only does the Git server not know who accesses the repositories, but also that the client must trust the network to not spoof the address while accessing the server. This transportation is not encrypted.

The smart HTTP(S) protocol

Setting up the so-called "smart" HTTP(S) protocol consists basically of enabling a server script that would invoke `git receive-pack` and `git upload-pack` on the server. Git provides a CGI script named `git-http-backend` for this task. This CGI script can detect whether the client understands the smart HTTP protocol; if not, it will fall back on the "dumb" behavior (a backward compatibility feature).

To use this protocol, you need a CGI server – for example, Apache (with this server , you would also need the `mod_cgi` module or its equivalent, and the `mod_env` and `mod_alias` modules). The parameters are passed using environment variables (hence the need for `mod_env` when using Apache) – `GIT_PROJECT_ROOT` to specify where repositories are and an optional `GIT_HTTP_EXPORT_ALL` if you want to have all the repositories exported, not only those with the `git-daemon-export-ok` file in them.

The authentication is done by the web server. In particular, you can set it up to allow unauthenticated anonymous read-only access, while requiring authentication for a push. Utilizing HTTPS gives encryption and server authentication, like with the SSH protocol. The URL for fetching and pushing is the same when using HTTP(S); you can also configure it so that the web interface to browse Git repositories uses the same URL for fetching.

> **Note**
>
> The documentation of `git-http-backend` includes a setup for Apache for different situations, including unauthenticated read and authenticated write. The setup presented there is a bit involved because initial ref advertisements use the query string, while the `receive-pack` service invocation uses path info.
>
> Conversely, requiring authentication with any valid account for reads and writes, and leaving the restriction of writes to the server-side hook, is a simpler and often acceptable solution.

If you try to push to the repository that requires authentication, the server can prompt for credentials. Because the HTTP protocol is stateless and involves more than one connection sometimes, it is useful to utilize credential helpers (see *Chapter 13, Customizing and Extending Git*) to avoid either having to give the password more than once for a single operation, or having to save the password somewhere on the disk (for example, in the remote URL).

> **Gitolite for smart HTTPS access control**
>
> While Gitolite (`https://gitolite.com/`) provides an access control layer on top of Git for access over SSH, it can be configured to perform authorization for smart HTTP mode.

Dumb protocols

If you cannot run Git on the server, you can still use the dumb protocol, which does not require it. The dumb HTTP(S) protocol expects the Git repository to be served like normal static files from the web server. However, to be able to use this kind of protocol, Git requires the extra `objects/info/packs` and `info/refs` files to be present on the server and kept up to date with `git update-server-info`. This command is usually run on a push via one of the earlier mentioned smart protocols (the default `post-update` hook does that, and so does `git-receive-pack` if `receive.updateServerInfo` is set to `true`).

It is possible to push with the dumb protocol, but this requires a setup that allows you to update files using a specified transport; for the dumb HTTP(S) transport protocol, this means configuring WebDAV.

Authentication, in this case, is done by the web server for static files. Obviously, for this kind of transport, Git's server-side hooks are not invoked, and thus they cannot be used to further restrict access.

> **Historical note**
>
> Note that, for modern Git, the dumb transport is implemented using the curl family of remote helpers, which may be not installed by default.

This transport works (for fetching) by downloading requested refs (as plain files), examining where to find files containing the referenced commit objects (hence the need for server information files, at least for objects in packfiles), getting them, and then walking through the chain of revisions, examining each object needed, and downloading new files if the object is not present yet in the local repository. This walker method can be horrendously inefficient if the repository is not packed well with respect to the requested revision range. It requires a large number of connections and always downloads the whole pack, even if only one object from it is needed.

With smart protocols, Git on the client side and Git on the server side negotiate between themselves which objects need to be sent (a want/have negotiation). Git then creates a customized packfile, utilizing the knowledge of what objects are already present on the other side, and usually includes only deltas – that is, the difference from what the other side has (a thin packfile). The other side rewrites the received packfile to be self-contained.

Remote helpers

Git allows us to create support for new transport protocols by writing remote helper programs. This mechanism can be also used to support foreign repositories. Git interacts with a repository requiring a remote helper by spawning the helper as an independent child process, communicating with this process through its standard input and output with a set of commands. The use of remote transport helpers is described in *Chapter 6, Collaborative Development with Git*.

You can find third-party remote helpers to add support to the new ways of accessing repositories – for example, there is `git-remote-dropbox` to use Dropbox to store the remote Git repository. Note, however, that remote helpers are (possibly yet) limited in features compared to built-in transport support.

Tools to manage Git repositories

Nowadays, there is no need to write a Git repository management solution yourself. There is a wide range of various third-party solutions that you can use. It is impossible to list them all, and even giving recommendations is risky. The Git ecosystem is actively developed; which tool is the best could have changed since the time of writing.

I'd like to focus here just on the types of tools for administrators, just as I did for GUIs in *Chapter 13, Customizing and Extending Git*.

First, there are **Git repository management** solutions (we have seen one example of such in the form of the `update-paranoid` script in the `contrib/` area). These tools focus on access control, usually the authorization part, making it easy to add repositories and manage their permissions. An example of such a tool is *Gitolite*.

They often support some mechanism to add your own additional access constraints.

Then, there are **web interfaces** that allow us to view Git repositories using a web browser. Some make it even possible to create new revisions using a web interface. They differ in capabilities, but they usually offer at least a list of available Git repositories, a summary view for each one, an equivalent of the `git log` and `git show` commands, and a view with a list of files in the repository. An example of such tools is the `gitweb` script in Perl that is distributed with Git; another is `cgit`, used by `git.kernel.org`.

Also useful are the **code review** (**code collaboration**) tools. These make it possible for developers in a team to review each other's proposed changes using a web interface. These tools often allow the creation of new projects and the handling of access management. An example of such a tool is Gerrit Code Review.

Finally, there are **Git hosting** solutions, also called **software forges**, usually with a web interface for the administrative side of managing repositories, allowing us to add users, create repositories, manage their access, and often work from the web browser on Git repositories. Examples of such tools are GitLab and Gitea. There are also similar **source code management** systems, which provide (among other web-based interfaces) repository hosting services, together with the features to collaborate and manage development. One example of such a system is Kallithea; however, nowadays, many software forges include some source code management features, such as issue tracking, and **CI/CD** (**Continuous Integration/Continuous Delivery**) pipelines.

Of course, you don't need to self-host your code. There is a plethora of third-party hosted options – GitHub, Bitbucket, and so on. There are even hosted solutions using open source hosting management tools, such as GitLab and Codeberg.

Tips and tricks to host repositories

If you want to self-host Git repositories, there are a few things that may help you with server performance and user satisfaction.

Reducing the size taken by repositories

If you are hosting many forks (clones) of the same repository, you might want to reduce disk usage by somehow sharing common objects. One solution is to use **alternates** (for example, with `git clone --reference`) while creating a fork. In this case, the derived repository would look to its parent object storage if the object is not found on its own.

There are, however, two problems with this approach. First, you need to ensure that the object the borrowing repository relies on does not vanish from the repository set as the alternate object storage (the repository you borrow from). This can be done, for example, by linking the borrowing repository refs in the repository lending the objects, (e.g., in the `refs/borrowed/` namespace). Second is that the objects entering the borrowing repository are not automatically de-duplicated; you need to run `git repack -a -d -l`, which internally passes the `--local` option to `git pack-objects`.

An alternate solution would be to keep every fork together in a single repository and use **git namespaces** to manage separate views into the DAG of revisions, one for each fork. With plain Git, this solution means that the repository is addressed by the URL of the common object storage and the namespace to select a particular fork. Usually, this is managed by a server configuration or a repository management tool; such a mechanism translates the address of the repository into a common repository and the namespace. The `git-http-backend` manpage includes an example configuration to serve multiple repositories from different namespaces in a single repository. Gitolite also has some support for namespaces in the form of logical and backing repositories and `option namespace.pattern`, although not every feature works for logical repositories.

Storing multiple repositories as the namespace of a single repository avoids storing duplicated copies of the same objects. It automatically prevents duplication between new objects without the need for ongoing maintenance, as opposed to the alternate solution. Conversely, security is weaker; you need to treat anyone with access to the single namespace, which is within the repository, as if they had access to all the other namespaces (although this might not be a problem for your case).

Speeding up smart protocols with pack bitmaps

Another issue that you can stumble upon while self-hosting repositories is the performance of smart protocols. For the clients of your server, it is important that operations finish quickly; as an administrator, you would not want to generate a high CPU load on the server due to serving Git repositories.

One feature, ported from JGit, should significantly improve the performance of the counting objects phase, while serving objects from a repository that uses it. This feature is a **bitmap-index file**, available since Git 2.0.

> **The bitmap-index file**
>
> The major function of the `bitmap-index` file is providing for a selected subset of commits, including the most recent ones, bit vectors (bitmaps) that store reachability information for a set of objects in a packfile, or in a multi-pack index. In each bit vector, the value of 1 at index i means that the i-th object (in the order defined by a packfile or a multi-pack index file) is reachable from the commit that the given bit vector belongs to.

This file is stored alongside packfiles and their indexes. It can be generated manually by running `git repack -A -d --write-bitmap-index`, or it can be generated automatically together with the packfile by setting the `repack.writeBitmaps` configuration variable to `true`. The disadvantage of this solution is that bitmaps take additional disk space, and the initial repack requires extra time to create bitmap-index. With modern Git, thanks to the multi-pack index, you no longer need to repack everything into a single packfile to be able to use the bitmap file. This feature also makes it faster to update the bitmap.

Nowadays, this feature is turned on by default for bare repositories.

Solving the large non-resumable initial clone problem

Repositories with a large code base and a long history can get quite large. The problem is that the initial clone, where you need to get everything in a possibly large repository, is an all-or-nothing operation, at least for modern (safe and effective) smart transfer protocols – SSH, `git://`, and smart HTTP(S). This might be a problem if a network connection is not very reliable. There is no support for a resumable clone, and it unfortunately looks like it is a fundamentally hard problem to solve for Git developers. This does not mean, however, that you, as a hosting administrator, can do nothing to help users get this initial clone.

One solution is to create, with the `git bundle` command, a static file that can be used for the initial clone, or as a reference repository for the initial clone (the latter can be done with the `git clone --reference=<bundle> --dissociate` command after downloading the bundle). This bundle file can be distributed using any transport – in particular, one that can be resumed if interrupted, be it HTTP(S), FTP, rsync, or BitTorrent. The convention that people use, besides explaining how to get such a bundle in the developer documentation, is to use the same URL as that used for the repository but with the `.bundle` extension (instead of an empty extension or a `.git` suffix). If the bundle is available via the HTTP(S) or SSH protocols, it can be used without explicitly downloading it first with `git clone --bundle-uri=<bundle uri>`.

There is also the **bundle-uri** capability of Git, where the server suggests where you can download such a bundle from the client, which in turn can use the bundle to speed up the initial clone. At the time of writing, no software forge supports this feature, but there is the **git bundle-server** (`https://github.com/git-ecosystem/git-bundle-server`) web server and management interface for use with this feature.

There are also more esoteric approaches to solving the problem of the initial clone cost, such as a step-by-step deepening of a shallow clone (or perhaps just using a shallow clone with `git clone --depth` is all that's needed), starting with a partial clone, or using approaches such as GitTorrent.

Augmenting development workflows

Handling version control is only a part of the development workflow. There is also work management, code review and audit, running automated tests, and generating builds.

Many of these steps can be aided by specialized tools. Many of them offer Git integration. For example, code review can be managed using Gerrit, requiring that each change passes a review before being made public. Another example is setting up development environments so that pushing changes to the public repository can automatically close tickets in the issue tracker, based on the patterns in the commit messages. This can be done with server-side hooks or with the hosting service's Webhooks.

A repository can serve as a gateway, running automated tests (for example, with the help of Jenkins' or Hudson's continuous integration service) and deploying changes to ensure quality environments only after passing all of these tests. Another repository can be configured to trigger builds for various supported systems. Many tools and services support push-to-deploy mechanisms (for example, Heroku or Google's App Engine).

Git can automatically notify users and developers about published changes. This can be done via email, a mailing list, an IRC/Discord/Slack channel, or a web-based dashboard application. The possibilities are plentiful; you only need to find them.

Defining development workflows in the repository

Many software forges allow you to automate, customize, and execute software development workflows right from the repository. Those solutions, such as *GitHub Actions* and *GitLab CI/CD*, let you run various workflows (for example, to run tests or to deploy an application) when other events happen in your repository at the software forge. Those workflows are run using runners, either virtual machines or containers. They are usually defined by a YAML file checked into your repository.

While the specific dialect of the YAML markup language, the pathname of the file, and the available pre-defined actions differ from service to service, they are similar enough that you should be able to migrate from one solution to the other.

GitOps – using Git for operational procedures

The natural extension of defining software development workflows in the Git repository is to use Git to automatically manage deployment infrastructure, especially for cloud-native applications. This is called **GitOps** – an operational framework that uses the Git repository to store **infrastructure as code (IoC)** files and application configuration files. This data can be stored in the same repository as the application code, or in a separate repository.

GitOps ensures that the infrastructure (including the development, testing, and deployment environments) is immediately reproducible, based on the state of the Git repository. This provides version control for operations should a rollback be needed.

Often, the infrastructure configuration is defined declaratively, and a specialized software agent (such as Argo CD, Flux, or Gitkube) running in the cloud pulls from the Git repository at regular intervals and checks the configuration against the live state, adjusting the state as necessary.

Summary

This chapter covered various issues related to the administrative side of working with Git. You learned the basics of maintenance, data recovery, and repository troubleshooting. You also learned how to set up Git on a server, how to use server-side hooks, and how to manage remote repositories. The chapter covered tips and tricks for a better remote performance. It described how you can use Git (with the help of third-party tools) to augment development workflows. The information in this chapter should help you to choose a Git repository management solution, or even write your own.

The next chapter will include a set of recommendations and best practices, both specific to Git and those that are version control-agnostic. A policy based on these suggestions can be enforced and encouraged with the help of the tools explored in this chapter.

Questions

Answer the following questions to test your knowledge of this chapter:

1. How do you set up automatic repository maintenance to ensure that Git operations will not slow down?

2. How you can try to recover a lost commit?

3. How do you find out why some Git commands started to perform badly and took too much time to execute?

4. How you can ensure that development follows a given defined policy?

5. What is the simplest solution to sharing the repository privately, where all developers work on a single computer (on a single machine)?

Answers

Here are the answers to the questions given above:

1. Use the `git maintenance` command.

2. First, check the branch and HEAD reflogs if the lost committing question is not readily available from there. If this fails, you can try to browse through unreachable commits with `git fsck`.

3. You can use the "Git trace" mechanism – for example, with the `GIT_TRACE2_PERF` or `GIT_TRACE_PERFORMANCE` environment variables.

4. Use your software forge features, if possible (for example, to protect a branch against changes or deletion), or use server-side hooks. Enforcing the policy can be helped, but not ensured, with client-side hooks.

5. Simply create the bare repository with `git init --bare --shared`, while ensuring that all developers that need access to it have appropriate filesystem permissions. If necessary, push to that repository.

Further reading

To learn more about the topics that were covered in this chapter, take a look at the following resources:

- Scott Chacon, Ben Straub: *Pro Git, 2nd Edition*, Apress (2014) *Chapter 4, Git on the Server* `https://git-scm.com/book/en/v2/Git-on-the-Server-The-Protocols`

- Scott Chacon: *Git Tips 2: New Stuff in Git* (2024) `https://blog.gitbutler.com/git-tips-2-new-stuff-in-git/#git-maintenance`

- Konstantin Ryabitsev: *Signed git pushes* (2020) `https://people.kernel.org/monsieuricon/signed-git-pushes`

- Vicent Martí: *Counting Objects* (2015) `https://github.blog/2015-09-22-counting-objects/`

- Sitaram Chamarty: *Gitolite Essentials*, Packt (2014) `https://subscription.packtpub.com/book/programming/9781783282371`

- Derrick Stolee: *Exploring new frontiers for Git push performance* (2019) `https://devblogs.microsoft.com/devops/exploring-new-frontiers-for-git-push-performance/`

- Taylor Blau: *Scaling monorepo maintenance* (2021) `https://github.blog/2021-04-29-scaling-monorepo-maintenance/`

Git Best Practices

The last chapter of *Mastering Git* presents a collection of generic and Git-specific version control recommendations and best practices. You have encountered many of these recommendations already in the earlier chapters; they are here as a summary and as a reminder. For details and the reasoning behind each best practice, refer to the specific chapters.

This chapter will cover the issues of managing the working directory, creating commits and series of commits (pull requests), submitting changes for inclusion, and the peer review of changes.

In this chapter, we will cover the following topics:

- How to separate projects into repositories
- What types of data to store in a repository and which files Git should ignore
- What to check before creating a new commit
- How to create a good commit and a good commit series (or, in other words, how to create a good pull request)
- How to choose an effective branching strategy, and how to name branches and tags
- How to review changes and how to respond to the review

Starting a project

When starting a project, you should choose and clearly define a project governance model (who manages work, who integrates changes, and who is responsible for what). You should decide about the license and the copyright of the code: whether it is work for hire and whether contributions require a copyright assignment, a contributor agreement, a contributor license agreement, or simply a digital certificate of origin.

Dividing work into repositories

In centralized version control systems, often everything is put under the same project tree. With distributed version control systems such as Git, it very much depends on the nature of the project. Often, it is better to split separate projects into separate repositories, but if those projects are tightly coupled together it might be better to use a **monorepo** – all projects in a single large repository.

If some part of the code is needed by multiple separate projects, consider extracting it into its own project and then incorporating it as a submodule or subtree, grouping concepts into a superproject. See *Chapter 11*, *Managing Subprojects*, for the details.

Selecting the collaboration workflow

You need to make decisions on the collaboration structure, whether your project will use a dispersed contributor model, a "blessed" repository model, or a central repository, and so on (as found in *Chapter 6, Collaborative Development with Git*).

This often requires setting up an access control mechanism and deciding on the permission structure; see *Chapter 14, Git Administration*, for details on how one can set up this.

You also need to decide on the branching patterns to use. See *Chapter 8, Advanced Branching Techniques*, for examples of the most common patterns. You need to decide how to integrate changes, and how to isolate independent work. Those branching patterns are often grouped together into a single named branching workflow.

This decision about branching doesn't need to be cast in stone. As your project and your team experience grow, you might want to consider changing the branching model, for example, from the trunk-based workflow to a plain branch-per-feature model, a GitHub flow, or any of the other derivatives.

The decisions about licensing, the collaboration structure, and the branching model should all be stated explicitly in the developer documentation (at a minimum in the README and LICENSE/ COPYRIGHT files, and perhaps also in CodingGuidelines and CodeOfConduct). You need to remember that if the way in which the project is developed changes, this documentation needs to be updated to reflect the changes. This can happen, for example, because the project has grown beyond its initial stage.

Choosing which files to keep under version control

In most cases, you should not include any of the **generated files** in the version control system (though there are some very rare exceptions). You should track only the sources (the original resources); Git works best if these sources are plain text files, but it also works well with binary files.

To avoid accidentally including unwanted files in a repository, you should use the **gitignore patterns**. These ignore patterns that are specific to a project (for example, results and by-products of a build system) should go into the .gitignore file in the project tree; those specific to the developer (for example, backup files created by the editor one uses or the operating system-specific helper files) should go into their per-user `core.excludesFile` (which, in modern Git, is the `~/.config/git/ignore` file), or into a local configuration of the specific clone of the repository, that is, `.git/info/excludes`. See *Chapter 3*, *Managing Your Worktrees*, for details.

A good start for ignore patterns is the `https://gitignore.io` site with its `.gitignore` templates for various operating systems, IDEs, and programming languages.

Another suggestion is to not add to Git the configuration files that might change from environment to environment (for example, those that are different for MS Windows and Linux).

Working on a project

Here are some guidelines on how to create changes and develop new revisions. These guidelines can be used either for your work on your own project, or to help contribute your code to a project maintained by somebody else.

Different projects can use different development workflows; therefore, some of the recommendations presented here might not make sense depending on the given workflow in use.

Working on a topic branch

Branching in Git has two functions (*Chapter 8*, *Advanced Branching Techniques*): as a mediator for the code contributed by developers keeping to the **specified level of code stability and maturity** (long-running public branches), providing the road to integration and deployment, and as a **sandbox for the development of a new idea** (short-lived private branches).

The ability to sandbox changes is why it is considered a good practice to create a separate branch for each new task you work on. Such a branch is called a topic branch or a feature branch. Using separate branches makes it possible to switch between tasks easily, and to keep disparate pieces of work in progress from interfering with each other. On the other hand, if such branches are long-lived, it would go against **continuous integration** (**CI**) practices, reduce changeset visibility, and lead to more difficult integration because of larger divergence.

You should choose short and descriptive names for branches. There are different naming conventions for topic branches; the convention your project uses should be specified in the developer documentation. In general, branches are usually named with a summary of a topic they host, usually in all-lowercase and with the spaces between words replaced by hyphens or underscores (see the `git-check-ref-format` manpage to know what is forbidden in branch names). Branch names can include slashes (be hierarchical).

If you are using an issue tracker, then a branch that fixes a bug or implements an issue can have its name prefixed with the identifier (the number) of the ticket describing the issue, for example, `1234-doc_spellcheck`. On the other hand, the maintainer, while gathering submissions from other developers, could put these submissions in topic branches named after the initials of the developer and the name of the topic, for example, `ad/whitespace-cleanup` (this is an example of a **hierarchical branch name**).

It is considered a good practice to delete your topic branch from your local repository, and also from the upstream repository after you are done with the branch in question, to reduce clutter.

Deciding what to base your work on

As a developer, you are usually working on some specific issue at a given time, be it a bug fix, an enhancement, a correction to some topic, or a new feature.

Where to start your work on a given topic, and what branch to base your work on, both depend on the branching workflow chosen for the project (see *Chapter 8, Advanced Branching Techniques*, for a selection of branching workflows). This decision also depends on the type of work you do.

For a topic branch workflow (or a branch-per-feature workflow), you would want to base your work on the oldest and most stable long-running branch that your change is relevant to, and for which you plan to merge your changes into. This is because, as described in *Chapter 8, Advanced Branching Techniques*, you should **never merge** a **less stable branch** into a **more stable branch**. The reason behind this best practice rule is to avoid destabilizing the branch as merging carries over all the changes.

Different types of changes require a different long-lived branch to be used as a base for a topic branch with those changes, or to put those changes onto. In general, to help developers working on a project, this information should be described in the developer documentation; not everybody needs to be knowledgeable about the branching workflow used by the project.

The following describes what is usually used as a base branch, depending on the purpose of the changes:

- **Bugfix**: In this case, the topic branch (the bugfix branch) should be based on the oldest and the most stable branch in which the bug is present. This means, in general, starting with the maintenance branch. If the bug is not present in the maintenance branch, then base the bugfix branch on the stable branch. For a bug that is not present in the stable branch, find the topic branch that introduced it and base your work on top of that topic branch.

- **New feature**: In this case, the topic branch (the feature branch) should be based on the stable branch, if possible. If the new feature depends on some topic that is not ready for the stable branch, then base your work on that topic (from a topic branch).

- Corrections and enhancements to a topic that didn't get merged into the stable branch should be based on the tip of the topic branch being corrected. If the topic in question is not considered published, it's alright to make changes to the steps of the topic, squashing minor corrections in the series (see the section about rewriting history in *Chapter 10, Keeping History Clean*).

If the project you are contributing to is large enough to have dedicated maintainers for selected parts (subsystems) of the system, you first need to decide which repository and fork (sometimes named "a tree") to base your work on.

Splitting changes into logically separate steps

Unless your work is really simple and can be done in a single step (a single commit)—as is the case with many bugfixes—you should make separate commits for logically separate changes, one commit per single step. Those commits should be ordered logically.

Following good practice for commit messages (with an explanation of what you have done—see the next section) could help in deciding when to commit. If your description gets too long and you begin to see that you have two independent changes squashed together, that's a sign that you probably need to split your commit into finer-grained pieces and use smaller steps.

Remember, however, that it is a matter of balance between the project conventions and the development workflow chosen. Changes should, at a minimum, stand on their own. At each step (at each commit) of the implementation of a feature, the code compiles and the program passes the test suite. You should **commit early and often**. Smaller self-contained revisions are easier to review, and with smaller (but still complete) changes, it is easier to find regression bugs with `git bisect` (which is described in *Chapter 4, Exploring Project History*).

Note that you don't necessarily need to come up with the perfect sequence of steps from the start. If you notice that you have entangled the work directory's state, you can make use of the staging area, using an interactive add to disentangle it (this is described in *Chapter 2, Developing with Git,* and *Chapter 3, Managing Your Worktrees*). You can also use an interactive rebase or a similar technique, as shown in *Chapter 10, Keeping History Clean*, to curate commits into an easy-to-read (and easy-to-bisect) history before publishing.

> **Important note**
> You should remember that a commit is a place to record your result (or a particular step towards the result), not a place to save the temporary state of your work. If you need to temporarily save the current state before going back to it, use `git stash`.

Writing a good commit message

A good commit message should include an explanation for the change with sufficient detail so that other developers on the team (including reviewers and the maintainer) can judge whether it is a good idea to include the change in the codebase. This good-or-not decision should not require them to read the actual changes to find out what the commit intends to do.

The first line of the commit message should be a short, terse description (from around 50 to 72 characters) with a summary of the changes. It should be separated by an empty line from the rest of the commit message, if there is one. This is partly because, in many places, such as in the `git log --oneline` command output, in a graphical history viewer such as `gitk`, and in the instruction sheet of `git rebase --interactive`, you will see only this one line of the commit message and have to decide the action with respect to that commit on the basis of this one line. If you have trouble with coming up with a good summary of changes, this might mean that these changes need to be split into smaller steps.

There are various conventions for this summary line of changes. One convention is to prefix the first summary line with **area:**, which is an identifier for the general area of the code being modified: the name of the subsystem, of an affected subdirectory, or a filename of a file being changed. If the development is managed via an issue tracker, this summary line can start with something like the **[#1234]** prefix, where **1234** is the identifier of an issue or task implemented in the commit. In general, when not sure about what information to include in the commit message, refer to the development documentation or fall back to the current convention used by other commits in the history.

> **Tip**
> If you are using agile development methods, you can look for especially good commit messages during retrospectives and add them as examples to the developer documentation for the future.

For all but trivial changes, there should be a longer meaningful description, the body of the commit message. There is something that people coming from other version control systems might need to unlearn: namely, not writing a commit message at all or writing it all on one long line. Note that Git will not allow the creation of a commit with an empty commit message unless forced to with `--allow-empty`.

The commit message should do the following:

- Include the rationale for the commit, explaining the problem that the commit tries to solve – the *why*, in other words. It should include a description of what is wrong with the current code or the current behavior of the project without the change. This should be self-contained, but it can refer to other sources including the issue tracker (the bug tracker) or other external documents such as articles, wikis, or **Common Vulnerabilities and Exposures (CVEs)**.

- Include a quick summary. In most cases, it should also explain the *how* and justify the way the commit solves the problem.

- Describe why you think the result with the change is better; this part of the description does not need to explain what the code does, as that is largely a task for the code comments.

- If there was more than one possible solution, include a description of the alternative solutions that were considered but ultimately discarded, perhaps with links to the discussion or review(s).

It's a good idea to try to make sure that your explanation of the changes can be understood without access to any external resources (that is, without accessing the issue tracker, the internet, or a mailing list archive). Instead of just referring to the discussion, or in addition to giving a URL or an issue number, write a summary of the relevant points in the commit message.

One of the possible recommendations when writing a commit message is to describe changes in the imperative mood, for example, **make foo do bar**, as if you are giving orders to the codebase to change its behavior, instead of writing **This commit makes...** or **[I] changed ...**.

Here, `commit.template` and commit message hooks can help in following these practices. see *Chapter 13*, *Customizing and Extending Git*, for details (and *Chapter 14*, *Git Administration*, for a description of the way to enforce this recommendation).

Preparing changes for submission

If the topic branch was started a long time ago, consider rebasing the branch to be submitted on top of the current tip of the base branch. This should make it easier to integrate changes in the future. If your topic branch was based on the development version, or on the other in-flight topic branch (perhaps because it depended on some specific feature), and the branch it was based on got merged into a stable line of development, you should rebase your changes on top of the stable integration branch instead.

Rebasing is also a chance for a final clean-up of the history; the chance to make submitted changes easier to review. Simply run an interactive rebase with `git rebase --interactive`, or a patch management tool if you prefer (see *Chapter 10*, *Keeping History Clean*). One caveat: **do not rewrite the published history**.

Consider testing that your changes merge cleanly, and fix it if they don't (if possible). Make sure that they apply or merge cleanly into the appropriate integration branch.

Take a last look at your commits to be submitted. Make sure that your changes do not add the commented-out (or the ifdef-ed-out) code, nor any extra files not related to the purpose of the patch (for example, changes in an upcoming feature). Review your commit series before submission to ensure accuracy.

Integrating changes

The exact details on how to submit changes for merging depends, of course, on the development workflow that the project is using. Various classes of possible workflows are described in *Chapter 6*, *Collaborative Development with Git*.

Submitting and describing changes

If the project has a dedicated maintainer or, at least someone responsible for merging the proposed changes into the official version, you also need to describe the submitted changes as a whole (in addition to describing each commit in the series). This can be done in the form of a cover letter for the patch series while sending changes as patches via email. It can also be done with comments in the pull request while using the collocated contributor repositories model, or it can be the description in an email with a pull request, which already includes the URL and the branch in your public repository with changes (generated with `git request-pull`).

This cover letter or pull request should include a description of the purpose of the patch series or the pull request. Consider also providing an overview of why the work is taking place (with any relevant links and a summary of the discussion). Be explicit in stating that it is a work in progress in the description of changes.

In the dispersed contributor model, where changes are submitted for review as patches or patch series, usually to a mailing list, you should use Git-based tools such as `git format-patch` and, if possible, `git send-email`. Multiple related patches should be grouped together, for example, in their own email thread. The convention is to send them as replies to an additional cover letter message, which should describe the feature as a whole.

If the changes are sent to the mailing list, it is a common convention to prefix your subject line with `[PATCH]` or `[PATCH m/n]` (where m is the patch number in the series of the n patches). This lets people easily distinguish patch submissions from other emails. This part can be done with `git format-patch`. What you need to decide yourself is whether to use additional markers after PATCH to mark the nature of the series, for example, PATCH/RFC. (**RFC** here means **Request For Comments**, i.e., an idea for a feature with an example of its implementation. Such a patch series should only be examined if the idea is worthy; it is not ready to be applied/merged yet and is provided only for discussion among the developers.)

In the collocated contributor repositories model, where all the developers use the same Git hosting website or software (for example, GitHub, Bitbucket, GitLab, or a private instance of it), you would push changes to your own public repository, a fork of the official version. Then, you would create a merge request or pull request, usually via the web interface of the hosting service, again describing the changes as a whole there.

In the case of using the central repository (perhaps in a shared maintenance model), you would push changes to a separate and possibly new branch in the integration repository, and then send an announcement to the maintainer so that they can find the changes to merge. The details of this step depend on the exact setup; sending announcements might be done via email, some kind of internal messaging mechanism, or even via tickets (or the comments in the tickets).

The development documentation might include rules specifying where to send announcements and/or changes. It is considered a courtesy to notify the people who are involved in the area of code you are changing about the new changes (you can use `git blame` and `git shortlog` to identify these people; see *Chapter 4, Exploring Project History*). These people are important; they can write comments about the change and help review it.

Crediting people and signing your work

Some open source projects, in order to improve the tracking provenance of the code, use the sign-off procedure borrowed from the Linux kernel called **Digital Certificate of Origin**. The sign-off is a simple line at the end of the commit message, like the following example:

Signed-off-by: Random Developer <rdeveloper@company.com>

By adding this line, you certify that the contribution is either created as a whole or in part by you, or is based on previous work, or was provided directly to you, and that everybody in the chain has the right to submit it under the appropriate license. If your work is based on work by somebody else, or if you are just passing somebody's work, then there can be multiple sign-off lines forming a chain of provenance.

In order to credit people who helped with creating the commit, you can append to the commit message other trailers, such as **Reported-by:**, **Reviewed-by:**, **Acked-by:** (this one states that it was liked by the person responsible for the area covered by the change), or **Tested-by:**.

The art of the change review

Completing a peer review of changes is time-consuming (although so is using version control), but the benefits are huge: better code quality, a reduction in the time needed for quality assurance testing, transfer of knowledge, and so on. The change can be reviewed by a peer developer, reviewed by a community (requiring consensus), or reviewed by the maintainer or one of their lieutenants.

Before beginning the code review process, you should read through the description of the proposed changes to discover why the change was proposed and decide whether you are the correct person to perform the review (that is one of the reasons why good commit messages are so important). You need to understand the problem that the change tries to solve. You should familiarize yourself with the context of the issue, and with the code in the area of changes.

The first step is to reproduce the state before the change and check whether the program works as described (for example, that the bug in a bugfix can be reproduced). Then, you need to check out the topic branch with the proposed changes and verify that the result works correctly. If it works, review the proposed changes, creating a comprehensive list of everything wrong (though if there are errors early in the process, it might be unnecessary to go deeper), as follows:

- Are the commit messages descriptive enough? Is the code easily understood?
- Is the contribution architected correctly? is it architecturally sound?
- Does the code comply with the project's coding standards and with the agreed-upon coding conventions?
- Are the changes limited to the scope described in the commit message?
- Does the code follow the industry's best practices? Is it safe and efficient?
- Is there any redundant or duplicate code? Is the code as modular as possible?

- Does the code introduce any regressions in the test suite? If it is a new feature, does the change include the tests for the new feature, both positive and negative?

- Is the new code performing the way it did before the change (within the project's tolerances)?

- Are all the words spelled correctly, and does the new version follow the formatting guidelines for the content?

This is only one possible proposal for such a code review checklist. Depending on the specifics of the project, there might be more questions that need to be asked as a part of the review; make the team write their own checklist. You can find good examples online.

Divide the problems that you have found during reviews into the following categories:

- **Wrong problem**: This feature does not lie within the scope of the project. It is sometimes used for a bug that cannot be reproduced. Is the idea behind the contribution sound? If so, eject the changes with or without prejudice and do not continue the analysis for the review.

- **Does not work**: This does not compile, introduces a regression, doesn't pass the test suite, doesn't fix the bug, and so on. These problems absolutely must be fixed.

- **Fails best practices**: This does not follow the industry guidelines or the project's coding conventions. Is the contribution polished? These are pretty important to fix, but there might be some nuances as to why it is written the way it is.

- **Does not match reviewer preferences**. In this case, you should suggest modifications, or alternatively ask for clarification.

Minor problems, for example, typos or spelling errors, can be fixed immediately by the reviewer. If the exact problem repeats itself, however, consider asking the original author for a fix and resubmission; this is done to spread knowledge. You should not be making any substantive edits in the review process (barring extenuating circumstances).

Ask, don't tell. Explain your reasoning about why the code should be changed. Offer ways to improve the code. Distinguish between facts and opinions.

Responding to reviews and comments

Changes are not always accepted on the first try. You can and will get suggestions for improvement (and other comments) from the maintainer, the code reviewer, and other developers. You might even get these comments in the patch form or a fixup commit form.

First, consider leading your response with an expression of appreciation for the commenter having taken the time to perform a review. If anything in the review is unclear, do ask for clarification; if there is a lack of understanding between you and the reviewer, offer clarification.

The next step is often to polish and refine the changes. Then, you should resubmit them (perhaps, marking them as **v2**). You should respond to the review for each commit and for the whole series.

If you are responding to comments in a pull request, reply in the same way. In the case of patch submissions via email, you can put the comments for a new version (with a response to the review or a description of the difference from the previous attempt), either between three dashes --- and the `diffstat`, or at the top of an email separated from what is to be in the commit message by a "scissors" line, for example, --- >8 ---. An explanation of the changes that stays constant between iterations, but nevertheless should be not included in the commit message, can be kept in the Git notes (see *Chapter 10*, *Keeping History Clean*) and inserted automatically via `git format-patch --notes`.

Depending on the project's governance structure, you will likely have to wait for the changes to be considered good and ready for inclusion. This can be the decision of a benevolent dictator for life in open source projects, or the decision of the team leader, a committee, or a consensus. It is considered a good practice to summarize the discussion while submitting a final version of a feature.

Note that changes that have been accepted might nevertheless go through a few more stages before finally graduating to the stable branch and being present in the project.

Other recommendations

In this section, you will find the best practices and recommendations that do not fit cleanly into one of the areas described so far, namely starting a project, working on a project, and integrating changes.

Don't panic, recovery is almost always possible

As long as you have committed your work and stored your changes in the repository, it will not be lost. It could only perhaps be misplaced. Git also tries to preserve your current uncommitted (unsaved) work, but it cannot distinguish for example between the accidental and the conscious removal of all the changes to the working directory with `git reset --hard`. Therefore, make sure to commit or stash your current work before trying to recover lost commits.

Thanks to the reflog (both for the specific branch and for the HEAD ref), it is easy to undo most operations. Then, there is the list of stashed changes (see *Chapter 3*, *Managing Your Worktrees*), where your changes might be hiding. And there is `git fsck` as the last resort. See *Chapter 14*, *Git Administration*, for some further information about data recovery.

If the problem is that you have made a mess of the working directory, stop and think. Do not drop your changes needlessly. With the help of interactive add, interactive reset (the `--patch` option), and interactive checkout (the same), you can usually disentangle the mess.

Running `git status` and carefully reading its output helps in many cases where you are stuck after doing some lesser-known `git` operation.

If you have a problem with a rebase or merge, and you cannot pass the responsibility to another developer, there is always the third-party `git-imerge` tool.

Don't change the published history

Once you have made your changes public, you should ideally consider those revisions to be etched in stone, immutable, and unchanging. If you find problems with commits, create a fix (perhaps by undoing the effect of the changes with `git revert`). This is all described in *Chapter 10, Keeping History Clean*: that is, unless it is stated explicitly in the development documentation that these specific branches can be rewritten or redone; but it is nevertheless better to avoid creating such branches.

In some rare cases, you might really need to change the history: remove a file, clean up an unencrypted stored password, remove accidentally added large files, and so on. If you need to do it, notify all the developers of the fact.

Numbering and tagging releases

Before you release a new version of your project, mark the version to be released with a signed tag. This ensures the integrity of the just-created revision.

There are various conventions for naming the release tags and using release numbering. One of the more common ones is tagging releases by using, for example, `1.0.2` or `v1.0.2` as a tag name.

> **Tip**
>
> If the integrity of the project is important, consider using signed merges for integration (that is, merging signed tags). See *Chapter 6, Collaborative Development with Git*, and for signed pushes, see *Chapter 14, Git Administration*.

There are different conventions for naming releases. For example, with time-based releases, there is the convention of naming releases after dates, such as `2015.04` (or `15.04`). Then, there is the common convention of **semantic versioning** (`http://semver.org/`) with the `MAJOR.MINOR.PATCH` numbering, where `PATCH` increases when you are making backward-compatible bug fixes, `MINOR` is increased when adding functionality that is backward compatible, and the `MAJOR` version is increased when making incompatible API changes. Even when not using full semantic versioning, it is common to add a third number for maintenance releases, for example, `v1.0` and `v1.0.3`.

Automate where possible

You should not only have the coding standards written down in the development documentation; you also need to enforce them. Following these standards can be facilitated with client-side hooks (*Chapter 13, Customizing and Extending Git*) and enforced with server-side hooks (*Chapter 14, Git Administration*).

Hooks can also help by automatically managing tickets in the issue tracker and selecting an operation based on given triggers (patterns) in the commit message. Hooks can also be used to protect against rewriting the history.

Consider using third-party solutions, such as Gitolite or GitLab, to enforce rules for access control. If you need to do a code review, use appropriate tools such as Gerrit or the pull requests of GitHub, Bitbucket, or GitLab.

Summary

These recommendations, based on the best practices of using Git as a version control system, can really help your development and your team. You have learned the steps along the road, starting from an idea, going all the way, and ending with the changes being integrated into the project. These checklists should help you develop better code.

Further reading

To learn more about the topics that were covered in this chapter, take a look at the following resources:

- Emma Jane Hogbin Westby: *Git for Teams* (2015), O'Reilly Media
- *Learn Git Branching* `https://learngitbranching.js.org/`
- *Conventional Commits: A specification for adding human and machine-readable meaning to commit messages* `https://www.conventionalcommits.org/`
- *Commitizen - a release management tool designed for teams* `https://commitizen-tools.github.io/commitizen/`
- Sage Sharp: *The Gentle Art Of Patch Review* (2014) `https://sage.thesharps.us/2014/09/01/the-gentle-art-of-patch-review/`
- *Dangit, Git!?!* `https://dangitgit.com/en`
- Julia Evans: *Oh shit, git!* Zine `https://wizardzines.com/zines/oh-shit-git/`
- *Semantic Versioning 2.0.0* `https://semver.org/`

Index

U

V

W

‹packt›

packtpub.com

Subscribe to our online digital library for full access to over 7,000 books and videos, as well as industry leading tools to help you plan your personal development and advance your career. For more information, please visit our website.

Why subscribe?

- Spend less time learning and more time coding with practical eBooks and Videos from over 4,000 industry professionals

- Improve your learning with Skill Plans built especially for you

- Get a free eBook or video every month

- Fully searchable for easy access to vital information

- Copy and paste, print, and bookmark content

Did you know that Packt offers eBook versions of every book published, with PDF and ePub files available? You can upgrade to the eBook version at packtpub.com and as a print book customer, you are entitled to a discount on the eBook copy. Get in touch with us at customercare@packtpub.com for more details.

At www.packtpub.com, you can also read a collection of free technical articles, sign up for a range of free newsletters, and receive exclusive discounts and offers on Packt books and eBooks.

Other Books You May Enjoy

If you enjoyed this book, you may be interested in these other books by Packt:

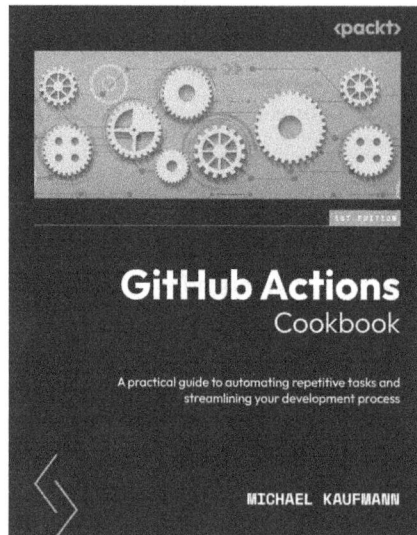

GitHub Actions Cookbook

Michael Kaufmann

ISBN: 978-1-83546-894-4

- Author and debug GitHub Actions workflows with VS Code and Copilot
- Run your workflows on GitHub-provided VMs (Linux, Windows, and macOS) or host your own runners in your infrastructure
- Understand how to secure your workflows with GitHub Actions
- Boost your productivity by automating workflows using GitHub's powerful tools, such as the CLI, APIs, SDKs, and access tokens
- Deploy to any cloud and platform in a secure and reliable way with staged or ring-based deployments

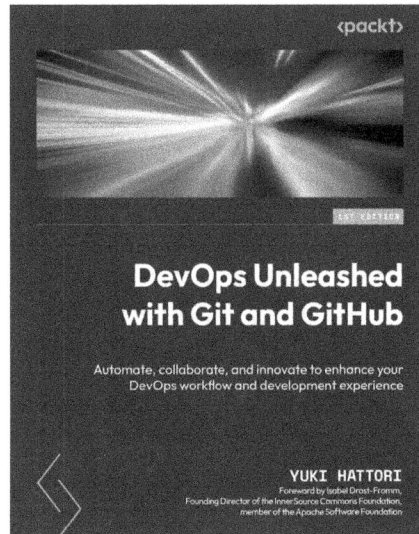

DevOps Unleashed with Git and GitHub

Yuki Hattori

ISBN: 978-1-83546-371-0

- Master the fundamentals of Git and GitHub
- Unlock DevOps principles that drive automation, continuous integration and continuous deployment (CI/ CD), and monitoring
- Facilitate seamless cross-team collaboration
- Boost productivity using GitHub Actions
- Measure and improve development velocity
- Leverage the GitHub Copilot AI tool to elevate your developer experience

Packt is searching for authors like you

If you're interested in becoming an author for Packt, please visit `authors.packtpub.com` and apply today. We have worked with thousands of developers and tech professionals, just like you, to help them share their insight with the global tech community. You can make a general application, apply for a specific hot topic that we are recruiting an author for, or submit your own idea.

Share Your Thoughts

Now you've finished *Mastering Git*, we'd love to hear your thoughts! Scan the QR code below to go straight to the Amazon review page for this book and share your feedback or leave a review on the site that you purchased it from.

`https://packt.link/r/1-835-08607-1`

Your review is important to us and the tech community and will help us make sure we're delivering excellent quality content.

Download a free PDF copy of this book

Thanks for purchasing this book!

Do you like to read on the go but are unable to carry your print books everywhere?

Is your eBook purchase not compatible with the device of your choice?

Don't worry, now with every Packt book you get a DRM-free PDF version of that book at no cost.

Read anywhere, any place, on any device. Search, copy, and paste code from your favorite technical books directly into your application.

The perks don't stop there, you can get exclusive access to discounts, newsletters, and great free content in your inbox daily

Follow these simple steps to get the benefits:

1. Scan the QR code or visit the link below

https://packt.link/free-ebook/978-1-83508-607-0

2. Submit your proof of purchase
3. That's it! We'll send your free PDF and other benefits to your email directly

www.ingramcontent.com/pod-product-compliance
Lightning Source LLC
Chambersburg PA
CBHW081037220326
41598CB00038B/6901